LINEAR-FRACTIONAL PROGRAMMING THEORY, METHODS, APPLICATIONS AND SOFTWARE

Applied Optimization
Volume 84

Series Editors:

Panos M. Pardalos
University of Florida, U.S.A.

Donald W. Hearn
University of Florida, U.S.A.

LINEAR-FRACTIONAL PROGRAMMING THEORY, METHODS, APPLICATIONS AND SOFTWARE

ERIK B.BAJALINOV
Senior Research Fellow
Department of Computer Science
Institute of Informatics
Debrecen University
HUNGARY

Kluwer Academic Publishers
Boston/Dordrecht/London

Distributors for North, Central and South America:
Kluwer Academic Publishers
101 Philip Drive
Assinippi Park
Norwell, Massachusetts 02061 USA
Telephone (781) 871-6600
Fax (781) 871-6528
E-Mail <kluwer@wkap.com>

Distributors for all other countries:
Kluwer Academic Publishers Group
Post Office Box 322
3300 AH Dordrecht, THE NETHERLANDS
Telephone 31 78 6576 000
Fax 31 78 6576 474
E-Mail <orderdept@wkap.nl>

Electronic Services <http://www.wkap.nl>

Library of Congress Cataloging-in-Publication

Bajalinov, Erik B.
Linear-Fractional Programming: Theory, Methods, Applications and Software
ISBN 1-4020-7626-6

This book is dedicated to the memory of my parents Mihri Makhmutova and Bakish Bajalinov

Contents

List of Figures

List of Tables

Preface

This is a book on *Linear-Fractional Programming* (here and in what follows we will refer to it as "LFP"). The field of LFP, largely developed by Hungarian mathematician B. Martos and his associates in the 1960's, is concerned with problems of optimization.

LFP problems deal with determining the best possible allocation of available resources to meet certain specifications. In particular, they may deal with situations where a number of resources, such as people, materials, machines, and land, are available and are to be combined to yield several products. In linear-fractional programming, the goal is to determine a permissible allocation of resources that will maximize or minimize some specific showing, such as profit gained per unit of cost, or cost of unit of product produced, etc.

Strictly speaking, linear-fractional programming is a special case of the broader field of *Mathematical Programming*. LFP deals with that class of mathematical programming problems in which the relations among the variables are linear: the constraint relations (i.e. the *restrictions*) must be in linear form and the function to be optimized (i.e. the *objective function*) must be a ratio of two linear functions.

At the same time LFP includes as a special case the well known and widespread *Linear Programming* (LP). In the problems of LP both the restrictions and the objective function must be linear in form. If in an LFP problem the denominator of the objective function is constant, which equals to 1, then we have an LP problem. Conversely, any problem of LP may be considered as an LFP one with the constant denominator of the objective function.

In a typical maximum problem, a manufacturer may wish to use available resources to produce several products. The manufacturer, knowing how much profit and cost are made for each unit of product produced, would wish to produce that particular combination of products that would maximize the profit gained per unit of cost.

The example of a minimum problem is as follows: A company owning several mines with varying grades of ore is given an order to supply certain quantities of each grade; how can the company satisfy the requirements in such a way that the cost of unit of the ore is minimized?

Transportation problems comprise a special class of linear-fractional programming. In a typical problem of this type the trucking company may be interested in finding the least expensive (minimum total cost in LP or minimum cost per unit of transported product in LFP) way of transporting each unit of large quantities of a product from a number of warehouses to a number of stores.

Assignment problems are related to transportation problems. A typical example of this type of problem is finding the best way to assign n applications to n jobs, given ratings of the applicants with respect to the different jobs.

This book will deal with the study of the types of problems described above. The emphasis will be on formulating the problem, mathematically analyzing and finally solving it, and then interpreting the solution. Some special advanced topics of LFP will be considered too.

The main computational technique in linear-fractional programming is the simplex method developed by George B. Dantzig in the 1940's for solving linear programming problems and later, in 1960 upgraded by Béla Martos for solving LFP problems.

This book is completely self-contained, with all the necessary mathematical background given in Chapter 2. Readers who are familiar with linear algebra may omit this chapter. Knowledge of LP is desirable but not necessary.

ERIK B.BAJALINOV

Acknowledgments

Preliminary versions of some parts of the book were included about two years ago in my previous book written in cooperation with Balázs Imreh (Szeged University, Hungary) and published in Hungary in 2001. The author is grateful to many students and colleagues from the Hungarian Operations Research community for their encouragement and useful comments and criticism.

My special thanks are to:

Pál Dömösi (Department of Computer Science, Institute of Mathematics and Informatics, Debrecen University, Hungary) for friendly support and wisdom advices;

my colleagues Katalin Bognár, Zoltán Papp, Attila Pethõ, Magda Várterész for their warmest support and administrative assistance;

Jacek Gondzio and Julian Hall (Department of Mathematics and Statistics, Edinburgh University, Scotland) for assistance and support during my visit to Edinburgh Parallel Computing Centre (EPCC, Edinburgh University, Scotland);

my students Tamás Barta, Ádám Benedek, Csaba Kertész, József Kiss, and others for their assistance in developing and debugging software tools necessary to check numerous numeric examples included in the book;

my teachers and mentors Juriy P. Chernov (former State Planning Committee of USSR, Russia) and Jozef V. Romanovsky (Department of Operations Research, State University of Sent-Petersburg, Russia).

Finally, my thanks are also due to the staff of Kluwer Academic Publishers for their interest in my book, encouragement, and cooperation.

Chapter 1

INTRODUCTION

1. Subject of the book

This book deals with linear-fractional programming (LFP). The object of LFP is to find the optimal (maximal or minimal) value of a *linear-fractional objective function* subject to *linear constraints* on the given variables. If all unknown variables are real valued then we say that the problem is *real* or *continuous*. In the case of one or more integer-valued variables we usually say that the problem is *integer* or *IP*. The IP problem may be *pure*, if all the variables must have in optimal solution an integer value, or *mixed* in the other case.

The constraints in the problem may be either equality or inequality constraints[1]. From the point of view of real-world applications, LFP possesses as many nice and extremely useful features, as *linear programming* (LP). If we have a problem formulated as an LP one, we can re-formulate this problem as LFP by replacing an original linear objective function with a ratio (fraction) of two linear functions. If in the original LP problem the objective function expresses, for example, the profit of some company, in the case of the LFP problem we can optimize the activity of the company in accordance with such fractional criteria as *profit/cost* or *profit/manpower requirement*, and so on. Moreover, from the point of view of applications such an optimal solution is often more preferable and attractive than obtained from the LP problem because of higher efficiency.

[1]The more general optimization problem arising when the objective function and/or the constraints contain non-linear expressions is beyond the scope of this book.

Problems of LFP arise when there appears a necessity to optimize the efficiency of some activity: profit gained by company per unit of expenditure of labor, cost of production per unit of produced goods, nutritiousness of ration per unit of cost, etc. Nowadays because of a deficit of natural resources the use of such specific criteria becomes more and more topical and relevant. So an application of LFP to solving real-world problems connected with optimizing efficiency could be as useful as in the case of LP. The only problem is that until now there has been no well-made software package developed especially for using LFP and teaching it. The matter might be explained by the following two reasons.

First, in 1962 A.Charnes and W.W.Cooper [38] showed that by a simple transformation the original LFP problem can be reduced to an LP problem that can therefore be solved using a regular simplex method for linear programming. It was found that this approach is very useful for mathematicians because most theoretical results developed in LP could be relatively easily expanded to include LFP problems. But from the point of view of users, this approach is by far the best because this transformation increases the number of constraints and variables and leads to changes in the structure of the original constraints. So if we want to find a production plan that optimizes the specific profit of a company, we must be very familiar with the technique of the transformation. But even if we have performed this transformation and got an LP problem, we are often unable to use our special software tools and methods, for example in the case of the transportation problem, because of changes in the structure of constraints and the objective function.

The second reason is that in the English-language special literature the discussions of a possible economic interpretation of dual variables in LFP had been completed with incorrect interpretation [109], criticism and negative results [122]. The only constructive result was given by J.S.H.Kornbluth and G.R.Salkin [115] in terms of derivatives and quite complicated formulas without trying at all to explain the results obtained in terms understandable for non-mathematicians. Later, in [9], [10], [11], [12], [13] the economic interpretation of dual variables in LFP was explained in economic terms and possible ways to use the obtained results in applications were shown there.

So it may be very useful to be able to solve a linear-fractional programming problem and utilize the information obtained with the optimal solution. Our interest will be mainly in the basic theory of LFP, the simplex technique[2], duality theory in LFP, and special problems of LFP.

[2]However, we will not deal in detail with such recent directions of research in LFP as the applicability of *Interior Point Methods* (IPM) in LFP. Some short overview of advanced methods will be given in Chapter 10. Such highly useful research arose after the seminal paper of N.Karmarkar [108] in 1984.

From the practical point of view it is necessary not only to know how the LFP problem may be solved, but to have such software tools which allow to solve it in reasonable time, and then perform *Sensitivity* or *Post-optimal Analysis.* This is why we include in this book relatively detailed information on *WinGULF*[3], a programming package for linear and linear-fractional programming developed by the author.

The aim of the present book is to describe the foundations of LFP and to provide readers with the basic knowledge necessary to solve LFP problems and utilize the optimal solution obtained.

2. Description of the content

As stated in the previous section, we intend to present an LFP approach to optimization, algorithms of LFP, duality, post-optimal analysis and software tools.

Chapter 2 is a self-contained treatment of the basics of linear algebra. It provides the main building blocks that will be needed in the rest of the book.

In Chapter 3 we begin our study of linear-fractional programming by describing the problem and its possible applications. We learn here how to solve graphically those LFP problems that involve only two variables. In the rest of this chapter we deal with the so-called *Charnes and Cooper transformation* and *Dinkelbach's algorithm.* In addition, several LFP models of real-world problems are briefly described at the end of the chapter.

The main computational technique used to solve LFP problems is the *Simplex Method.* It is presented in Chapter 4. After a detailed consideration of the theoretical background of the method, and an introduction to such special tools as the *Big M method* and the *Two-Phase Simplex Method,* we close this chapter by considering degeneracy and variables with special forms of restrictions.

In Chapter 5, "Duality Theory", we present the main theoretical and practical aspects of duality in linear-fractional programming. The natural approach to construct the dual problem for LFP is to transform the original fractional problem into a non-fractional form and then to formulate the dual of the latter using the classical way. We show in Section 1 that this approach is not of practical interest. After a brief description of different approaches to the duality in LFP, we concentrate on the approach based on the fractional *Lagrangian* introduced by E.G.Gol'stein (Section 2). In Section 3 we formulate and prove the main statements of duality theory in LFP. The rest of the chapter deals with the prac-

[3]WinGULF is a registered trademark of Optimum 95 Bt.

tical aspects of duality and the comparative analysis of dual variables in LP and LFP.

In Chapter 6 we discuss the Sensitivity (or Post-optimal) Analysis. We study here how an optimal solution or, in general, the output of a LFP model changes with (usually) small changes in the given data. In LFP, as well as in LP, a sensitivity is a basic part of the problem solution. In different sections of this chapter we show how much the coefficients of the objective function or right-hand side elements can vary before the optimal basis is either no longer optimal or feasible.

Chapter 7 deals with the interconnection between problems of LFP and LP, and their dual variables.

Topics connected with integer LFP are covered in Chapter 8. Here we formulate several practical LFP problems with integer variables and discuss methods used to solve integer LFP problems.

Other special LFP problems, such as the transportation problem and the assignment problem, are studied in Chapter 9. We formulate these problems and discuss some special methods, which allow to solve these special cases of LFP.

Topics connected with advanced methods and algorithms in LFP are covered in Chapter 10. Here we study such special modifications of the simplex method as *The Dual Simplex Method* (Section 1), *The Criss-Cross Method* (Section 2), and give a brief overview of new techniques and recent theoretical developments.

Some special extensions and generalizations of LFP are covered in Chapter 11, "Advanced Topics in LFP".

In Chapter 12 we discuss the computational issues of linear-fractional programming. Here we consider special techniques used to solve real-world large-scale LFP problems.

Using the program package WinGULF to solve LFP problems is discussed in Chapter 13.

The common thread through the various parts of the book will be the prominent role of linear-fractional programming as a generalization of LP - everywhere, if it is reasonable, we will show how the given LFP statement relates to linear programming.

It may be worthwhile devoting some words to the positioning of footnotes and exercises in this book. The footnotes are used to related references, or to make a small digression from the main thrust of reasoning. So we preferred to place the footnotes not at the end of each chapter (section) but at the bottom of the page they refer to. The exercises are grouped by chapters, not by sections, and are given at the end of each chapter.

3. What is new in this book?

The book offers an approach to LFP and to duality in LFP that is new in many aspects. First, we use in the book a "direct" approach to LFP problems, that is we consider the original LFP problem as it is, without reducing it to an LP problem. Moreover, we consider LFP as a generalization of linear programming and so most of the results are formulated in such a way that appropriate results of LP may be obtained as a special case of LFP. On the other hand, this approach allows us to compare dual variables in LP and LFP and describe the relation between them. In this respect important (and new) application possibilities of duality are indicated in Chapter 5, Chapter 7 and Chapter 11. Finally, we present a new software tool, WinGULF, developed especially for linear-fractional programming and so utilizing advances of LFP theory. The special "Student Edition" version of the package may be freely downloaded[4] from the authors's Web-page: *http:\\www.math.klte.hu\ ~bajalinov*.

4. Required knowledge and skills

We wanted to write such a book that presents the most important and prominent results on LFP in a unified and comprehensive way, with a full development of the most important items.

Chapter 2 can be considered as an elementary introduction to linear algebra, containing the most basically necessary definitions and results. Some basic knowledge of differential calculus is required in Chapters 3 and 10.

The mathematical tools we used in the book do not go beyond standard calculus and linear algebra. People educated in the Simplex method and Duality theory in linear programming will easily understand the material. So we expect that the people will quickly get acquainted with the formalism and mathematical manipulations, just as many generations of students have become familiar with linear programming.

In general, the level of the book will be accessible to any student (reader) with 2 to 3 years of basic training in calculus and linear algebra.

[4]If you have any questions, remarks, suggestions, bug reports, please feel free to contact me. I would appreciate it if you sent me your comments about this software. Be sure to check my Web-pages for updates. My e-mail: Bajalinov@math.klte.hu

To use the software tools described in the book the reader must be familiar with the basics of working with the operating systems Microsoft Windows 9*, NT, ME, 2000 or XP[5] and have the necessary skills to install and run the Windows application WinGULF[6].

5. How to use the book for courses

Recognizing the importance of the use of ratio-type criteria of optimality in real-world applications we may expect LFP soon to become more a popular topic in Operations Research and other related fields where methods of optimization are used, such as Business, Economics, and Engineering. This is why when writing this book one of our goals was to collect such relevant materials from research papers that cover the most important topics of the theory, methods and applications of LFP. Our second aim was to introduce those software tools that allow to solve a linear-fractional programming problem.

The author hopes and believes that different sections of the book may be used in the frame of a basic course of Operations Research, or may be considered as the basis for a special course of Linear-Fractional Programming.

Of course, a well-motivated reader can also use this book for self-study to learn LFP. Operations researchers and people who use optimization in their professional work can use this book as a reference for details, or for developing real-world LFP models, or even for developing improved methods, algorithms and special techniques.

[5]Windows 95, 98, NT, ME, 2000, and XP are registered trademarks of Microsoft Inc.

[6]At the moment there is no Mac version of the WinGULF package. Special versions for high performance computing including parallel environments for Linux/Unix/Solaris are available from the author

Chapter 2

BASIC LINEAR ALGEBRA

In this chapter, we begin by giving some familiar definitions for the sake of completeness and to refresh readers' memory. We survey the topics of linear algebra that will be needed in the rest of the book. First, we discuss the building blocks of linear algebra: vectors, matrices, linear dependence and independence, determinants, etc. We continue the chapter with an introduction to inverse of matrix, then we use our knowledge of matrices and vectors to develop a systematic procedure (Gaussian elimination method) for solving linear equations, which we then use to invert matrices. Finally, we close the chapter with a short description of the Gauss-Jordan method for solving systems of linear equations.

The material covered in this chapter will be used in our study of linear-fractional programming.

1. Matrices and their Properties

DEFINITION 2.1 *A **matrix** is any rectangular array of numbers. The general form of a matrix with m rows and n columns is*

$$A = \begin{pmatrix} a_{11} & a_{12} & \dots & a_{1n} \\ a_{21} & a_{22} & \dots & a_{2n} \\ \vdots & \vdots & \vdots & \vdots \\ a_{m1} & a_{m2} & \dots & a_{mn} \end{pmatrix}$$

For example,

$$A = \begin{pmatrix} 12 & 10 \\ 21 & 2 \end{pmatrix}, \quad B = \begin{pmatrix} 3 & 0 & 13 \\ 2 & 6 & 3 \end{pmatrix}, \quad C = \begin{pmatrix} 7 & 16 \\ 1 & 32 \\ 14 & 8 \end{pmatrix}$$

are all matrices.

If a matrix A has m rows and n columns, we call A an $m \times n$ matrix.

We will usually denote matrices by uppercase letters (A, B, etc.). Lowercase letters, such as a, b, c, etc., denote real numbers. Another name for a real number (frequently used during discussions of matrices and vectors) is**scalar**.

The numbers in the array, such as $a_{11}, a_{12}, \ldots, a_{mn}$, are called the **elements** of the matrix. The double subscript is used to denote the location of an element in a matrix. The first subscript gives the row and the second one gives the column in which the element is located. So subscripts $a_{i,j}$ are always in row i, column j order. Often the comma is omitted, so notation a_{11} usually means "$a_{\mathit{one\ one}}$" not "a_{eleven}". For example, a_{34} refers to the element in row 3 and column 4.

DEFINITION 2.2 *The element in the ith row and jth column of matrix A is called the ij-**th element** of A and is written a_{ij}.*

For example, if

$$A = \begin{pmatrix} 11 & 12 & 14 \\ 21 & 22 & 24 \\ 31 & 32 & 34 \end{pmatrix},$$

then $a_{11} = 11$, $a_{22} = 22$, and $a_{31} = 31$.

Sometimes we will use the notation $A = \|a_{ij}\|_{m \times n}$ to indicate that A is the matrix which consists of m rows and n columns, and whose ijth element is a_{ij}.

DEFINITION 2.3 *Two matrices A and B are said to have the same **shape** or **order** if they have the same respective number of rows and columns.*

DEFINITION 2.4 *Two matrices A and B are **equal** if they have the same shape and if the corresponding elements are equal.*

For example, if

$$A = \begin{pmatrix} 1 & 2 \\ 3 & 4 \end{pmatrix}, \quad B = \begin{pmatrix} x & y \\ w & z \end{pmatrix},$$

then $A = B$ if and only if $x = 1$, $y = 2$, $w = 3$, and $z = 4$.

DEFINITION 2.5 *A matrix with the same number of rows and columns is called a* ***square matrix****. If a square matrix has m rows and m columns, it is said to be of order m.*

The square matrix having nonzero entries along the **main diagonal** (the diagonal running from the upper left corner to the lower right corner) and zeros elsewhere is called the **diagonal matrix** and is denoted by D. For example,

$$D = \begin{pmatrix} 3 & 0 & 0 \\ 0 & 11 & 0 \\ 0 & 0 & -8 \end{pmatrix}.$$

The square matrix having ones along the main diagonal and zeros elsewhere is called the **identity matrix** or **unit matrix** and is denoted by I or I_n. For example, the third-order unit matrix is

$$I_3 = \begin{pmatrix} 1 & 0 & 0 \\ 0 & 1 & 0 \\ 0 & 0 & 1 \end{pmatrix}.$$

The notation E (an abbreviation for the German term, "*Einheitsmatrix*") is sometimes also used. The standard notation for the jth column of identity matrix I_n is e_j, so $I_n = (e_1, e_2, \ldots, e_n)$. In the case of I_3 we have:

$$e_1 = \begin{pmatrix} 1 \\ 0 \\ 0 \end{pmatrix}, \quad e_2 = \begin{pmatrix} 0 \\ 1 \\ 0 \end{pmatrix}, \quad e_3 = \begin{pmatrix} 0 \\ 0 \\ 1 \end{pmatrix}.$$

DEFINITION 2.6 *A* ***permutation matrix*** *is formed by reordering the columns of identity matrix I. The standard notation of permutation matrices is P.*

For example,

$$P = (e_3, e_1, e_2) = \begin{pmatrix} 0 & 1 & 0 \\ 0 & 0 & 1 \\ 1 & 0 & 0 \end{pmatrix}.$$

These permutation matrices are usually used to interchange rows (multiplying from the left, or pre-multiplying) or columns (multiplying from the right, or post-multiplying) of a matrix. For example, if

$$A = \begin{pmatrix} 11 & 12 & 14 \\ 21 & 22 & 24 \\ 31 & 32 & 34 \end{pmatrix} \quad \text{and} \quad P = \begin{pmatrix} 0 & 1 & 0 \\ 0 & 0 & 1 \\ 1 & 0 & 0 \end{pmatrix}$$

then

$$PA = \begin{pmatrix} 0 & 1 & 0 \\ 0 & 0 & 1 \\ 1 & 0 & 0 \end{pmatrix} \begin{pmatrix} 11 & 12 & 14 \\ 21 & 22 & 24 \\ 31 & 32 & 34 \end{pmatrix} = \begin{pmatrix} 21 & 22 & 24 \\ 31 & 32 & 34 \\ 11 & 12 & 14 \end{pmatrix}$$

and

$$AP = \begin{pmatrix} 11 & 12 & 14 \\ 21 & 22 & 24 \\ 31 & 32 & 34 \end{pmatrix} \begin{pmatrix} 0 & 1 & 0 \\ 0 & 0 & 1 \\ 1 & 0 & 0 \end{pmatrix} = \begin{pmatrix} 14 & 11 & 12 \\ 24 & 21 & 22 \\ 34 & 31 & 32 \end{pmatrix}.$$

Note that permutation matrices have the following very surprising and useful property $P^T = P$.

DEFINITION 2.7 *An $m \times m$ matrix A is said to be an* ***upper triangular matrix*** *if $a_{ij} = 0, \ \forall i > j, \ j = 1, 2, \dots, m$.*

DEFINITION 2.8 *An $m \times m$ matrix A is said to be a* ***lower triangular matrix*** *if $a_{ij} = 0, \ \forall i < j, \ j = 1, 2, \dots, m$.*

The standard notation for an upper triangular matrix is U and for a lower triangular matrix is L. For example,

$$U = \begin{pmatrix} 1 & 1 & 0 & 2 \\ 0 & 5 & 1 & 6 \\ 0 & 0 & 7 & 1 \\ 0 & 0 & 0 & 3 \end{pmatrix}$$ is an upper triangular matrix.

$$L = \begin{pmatrix} 1 & 0 & 0 & 0 \\ 3 & 5 & 0 & 0 \\ 1 & 4 & 7 & 0 \\ 8 & 2 & 4 & 3 \end{pmatrix}$$ is a lower triangular matrix.

DEFINITION 2.9 *An upper (lower) triangular matrix is said to be a* ***unit upper (lower) triangular matrix*** *if it has ones on the main diagonal.*

Examples:

$$U = \begin{pmatrix} 1 & 1 & 0 & 2 \\ 0 & 1 & 1 & 6 \\ 0 & 0 & 1 & 1 \\ 0 & 0 & 0 & 1 \end{pmatrix}$$ is a unit upper triangular matrix.

$$L = \begin{pmatrix} 1 & 0 & 0 & 0 \\ 3 & 1 & 0 & 0 \\ 1 & 4 & 1 & 0 \\ 8 & 2 & 4 & 1 \end{pmatrix} \quad \text{is a unit lower triangular matrix.}$$

DEFINITION 2.10 *The* ***zero matrix*** *is any $m \times n$ matrix with entries all 0, i.e.* $a_{ij} = 0,\ i = 1, 2, \ldots, m;\ j = 1, 2, \ldots, n.$

For example,

$$0 = \begin{pmatrix} 0 & 0 & 0 & 0 \\ 0 & 0 & 0 & 0 \\ 0 & 0 & 0 & 0 \end{pmatrix} \quad \text{is a zero matrix.}$$

Now we describe the operations on matrices that are used later in this book.

The scalar multiple of a matrix. Given any matrix A and any scalar k, matrix kA is obtained from matrix A by multiplying each element of A by k. For example,

$$\text{if } A = \begin{pmatrix} -1 & 2 \\ 6 & 0 \end{pmatrix}, \text{ then } 5A = \begin{pmatrix} -5 & 10 \\ 30 & 0 \end{pmatrix}.$$

Note that for any k A and kA have the same order. For $k = -1$, scalar multiplication of the matrix A is sometimes written as $-A$.

Addition of two matrices. First of all, addition of matrices A and B is defined only if A and B have the same order (say, $m \times n$). Let $A = \|a_{ij}\|_{m \times n}$ and $B = \|b_{ij}\|_{m \times n}$ be the given matrices. Then the matrix $C = A + B$ is defined to be the $m \times n$ matrix whose ijth element c_{ij} is $a_{ij} + b_{ij}$. Thus, to obtain the **sum** of two matrices A and B, we add the corresponding elements of A and B. For example, if

$$A = \begin{pmatrix} 12 & 10 \\ 21 & 2 \end{pmatrix}, \quad B = \begin{pmatrix} 3 & 13 \\ 2 & 3 \end{pmatrix},$$

then

$$A + B = \begin{pmatrix} 12+3 & 10+13 \\ 21+2 & 2+3 \end{pmatrix} = \begin{pmatrix} 15 & 23 \\ 23 & 5 \end{pmatrix}.$$

The transpose of a matrix. Given any $m \times n$ matrix

$$A = \begin{pmatrix} a_{11} & a_{12} & \ldots & a_{1n} \\ a_{21} & a_{22} & \ldots & a_{2n} \\ \vdots & \vdots & \vdots & \vdots \\ a_{m1} & a_{m2} & \ldots & a_{mn} \end{pmatrix},$$

the **transpose** of A (written A^T) is the $n \times m$ matrix

$$A^T = \begin{pmatrix} a_{11} & a_{21} & \dots & a_{m1} \\ a_{12} & a_{22} & \dots & a_{m2} \\ \vdots & \vdots & \vdots & \vdots \\ a_{1n} & a_{2n} & \dots & a_{mn} \end{pmatrix}.$$

Thus, A^T is obtained from A by letting row 1 of A be column 1 of A^T, letting row 2 of A be column 2 of A^T, and so on. For example,

$$\text{if } A = \begin{pmatrix} 1 & 2 & 3 \\ 4 & 5 & 6 \end{pmatrix}, \text{ then } A^T = \begin{pmatrix} 1 & 4 \\ 2 & 5 \\ 3 & 6 \end{pmatrix}.$$

The transpose operation on matrices obeys the following rules:

$$\begin{array}{lll} a. & (A+B)^T & = & A^T + B^T \\ b. & (A^T)^T & = & A \\ c. & (kA)^T & = & kA^T \\ d. & (AB)^T & = & B^T A^T. \end{array}$$

Note: Property (d), which is fairly important, states that transpose of a product equals the product of transposes, but *in the opposite order.*

DEFINITION 2.11 *Matrix A is* ***symmetric*** *if $A = A^T$.*

For example, if

$$A = \begin{pmatrix} 1 & 2 & 5 \\ 2 & 4 & -8 \\ 5 & -8 & 14 \end{pmatrix}, \quad \text{then} \quad A^T = \begin{pmatrix} 1 & 2 & 5 \\ 2 & 4 & -8 \\ 5 & -8 & 14 \end{pmatrix}.$$

So matrix A is a symmetric one.

DEFINITION 2.12 *A square matrix A is said to be* ***orthogonal*** *if $AA^T = I$.*

For example,

$$A = \frac{1}{\sqrt{2}} \begin{pmatrix} 1 & 1 \\ 1 & -1 \end{pmatrix}, \quad \text{and} \quad B = \frac{1}{3} \begin{pmatrix} 2 & -2 & 1 \\ 1 & 2 & 2 \\ 2 & 1 & -2 \end{pmatrix}$$

are orthogonal matrices.

Matrix multiplication. First of all, the **matrix product** of two matrices A and B (written AB) is defined if and only if the number of columns in A is equal to the number of rows in B. Let $A = ||a_{ir}||_{m \times p}$ and $B = ||b_{rj}||_{p \times n}$ be two matrices to multiply. Then AB (read as A *times* B) is a matrix $C = ||c_{ij}||_{m \times n}$ that has the same number of rows as A and the same number of columns as B. The rule for computing a typical element c_{ij} of matrix C is as follows :

$$c_{ij} = a_{i1}b_{1j} + a_{i2}b_{2j} + \cdots + a_{ip}b_{pj} = \sum_{r=1}^{p} a_{ir}b_{rj}.$$

There are p terms in this sum. For example, if

$$A = \begin{pmatrix} 0 & 1 \\ 2 & 3 \\ 4 & 5 \end{pmatrix}, \text{ and } B = \begin{pmatrix} 6 & 7 \\ 8 & 9 \end{pmatrix},$$

then the matrix product $C = ||c_{ij}||_{3 \times 2}$ of these two matrix A and B is as follows :

$$C = \begin{pmatrix} 8 & 9 \\ 36 & 41 \\ 64 & 73 \end{pmatrix},$$

because

$$\begin{aligned}
c_{11} &= a_{11}b_{11} + a_{12}b_{21} &= 0 \times 6 + 1 \times 8 &= 8\,, \\
c_{12} &= a_{11}b_{12} + a_{12}b_{22} &= 0 \times 7 + 1 \times 9 &= 9\,, \\
c_{21} &= a_{21}b_{11} + a_{22}b_{21} &= 2 \times 6 + 3 \times 8 &= 36\,, \\
c_{22} &= a_{21}b_{12} + a_{22}b_{22} &= 2 \times 7 + 3 \times 9 &= 41\,, \\
c_{31} &= a_{31}b_{11} + a_{32}b_{21} &= 4 \times 6 + 5 \times 8 &= 64\,, \\
c_{32} &= a_{31}b_{12} + a_{32}b_{22} &= 4 \times 7 + 5 \times 9 &= 73\,.
\end{aligned}$$

Algorithmically, if $A = ||a_{ir}||_{m \times p}$ and $B = ||b_{rj}||_{p \times n}$ the multiplication of matrices A and B may be implemented as shown in Figure 2.1.

Note that for given two matrices A and B matrix product BA is not defined.

To close this section, we discuss some important properties of matrix multiplication. In what follows, we assume that all matrix products are defined.

1 Matrix multiplication is **associative**. That is, $A(BC) = (AB)C$. To illustrate, let

$$A = \begin{pmatrix} 1 & 2 \end{pmatrix}, \quad B = \begin{pmatrix} 2 & 3 \\ 4 & 5 \end{pmatrix}, \quad c = \begin{pmatrix} 2 \\ 1 \end{pmatrix}.$$

Matrix-Matrix Multiplication

```
For i := 1 To m Do                      {Loop over rows}
   For j := 1 To n Do                   {Loop over columns}
      Begin
      c_ij := 0;
      For k := 1 To p Do
         c_ij := c_ij + a_ik * b_kj;
      End
```

Figure 2.1. Algorithm – Matrix-matrix multiplication.

Then $AB = (10 \ \ 13)$ and $(AB)C = 10 \times 2 + 13 \times 1 = (33)$. On the other hand,

$$BC = \begin{pmatrix} 7 \\ 13 \end{pmatrix}$$

so $A(BC) = 1 \times 7 + 2 \times 13 = (33)$. In this case, $A(BC) = (AB)C$ does hold.

2 Matrix multiplication is **distributive**. That is, $A(B+C) = AB + AC$ and $(B+C)A = BA + CA$.

3 The product of two lower (upper) triangular matrices is also lower (upper) triangular.

2. Vectors and their Properties

Any matrix with only one column (that is, any $m \times 1$ matrix) may be thought of as a *column vector*. The number of rows in a column vector is the *dimension* of the column vector. Thus,

$$\begin{pmatrix} 1 \\ 2 \\ 3 \end{pmatrix}$$

may be thought of as a 3×1 matrix or a three-dimensional column vector.

In analogous fashion, we can think of any matrix with only one row (that is, any $1 \times n$ matrix) as a *row vector*. The dimension of a row vector is the number of columns in the vector. Thus, row vector $(1, \ 2, \ 3)$ may be viewed as a 1×3 matrix or a three-dimensional row vector. In this book, vectors appear in boldface type : for instance, vectors **a**, **b**, and **c**. An n-dimensional vector

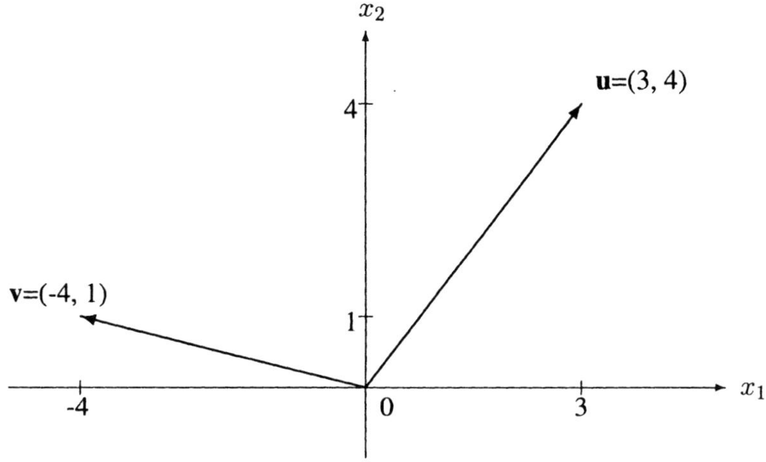

Figure 2.2. Vectors – As directed line segments.

(either column or row) in which all elements equal zero is called a *zero vector* (written 0). Thus,

$$\begin{pmatrix} 0, & 0 \end{pmatrix} \text{ and } \begin{pmatrix} 0 \\ 0 \end{pmatrix}$$

are two-dimensional zero vectors.

It is convenient to give vectors a geometrical interpretation. This is particularly simple if we are dealing with two- or three-dimensional spaces. For example, in the two-dimensional plane, the vector $\mathbf{u} = (3,\ 4)$ corresponds to the line segment joining the **origin** (i.e. point (0, 0)) to the point (3, 4). Thus, we associate the point (3, 4) with the vector $\mathbf{u} = (3,\ 4)$. Therefore, the vector **u** also determines a direction. The directed line segments corresponding to vectors

$$\mathbf{u} = \begin{pmatrix} 3, & 4 \end{pmatrix}, \quad \mathbf{v} = \begin{pmatrix} -4 \\ 1 \end{pmatrix}$$

are drawn in Figure 2.2. The situation is similar in three-dimensional space. For example, the three-component vector $\mathbf{b} = (b_1, b_2, b_3)$ can be represented as a point in three-dimensional space whose x_1, x_2, and x_3 coordinates are b_1, b_2, and b_3 respectively. The X_1, X_2, and X_3 axes are mutually perpendicular. Here, too, we can again think of the vector **b** as a directed line segment from the origin to the point that characterizes the vector.

Analogously, any m-dimensional vector corresponds to a point and a directed line segment in the m-dimensional space.

It should be noted that there is no geometric distinction between column and row vectors, i.e. they are geometrically equivalent.

Four operations pertaining to vectors are of importance. Three of them – transpose, addition (subtraction), and scalar multiplication – are defined in the same way as for general matrices. The fourth operation is forming the*scalar product.* To define the scalar product of two vectors, suppose we have a row vector $\mathbf{a} = (a_1, a_2, \ldots, a_n)$ and a column vector

$$\mathbf{b} = \begin{pmatrix} b_1 \\ b_2 \\ \vdots \\ b_n \end{pmatrix}$$

of the same dimension. The scalar product of $\mathbf{a}$ and $\mathbf{b}$ (written $\mathbf{a} \cdot \mathbf{b}$) is the number

$$a_1 b_1 + a_2 b_2 + \ldots + a_n b_n = \sum_{j=1}^{n} a_j b_j.$$

For the scalar product of two vectors to be defined, the first vector must be a row vector, the second one must be a column, and the both must be of the same dimension. For example, if

$$\mathbf{u} = (\ 1,\ \ 2,\ \ 3\ ) \quad \text{and} \quad \mathbf{v} = \begin{pmatrix} 2 \\ 1 \\ 2 \end{pmatrix},$$

then $\mathbf{u} \cdot \mathbf{v} = 1 \times 2 + 2 \times 1 + 3 \times 2 = 10$. By these rules for computing a scalar product, if

$$\mathbf{u} = (\ 1,\ \ 2,\ \ 3\ ) \quad \text{and} \quad \mathbf{v} = \begin{pmatrix} 1 \\ 2 \end{pmatrix},$$

then $\mathbf{u} \cdot \mathbf{v}$ is not defined. Also, if

$$\mathbf{u} = \begin{pmatrix} 1 \\ 2 \\ 3 \end{pmatrix} \quad \text{and} \quad \mathbf{v} = (\ 4,\ \ 5,\ \ 6\ ),$$

then $\mathbf{u} \cdot \mathbf{v}$ is not defined, but the following scalar products are correct : $\mathbf{u}^T \cdot \mathbf{v}^T$ and $\mathbf{v} \cdot \mathbf{u}$.

Such manipulations involving vectors with many components lead to the abstract concept known as n-dimensional Euclidean space. This space consists of all n-dimensional vectors and will be denoted by R^n.

3. Linear Independence and Dependence

In this section we study the concepts of a linearly independent set of vectors, a linearly dependent set of vectors, and the rank of matrix. Consider vectors $A_1, A_2, \ldots, A_n$, where $A_j = (a_{1j}, a_{2j}, \ldots, a_{mj})^T, j = 1, 2, \ldots, n.$

DEFINITION 2.13 *A* **linear combination** *of the vectors* $A_1, A_2, \ldots, A_n$ *is any vector of the form* $\lambda_1 A_1 + \lambda_2 A_2 + \ldots + \lambda_n A_n$, *where* $\lambda_j,\ j = 1, 2, \ldots, n$, *are arbitrary scalars.*

DEFINITION 2.14 *Vectors* $A_1, A_2, \ldots, A_n$ *are* **linearly dependent** *if there is at least one vector* $\lambda = (\lambda_1, \lambda_2, \ldots, \lambda_n)$ *such that* $\lambda_1 A_1 + \lambda_2 A_2 + \ldots + \lambda_n A_n = 0$, *and at least one* $\lambda_j \neq 0,\ 1 \leq j \leq n$. *In other cases we say that vectors* $A_1, A_2, \ldots, A_n$ *are* **linearly independent**

For example, vectors

$$A_1 = \begin{pmatrix} 0 \\ 0 \end{pmatrix}, \quad A_2 = \begin{pmatrix} 1 \\ 0 \end{pmatrix}, \quad A_3 = \begin{pmatrix} 0 \\ 1 \end{pmatrix}$$

are linearly dependent because

$$\lambda_1 \begin{pmatrix} 0 \\ 0 \end{pmatrix} + \lambda_2 \begin{pmatrix} 1 \\ 0 \end{pmatrix} + \lambda_3 \begin{pmatrix} 0 \\ 1 \end{pmatrix} = 0$$

for any $\lambda_1 \neq 0,\ \lambda_2 = 0,\ \lambda_3 = 0.$ At the same time vectors A_2 and A_3 are linearly independent because $\lambda_2 A_2 + \lambda_3 A_3 = 0$ only if $\lambda_2 = \lambda_3 = 0.$

Let A be any $m \times n$ matrix, and A_j denotes its j-th column, $j = 1, 2, \ldots, n.$

DEFINITION 2.15 *The* **rank** *of matrix A is the largest number of linearly independent vectors* A_j.

4. Determinants

Associated with any square matrix A is a number called the determinant of A, often abbreviated as $det(A)$ or $|A|$. Knowing how to compute the determinant of a square matrix will be useful in our study of linear-fractional programming.

Consider matrix $A = ||a_{ij}||_{n\times n}$. If $n = 1$, then $det(A) = a_{11}$. For $n = 2$ we have $det(A) = a_{11} \cdot a_{22} - a_{21} \cdot a_{12}$. Before computing the determinant for $n > 2$ we need to define the concept of minor of a matrix.

DEFINITION 2.16 *If A is an $n \times n$ matrix, then for any values of indexes i and j the ij-th* **minor** *of A (written A_{ij}) is the $(n-1) \times (n-1)$ submatrix of A obtained by deleting the i-th row and the j-th column of A.*

For example,

$$\text{If} \quad A = \begin{pmatrix} a_{11} & a_{12} & a_{13} \\ a_{21} & a_{22} & a_{23} \\ a_{31} & a_{32} & a_{33} \end{pmatrix}, \quad \text{then} \quad A_{32} = \begin{pmatrix} a_{11} & a_{13} \\ a_{21} & a_{23} \end{pmatrix}.$$

Now, to compute $det(A)$ for $n > 2$ we pick any value of index i, $1 \leq i \leq n$, and compute

$$det(A) = \sum_{j=1}^{n} (-1)^{(i+j)} \cdot a_{i,j} \cdot det(A_{ij}) \,. \tag{2.1}$$

Formula (2.1) reduces the computation of the determinant of $n \times n$ matrix to computations involving only $(n-1) \times (n-1)$ matrices. We should apply formula (2.1) until the determinant of A can be expressed in terms of 2×2 matrices.

For $n = 3$ we can use the following simplified formula:

$$\begin{aligned} det(A) &= \begin{vmatrix} a_{11} & a_{12} & a_{13} \\ a_{21} & a_{22} & a_{23} \\ a_{31} & a_{32} & a_{33} \end{vmatrix} = \\ &= a_{11} \cdot a_{22} \cdot a_{33} + a_{21} \cdot a_{32} \cdot a_{13} + a_{31} \cdot a_{12} \cdot a_{23} \; - \\ &- \; a_{31} \cdot a_{22} \cdot a_{13} - a_{11} \cdot a_{32} \cdot a_{23} - a_{33} \cdot a_{21} \cdot a_{12}. \end{aligned}$$

Finally, we list some useful facts about determinants:

1 If two rows of a matrix are identical, its determinant is zero;

2 If two rows of a matrix are linearly dependent, its determinant is zero;

3 Interchanging two rows (or columns) of a matrix changes the sign of its determinant;

4 The determinant of the transposed matrix is equal to the original determinant;

5 If each element of some row (or column) of a matrix is multiplied by a constant c, the determinant is multiplied by the same constant c.

6 If a matrix is a triangular one, that is all its elements above the main diagonal (or below it) are zero, the determinant of the matrix is the product of the elements on the main diagonal.

7 If each element of a row (or column) of a matrix is multiplied by a constant c and the results are added to another row (or column), the determinant is not changed.

8 If each element in the row (column) of the determinant is a sum of two summands, then the determinant expands into the sum of two determinants, where in the considered row (column) in the first of them there will be the first summands and in the second of them there will be the second summands, and all the remaining rows (columns) will be identical to those of the given original determinant. So, if

$$A = \begin{pmatrix} a_{11} & a_{12} & \dots & a_{1n} \\ \vdots & \vdots & \vdots & \vdots \\ a_{k1} + b_{k1} & a_{k2} + b_{k2} & \dots & a_{kn} + b_{kn} \\ \vdots & \vdots & \vdots & \vdots \\ a_{n1} & a_{n2} & \dots & a_{nn} \end{pmatrix}$$

then

$$det(A) = \begin{vmatrix} a_{11} & a_{12} & \dots & a_{1n} \\ \vdots & \vdots & \vdots & \vdots \\ a_{k1} & a_{k2} & \dots & a_{kn} \\ \vdots & \vdots & \vdots & \vdots \\ a_{n1} & a_{n2} & \dots & a_{nn} \end{vmatrix} + \begin{vmatrix} a_{11} & a_{12} & \dots & a_{1n} \\ \vdots & \vdots & \vdots & \vdots \\ b_{k1} & b_{k2} & \dots & b_{kn} \\ \vdots & \vdots & \vdots & \vdots \\ a_{n1} & a_{n2} & \dots & a_{nn} \end{vmatrix}.$$

In accordance with property 2, $det(A) = 0$, if rows (or columns) of the matrix A are linearly dependent. Using this property of determinants we can re-write the definition 2.15 of the rank of matrix A in the following way

DEFINITION 2.17 *The* **rank** *of matrix A is the largest order of its minor* A_{ij} *with non-zero value.*

5. The Inverse of Matrix

For a given square $(m \times m)$ matrix A we will say, that matrix B is **inverse** of A (written $B = A^{-1}$), if

$$BA = AB = I,$$

where I is an identity matrix.

DEFINITION 2.18 *An* $m \times m$ *matrix A is said to be* **singular** *if det(A)=0. In the other case, matrix A is called* **nonsingular**.

Some square matrices do not have inverses. Only nonsingular matrices do have inverses and may be called **invertible** matrices. For given nonsingular (invertible) matrix A the inverse matrix A^{-1} may be calculated by the following formula:

$$A^{-1} = \frac{1}{det(A)} \cdot \begin{pmatrix} \tilde{A}_{11} & \tilde{A}_{21} & \dots & \tilde{A}_{m1} \\ \tilde{A}_{12} & \tilde{A}_{22} & \dots & \tilde{A}_{m2} \\ \vdots & & & \vdots \\ \tilde{A}_{1m} & \tilde{A}_{2m} & \dots & \tilde{A}_{mm} \end{pmatrix}$$

where $\tilde{A}_{ij} = (-1)^{i+j} \cdot det(A_{ij})$ - is a **cofactor** of minor A_{ij}.

The most important property of inverse matrix is that

$$A^{-1} \cdot A = A \cdot A^{-1} = I.$$

For example, consider the following matrix

$$A = \begin{pmatrix} 2 & 0 & -1 \\ 3 & 1 & 2 \\ -1 & 0 & 1 \end{pmatrix}.$$

To determine its inverse we have to calculate its determinant:

$$\begin{aligned} det(A) &= \begin{vmatrix} 2 & 0 & -1 \\ 3 & 1 & 2 \\ -1 & 0 & 1 \end{vmatrix} = \\ &= 2 \cdot 1 \cdot 1 + (-1) \cdot 0 \cdot 3 + (-1) \cdot 0 \cdot 2 - \\ &- (-1) \cdot 1 \cdot (-1) - 2 \cdot 2 \cdot 0 - 3 \cdot 0 \cdot 1 = \\ &= 2 - 0 - 0 - 1 - 0 - 0 = 1, \end{aligned}$$

and then evaluate cofactors $\tilde{A}_{ij}$ $(i = 1, 2, 3;\ j = 1, 2, 3;)$:

$$\begin{aligned} \tilde{A}_{11} &= (-1)^{1+1} \cdot \begin{vmatrix} 1 & 2 \\ 0 & 1 \end{vmatrix} = (-1)^2 \cdot (1 \cdot 1 - 2 \cdot 0) = 1; \\ \tilde{A}_{12} &= (-1)^{1+2} \cdot \begin{vmatrix} 3 & 2 \\ -1 & 1 \end{vmatrix} = (-1)^3 \cdot (3 \cdot 1 - 2 \cdot (-1)) = -5; \\ \tilde{A}_{13} &= (-1)^{1+3} \cdot \begin{vmatrix} 3 & 1 \\ -1 & 0 \end{vmatrix} = (-1)^4 \cdot (3 \cdot 0 - 1 \cdot (-1)) = 1; \\ \tilde{A}_{21} &= (-1)^{2+1} \cdot \begin{vmatrix} 0 & -1 \\ 0 & 1 \end{vmatrix} = (-1)^3 \cdot (0 \cdot 1 - 0 \cdot (-1)) = 0; \\ \tilde{A}_{22} &= (-1)^{2+2} \cdot \begin{vmatrix} 2 & -1 \\ -1 & 1 \end{vmatrix} = (-1)^4 \cdot (2 \cdot 1 - (-1) \cdot (-1)) = 1; \\ \tilde{A}_{23} &= (-1)^{2+3} \cdot \begin{vmatrix} 2 & 0 \\ -1 & 0 \end{vmatrix} = (-1)^5 \cdot (2 \cdot 0 - (-1) \cdot 0) = 0; \end{aligned}$$

$$\tilde{A}_{31} = (-1)^{3+1} \cdot \begin{vmatrix} 0 & -1 \\ 1 & 2 \end{vmatrix} = (-1)^4 \cdot (0 \cdot 2 - 1 \cdot (-1)) = 1;$$

$$\tilde{A}_{32} = (-1)^{3+2} \cdot \begin{vmatrix} 2 & -1 \\ 3 & 2 \end{vmatrix} = (-1)^5 \cdot (2 \cdot 2 - 3 \cdot (-1)) = -7;$$

$$\tilde{A}_{33} = (-1)^{3+3} \cdot \begin{vmatrix} 2 & 0 \\ 3 & 1 \end{vmatrix} = (-1)^6 \cdot (2 \cdot 1 - 3 \cdot 0) = 2.$$

So,

$$A^{-1} = \frac{1}{det(A)} \cdot \begin{pmatrix} \tilde{A}_{11} & \tilde{A}_{21} & \tilde{A}_{31} \\ \tilde{A}_{12} & \tilde{A}_{22} & \tilde{A}_{32} \\ \tilde{A}_{13} & \tilde{A}_{23} & \tilde{A}_{33} \end{pmatrix} = \frac{1}{1} \cdot \begin{pmatrix} 1 & 0 & 1 \\ -5 & 1 & -7 \\ 1 & 0 & 2 \end{pmatrix}.$$

Now, we can verify that

$$A \cdot A^{-1} = \begin{pmatrix} 2 & 0 & -1 \\ 3 & 1 & 2 \\ -1 & 0 & 1 \end{pmatrix} \cdot \begin{pmatrix} 1 & 0 & 1 \\ -5 & 1 & -7 \\ 1 & 0 & 2 \end{pmatrix} = \begin{pmatrix} 1 & 0 & 0 \\ 0 & 1 & 0 \\ 0 & 0 & 1 \end{pmatrix} = I,$$

and

$$A^{-1} \cdot A = \begin{pmatrix} 1 & 0 & 1 \\ -5 & 1 & -7 \\ 1 & 0 & 2 \end{pmatrix} \cdot \begin{pmatrix} 2 & 0 & -1 \\ 3 & 1 & 2 \\ -1 & 0 & 1 \end{pmatrix} = \begin{pmatrix} 1 & 0 & 0 \\ 0 & 1 & 0 \\ 0 & 0 & 1 \end{pmatrix} = I.$$

Thus,

$$A^{-1} = \begin{pmatrix} 1 & 0 & 1 \\ -5 & 1 & -7 \\ 1 & 0 & 2 \end{pmatrix}.$$

Another important property of inverse matrix: for two given nonsingular square matrices A and B the following equality holds:

$$(A \cdot B)^{-1} = B^{-1} \cdot A^{-1}.$$

Note that for a permutation matrix P we have

$$P^{-1} = P^T.$$

We close our discussion by noting that determinants and invert square matrices may be used to solve linear equation systems.

6. Matrices and Systems of Linear Equations

Consider a system of linear equations given by system

$$\left.\begin{array}{ccccccccc} a_{11}x_1 & + & a_{12}x_2 & + & \dots & + & a_{1n}x_n & = & b_1 \\ a_{21}x_1 & + & a_{22}x_2 & + & \dots & + & a_{2n}x_n & = & b_2 \\ \vdots & & \vdots & & & & \vdots & & \vdots \\ a_{m1}x_1 & + & a_{m2}x_2 & + & \dots & + & a_{mn}x_n & = & b_m \end{array}\right\}. \tag{2.2}$$

Using matrix notation we can re-write system (2.2) in the following form

$$Ax = b, \tag{2.3}$$

where

$$A = \begin{pmatrix} a_{11} & a_{12} & \dots & a_{1n} \\ a_{21} & a_{22} & \dots & a_{2n} \\ \vdots & & & \vdots \\ a_{m1} & a_{m2} & \dots & a_{mn} \end{pmatrix}, \quad x = \begin{pmatrix} x_1 \\ x_2 \\ \vdots \\ x_n \end{pmatrix}, \quad b = \begin{pmatrix} b_1 \\ b_2 \\ \vdots \\ b_m \end{pmatrix}.$$

The **augmented form** of system (2.2) is

$$A\,|b = \left(\begin{array}{cccc|c} a_{11} & a_{12} & \dots & a_{1n} & b_1 \\ a_{21} & a_{22} & \dots & a_{2n} & b_2 \\ \vdots & & & \vdots & \vdots \\ a_{m1} & a_{m2} & \dots & a_{mn} & b_m \end{array}\right)$$

In this system (2.2), $x_1, x_2, \dots, x_n$ are referred to as **variables** or **unknowns**, and the a_{ij}'s and b_i's are **constants**. A set of equations like (2.2) or (2.3) is called **a linear system of** m **equations in** n **variables**.

DEFINITION 2.19 *A* **solution** *to a linear system of m equations in n variables is a set of values for the unknowns that satisfies each of the system's m equations.*

We denote a possible solution to (2.2) by n-dimensional column vector **x**, in which the jth element of **x** is the value of x_j.

Suppose that matrix A in the system (2.3) is a square matrix of size $n \times n$, column-vector $b = (b_1, b_2, \dots, b_n)^T$ and inverse matrix A^{-1} exists, then the solution of the system (2.3) may be expressed by inverse A^{-1}. Indeed, multiplying both sides of (2.3) by A^{-1}, and using the associative law and the definition of a matrix inverse, we obtain

$$(A^{-1}\,A)x \;=\; A^{-1}\,b$$

or

$$Ix = A^{-1}\,b$$

or

$$x = A^{-1}\,b$$

This shows that knowing inverse A^{-1} enables us to find the unique solution to a square linear system.

The general strategy for solving a linear system (2.3) suggests that we should transform the original system into another one whose solution is the same as that of the original system but is easier to compute. What type of transformation of a linear system leaves its solution unchanged? The answer is that we can pre-multiply (i.e. multiply from the left) both sides of the linear system $Ax = b$ by any nonsingular matrix M without affecting the solution. Indeed, note that the solution of the linear system $MAy = Mb$ is given by

$$y = (MA)^{-1}Mb = A^{-1}M^{-1}Mb = A^{-1}Ib = A^{-1}b = x.$$

An important example of such type transformation is the fact that any two rows of matrix A and corresponding elements of right-hand-side vector b may be interchanged (reordered) without changing the solution x. Intuitively it is obvious: all of the equations in the system $Ax = b$ must be satisfied simultaneously, so the order in which they have been written down in the system is irrelevant. Formally, such reordering of the equations is accomplished by pre-multiplying both sides of the system by a permutation matrix P (see section 1, definition 2.6). For example, if

$$P = \begin{pmatrix} 0 & 1 & 0 \\ 1 & 0 & 0 \\ 0 & 0 & 1 \end{pmatrix} \quad \text{and} \quad x = \begin{pmatrix} x_1 \\ x_2 \\ x_3 \end{pmatrix}$$

then

$$Px = \begin{pmatrix} 0 & 1 & 0 \\ 1 & 0 & 0 \\ 0 & 0 & 1 \end{pmatrix} \cdot \begin{pmatrix} x_1 \\ x_2 \\ x_3 \end{pmatrix} = \begin{pmatrix} x_2 \\ x_1 \\ x_3 \end{pmatrix}.$$

A permutation matrix is always nonsingular and its inverse matrix is simply its transpose: $P^{-1} = P^T$. Thus, the reordered system may be written in the form $PAx = Pb$, and the solution x is unchanged.

Post-multiplying (i.e. multiplying from the right) by a permutation matrix reorders the columns of the matrix instead of the rows. Such a transformation *does change* the solution, but only in that the components of the solution are permuted. To see this, observe that the solution of the system $APy = b$ is given by

$$y = (AP)^{-1}b = P^{-1}A^{-1}b = P^T A^{-1}b = P^T x$$

and hence the solution of the original system $Ax = b$ is given by $x = Py$.

In order to understand the most widespread method for solving problems of linear-fractional programming, we need to know a great deal about the properties of solutions of linear equation systems. With this in mind, we will pay great attention to studying such systems.

7. The Gaussian Elimination

The reader has undoubtedly solved linear systems of three equations in three unknowns by the method of elimination of variables or using Cramer's rule[1] and determinants. We now study in this section an efficient direct method (the *Gaussian elimination*) for solving a system of linear equations. The method is important for us also because its main operations are used in the simplex method, which we will use when solving an LFP problem.

7.1 Elementary Row Operations

Consider matrix $A = ||a_{ij}||_{m \times m}$. The method is based on the use of the following three *elementary row operations* or *ero*'s:

1 *Multiplication of any row of matrix A by a nonzero scalar;*
For example, if

$$A = \begin{pmatrix} 1 & 2 & 3 & 4 \\ 1 & 3 & 5 & 6 \\ 0 & 1 & 2 & 3 \end{pmatrix},$$

then multiplying row 2 of matrix A by 3 would yield

$$A' = \begin{pmatrix} 1 & 2 & 3 & 4 \\ 3 & 9 & 15 & 18 \\ 0 & 1 & 2 & 3 \end{pmatrix}.$$

2 *Adding to any row of matrix A of nonzero scalar product of other row;*
For example, we may multiply row 2 of matrix A by 4 and replace row 3 of

[1]Cramer's rule is beyond the scope of this book since in this method each component of the solution is computed as a ratio of determinants. Though often taught in elementary linear courses, this method is astronomically expensive for full matrices of nontrivial size. Cramer's rule is useful mostly as a theoretical tool and is not usually used in operations research.

A with expression 4*(row 2 of A)+ row 3 of A. Then row 3 becomes

$$4 * (1,\ 3,\ 5,\ 6) + (0,\ 1,\ 2,\ 3) = (4,\ 13,\ 22,\ 27)$$

and

$$A' = \begin{pmatrix} 1 & 2 & 3 & 4 \\ 1 & 3 & 5 & 6 \\ 4 & 13 & 22 & 27 \end{pmatrix}.$$

3 *Interchange of any rows of matrix A;*
For example, if we interchange row 1 and row 3 of matrix A, we obtain

$$A' = \begin{pmatrix} 0 & 1 & 2 & 3 \\ 1 & 3 & 5 & 6 \\ 1 & 2 & 3 & 4 \end{pmatrix}.$$

Gauss elimination uses one or more of the above elementary row operations in a systematic fashion to reduce the given square matrix $A = ||a_{ij}||_{m \times m}$ to the triangular matrix. The method may result in one of the following possible cases:

1 The system has a unique solution;

2 The system has no solution;

3 The system has an infinite number of solutions.

DEFINITION 2.20 *If systems of linear equations $Ax = b$ and $Bx = d$ have the same solution, they are said to be* **equivalent** *systems.*

THEOREM 2.1 *If system of linear equations $Bx = d$ is obtained from other system $Ax = b$ by a finite sequence of elementary row operations then $Ax = b$ and $Bx = d$ are equivalent.*

7.2 Main Steps

We will illustrate Gaussian elimination with the square system

$$\begin{pmatrix} a_{11} & a_{12} & a_{13} \\ a_{21} & a_{22} & a_{23} \\ a_{31} & a_{32} & a_{33} \end{pmatrix} \cdot \begin{pmatrix} x_1 \\ x_2 \\ x_3 \end{pmatrix} = \begin{pmatrix} b_1 \\ b_2 \\ b_3 \end{pmatrix}. \qquad (2.4)$$

Multiplying the first equation by a_{21}/a_{11} (assuming that the inequality $a_{11} \neq 0$ holds) and subtracting the product from the second equation produces the equivalent system

$$\begin{pmatrix} a_{11} & a_{12} & a_{13} \\ 0 & a_{22}^{(2)} & a_{23}^{(2)} \\ a_{31} & a_{32} & a_{33} \end{pmatrix} \cdot \begin{pmatrix} x_1 \\ x_2 \\ x_3 \end{pmatrix} = \begin{pmatrix} b_1 \\ b_2^{(2)} \\ b_3 \end{pmatrix}, \tag{2.5}$$

where

$$a_{22}^{(2)} = a_{22} - (a_{21}/a_{11})\, a_{12}, \tag{2.6}$$

$$a_{23}^{(2)} = a_{23} - (a_{21}/a_{11})\, a_{13}, \tag{2.7}$$

and

$$b_2^{(2)} = b_2 - (a_{21}/a_{11})\, b_1. \tag{2.8}$$

Similarly, multiplying the first equation by a_{31}/a_{11} and subtracting the product from the third equation produces the following equivalent system

$$\begin{pmatrix} a_{11} & a_{12} & a_{13} \\ 0 & a_{22}^{(2)} & a_{23}^{(2)} \\ 0 & a_{32}^{(2)} & a_{33}^{(2)} \end{pmatrix} \cdot \begin{pmatrix} x_1 \\ x_2 \\ x_3 \end{pmatrix} = \begin{pmatrix} b_1 \\ b_2^{(2)} \\ b_3^{(2)} \end{pmatrix}, \tag{2.9}$$

where

$$a_{32}^{(2)} = a_{32} - (a_{31}/a_{11})\, a_{12}, \tag{2.10}$$

$$a_{33}^{(2)} = a_{33} - (a_{31}/a_{11})\, a_{13}, \tag{2.11}$$

and

$$b_3^{(2)} = b_3 - (a_{31}/a_{11})\, b_1. \tag{2.12}$$

Finally, multiplying the new second equation of (2.9) by $a_{32}^{(2)}/a_{22}^{(2)}$ (assuming that the inequality $a_{22}^{(2)} \neq 0$ holds) and subtracting the product from the third equation of (2.9) produces the system

$$\begin{pmatrix} a_{11} & a_{12} & a_{13} \\ 0 & a_{22}^{(2)} & a_{23}^{(2)} \\ 0 & 0 & a_{33}^{(3)} \end{pmatrix} \cdot \begin{pmatrix} x_1 \\ x_2 \\ x_3 \end{pmatrix} = \begin{pmatrix} b_1 \\ b_2^{(2)} \\ b_3^{(3)} \end{pmatrix}, \tag{2.13}$$

where

$$a_{33}^{(3)} = a_{33}^{(2)} - (a_{32}^{(2)}/a_{22}^{(2)})\, a_{23}^{(2)}, \tag{2.14}$$

and

$$b_3^{(3)} = b_3^{(2)} - (a_{32}^{(2)}/a_{22}^{(2)})\, b_2^{(2)}. \tag{2.15}$$

Notice that equation (2.13) has the upper triangular form with the correspondence

$$U = \begin{pmatrix} a_{11} & a_{12} & a_{13} \\ 0 & a_{22}^{(2)} & a_{23}^{(2)} \\ 0 & 0 & a_{33}^{(3)} \end{pmatrix}$$

Once we have reduced our system of equations to upper triangular form, we can determine the solution by the so-called **backward substitution** procedure. Starting with the final (third) row of system (2.13) we have

$$x_3 = b_3^{(3)}/a_{33}^{(3)}.$$

Substituting for x_3 in the second row of (2.13) gives

$$x_2 = (b_2^{(2)} - a_{23}^{(2)}x_3)/a_{22}^{(2)}.$$

Similarly, substituting known x_2 and x_3 in the first equation of (2.13), gives

$$x_1 = (b_1 - a_{12}x_2 - a_{13}x_3)/a_{11}.$$

So, starting from the last variable x_3 and progressing in backward direction we have determined the solution of the original system (2.4).

This process may be performed in general by creating zeros in the first column, then in the second column, and so on. In the general case of a square system with m equations and m variables for $k = 1, 2, \ldots, m-1$ we use the following formulas

$$a_{ij}^{(k+1)} = a_{ij}^{(k)} - (a_{ik}^{(k)}/a_{kk}^{(k)})a_{kj}^{(k)}, \quad i, j > k, \tag{2.16}$$

and

$$b_i^{(k+1)} = b_i^{(k)} - (a_{ik}^{(k)}/a_{kk}^{(k)})b_k^{(k)}, \quad i > k, \tag{2.17}$$

where $a_{ij}^{(1)} = a_{ij}, \; i = 1, 2, \ldots, m; \; j = 1, 2, \ldots, m$. The only assumption required is that the inequalities $a_{kk}^{(k)} \neq 0, \; k = 1, 2, \ldots, m$, hold (the case when this assumption is not valid will be considered below in section 7.4). These entries are called **pivots** in Gaussian elimination. It is convenient to use the notation

$$A^{(k)}x = b^{(k)}$$

for the system obtained after $(k-1)$ steps, $k = 1, 2, \ldots, m$, with $A^{(1)} = A$ and $b^{(1)} = b$. The final matrix $A^{(n)}$ is upper triangular, so for solution x in general case we have

$$x_m = b_m^{(m)}/a_{mm}^{(m)} \tag{2.18}$$

$$x_j = (b_j^{(j)} - \sum_{i=j+1}^{m} a_{ji}^{(j)} x_i)/a_{jj}^{(j)}, \quad j = m-1, m-2, \dots, 1. \qquad (2.19)$$

We have thus expressed the Gaussian elimination method algebraically. Now we can summarize the main rules of the method in the following way:
To solve the system of m linear equations with m variables represented by the matrix equation $Ax = b$ we carry out the following procedure:

1 Reduce the system to upper triangular form using elementary row operations;

2 Use backward substitution defined by formulas (2.18) and (2.19) to determine unknown x.

One way to implement it is shown in Figure 2.3.

Observe that to reduce the original matrix A to triangular form using this method we have to perform $m^3/3$ multiplications and a similar number of additions. Solving the resulting triangular system for a single RHS vector by backward or forward substitution requires about $m^2/2$ multiplications and the same number of subtractions. Thus, as the order m of the matrix grows, the reducing phase of the Gaussian elimination becomes increasingly dominant in the cost of solving a linear system.

Consider another example. Given the following system

$$A\,|b = \left(\begin{array}{ccc|c} 1 & 0 & 0 & 4 \\ 0 & 1 & -1/3 & 1 \\ 0 & 0 & 1 & 3 \end{array}\right).$$

If we replace row 2 in this system with 1/3(row 3)+(row 2) we obtain the following system

$$A\,|b = \left(\begin{array}{ccc|c} 1 & 0 & 0 & 4 \\ 0 & 1 & 0 & 2 \\ 0 & 0 & 1 & 3 \end{array}\right),$$

which gives us solution $x_1 = 4,\ x_2 = 2,\ x_3 = 3.$

If after performing the method we obtain a row in the form

$$(0, 0, \dots, 0 \mid b_i),$$

where $b_i \neq 0,$ the original linear system has no solution. In the case of $b_i = 0,$ the system has an infinite number of solutions.

Gauss Algorithm

```
{Reducing to Upper Triangular Form}
For k := 1 To m − 1 Do                          {Loop over columns}
   Begin
   If a_kk = 0 Then Stop;                       {Stop if pivot is zero}
   For i := k + 1 To m Do
      Begin
      μ := a_ik/a_kk;                           {Compute multiplier}
      b_i := b_i − μb_k;                        {Update RHS entry}
      For j := k To m Do
         a_ij := a_ij − μa_kj;    {Apply transformation to submatrix}
      End
   End
{Backward Substitution}
If a_mm = 0 Then Stop;                          {Stop if matrix is singular}
x_m := b_m/a_mm;
For j := m − 1 Downto 1 Do
   Begin
   If a_jj = 0 Then Stop;                       {Stop if matrix is singular}
   For i := j + 1 To m Do b_i := b_i − a_ij x_i;
   x_j := b_j/a_jj;
   End
```

Figure 2.3. Algorithm – Gauss elimination with backward substitution

7.3 Forward Substitution

It is obvious that elementary row operations may be used for reducing original square matrix A to the lower triangular form

$$\begin{pmatrix} a_{11}^{(3)} & 0 & 0 \\ a_{21}^{(2)} & a_{22}^{(2)} & 0 \\ a_{31} & a_{32} & a_{33} \end{pmatrix} \cdot \begin{pmatrix} x_1 \\ x_2 \\ x_3 \end{pmatrix} = \begin{pmatrix} b_1^{(3)} \\ b_2^{(2)} \\ b_3 \end{pmatrix}. \tag{2.20}$$

Similarly to (2.18) and (2.19), system (2.20) in a general case may be solved in the forward direction by the following steps:

$$x_1 = b_1^{(m)}/a_{11}^{(m)} \tag{2.21}$$

$$x_j = (b_j^{(j)} - \sum_{i=1}^{j-1} a_{ji}^{(j)} x_i)/a_{jj}^{(j)}, \quad j = 2, 3, \ldots, m. \tag{2.22}$$

This is known as **forward substitution**. This algorithm may be implemented as shown in Figure 2.4.

Gauss Algorithm

```
{Reducing to Lower Triangular Form}
For k := m Downto 2 Do                          {Loop over columns}
   Begin
   If a_kk = 0 Then Stop;                       {Stop if pivot is zero}
   For i := k - 1 Downto 1 Do
      Begin
      μ := a_ik/a_kk;                           {Compute multiplier}
      b_i := b_i - μb_k;                        {Update RHS entry}
      For j := 1 To k Do
         a_ij := a_ij - μa_kj;        {Apply transformation to submatrix}
      End
   End
{Forward Substitution}
If a_11 = 0 Then Stop;                          {Stop if matrix is singular}
x_1 := b_1/a_11;
For j := 2 To m Do
   Begin
   If a_jj = 0 Then Stop;                       {Stop if matrix is singular}
   For i := 1 To j - 1 Do b_i := b_i - a_ij x_i;
   x_j := b_j/a_jj;
   End
```

Figure 2.4. Algorithm – Gauss elimination with forward substitution.

So the goal of the Gaussian elimination method is to transform the original system to upper or lower triangular form and then perform backward or forward substitution respectively. Note that if the obtained triangular system has the form of a unit triangular matrix, in this case formulas (2.18)-(2.19) and (2.21)-(2.22) simplify.

7.4 Pivoting

There is one obvious problem with the Gaussian elimination process as we have described it in the previous sections.

The obvious potential difficulty is that the Gaussian elimination breaks down if the leading diagonal entry of the remaining unreduced portion of the matrix is zero at any stage, as calculating multipliers for a given column requires division by the diagonal entry in that column.

Consider the following system

$$\begin{pmatrix} 0 & 2 & 5 \\ 4 & 6 & 8 \\ 5 & 4 & 12 \end{pmatrix} \cdot \begin{pmatrix} x_1 \\ x_2 \\ x_3 \end{pmatrix} = \begin{pmatrix} 15 \\ 25 \\ 30 \end{pmatrix}. \tag{2.23}$$

It is obvious that we cannot perform the multiplication of the first row by a_{21}/a_{11} because of zero entry $a_{11} = 0$. Exchanging the first and the second equations (both matrix and the right-hand side vector b) in the system (2.23) gives the following

$$\begin{pmatrix} 4 & 6 & 8 \\ 0 & 2 & 5 \\ 5 & 4 & 12 \end{pmatrix} \cdot \begin{pmatrix} x_1 \\ x_2 \\ x_3 \end{pmatrix} = \begin{pmatrix} 25 \\ 15 \\ 30 \end{pmatrix} \tag{2.24}$$

and completely avoids the difficulty. Moreover, this interchange of rows makes the first elementary row operation absolutely unnecessary because variable x_1 has been excluded from the second row of system (2.24) by simple interchanging rows. We know that such an interchange does not alter the solution of the system.

This simple observation is the basis for the solution of such type problems for a matrix of any order. We will illustrate these techniques by working with $m = 5$. The extension to the general case is obvious.

Let us suppose that we have performed the first two steps on a system $Ax = b$ with square matrix A of order 5 and at the moment the system has the following form

$$\begin{pmatrix} a_{11} & a_{12} & a_{13} & a_{14} & a_{15} \\ 0 & a_{22}^{(2)} & a_{23}^{(2)} & a_{24}^{(2)} & a_{25}^{(2)} \\ 0 & 0 & 0 & a_{34}^{(3)} & a_{35}^{(3)} \\ 0 & 0 & a_{43}^{(3)} & a_{44}^{(3)} & a_{45}^{(3)} \\ 0 & 0 & a_{53}^{(3)} & a_{54}^{(3)} & a_{55}^{(3)} \end{pmatrix} \cdot \begin{pmatrix} x_1 \\ x_2 \\ x_3 \\ x_4 \\ x_5 \end{pmatrix} = \begin{pmatrix} b_1 \\ b_2^{(2)} \\ b_3^{(3)} \\ b_4^{(3)} \\ b_5^{(3)} \end{pmatrix}. \tag{2.25}$$

In this situation we cannot proceed with excluding variable x_3 from the fourth and fifth equations of system (2.25) because of zero entry $a_{33}^{(3)} = 0$. If one of the inequalities $a_{43}^{(3)} \neq 0$ or $a_{45}^{(3)} \neq 0$ holds, first we interchange the third equation in (2.25) with either the forth or fifth one and then we proceed the elementary row operations excluding variable x_3. Interchanging rows (equations) in this manner during elimination to obtain a nonzero pivot is called **pivoting** or **partial pivoting**. The only way this pivoting can break down is the case when all other entries below the zero entry $a_{33}^{(3)}$ are equal to zero, that is

$$a_{33}^{(3)} = a_{43}^{(3)} = a_{53}^{(3)} = 0 .$$

In this case, equations 3, 4, and 5 of system (2.25) do not involve variables x_1, x_2, x_3. It means that variable x_3 may be given an arbitrary value and variables x_1, x_2 then can be determined by backward substitution in equations 1 and 2 of system (2.25) in accordance with formulas (2.18) and (2.19). In fact, in this case the matrix is singular, that is $det(A) = 0$ because

$$det(A) = det(A^{(3)}) = a_{11} \cdot a_{22}^{(2)} \cdot \begin{vmatrix} 0 & a_{34}^{(3)} & a_{35}^{(3)} \\ 0 & a_{44}^{(3)} & a_{45}^{(3)} \\ 0 & a_{54}^{(3)} & a_{55}^{(3)} \end{vmatrix} = 0 .$$

Such indeterminacy of a variable, in the given case x_3, is a typical facet of singularity.

Extending to the general case, we can see that as long as the given square matrix A is nonsingular, the equations may always be reordered through interchanging rows so that diagonal entry $a_{kk}^{(k)}$ is nonzero.

Sometimes instead of interchanging rows (partial pivoting) we can use so-called **full pivoting**, i.e. interchanging not only rows but columns too. Finally, we just have to note that partial pivoting is easier than the full pivoting, because we do not have to keep track of the permutation of the solution vector x. Partial pivoting makes available as pivots only the elements of the current column, so no permutations appear in this case.

8. The Gauss-Jordan Elimination

The motivation for the Gaussian elimination is to reduce a general matrix to triangular form, because the resulting linear system is easy to solve. Diagonal linear systems are even easier to solve, however, so diagonal form would appear to be an even more desirable aim. The *Gauss-Jordan elimination* is a variation of the standard Gaussian elimination in which the matrix is reduced to diagonal form rather than merely to upper or lower triangular form. The same type of elementary row operations are used to eliminate matrix entries as in the standard Gaussian elimination method. Moreover, the method allows to avoid such relatively expensive operations as backward (or forward) substitution. The backward substitution phase of Gaussian elimination in the case if we reduce the original square matrix A to an upper triangular form U may be avoided if we replace the formulas (2.16)-(2.17) with some other ones that make entries above the main diagonal zero. To produce zeros above the main diagonal the operations (2.16)-(2.17) must be replaced by the following operations ([57]):

$$a_{ij}^{(k+1)} = a_{ij}^{(k)} - (a_{ik}^{(k)}/a_{kk}^{(k)})\, a_{kj}^{(k)}, \quad i \neq k,\ j > k, \tag{2.26}$$

$$b_i^{(k+1)} = b_i^{(k)} - (a_{ik}^{(k)}/a_{kk}^{(k)})\, b_k^{(k)}, \quad i \neq k, \tag{2.27}$$

and

$$b_k^{(k+1)} = b_k^{(k)}. \tag{2.28}$$

In this case, the system is reduced to the diagonal form

$$a_{kk}^{(k)} x_k = b_k^{(m)}, \tag{2.29}$$

that is

$$\left(\begin{array}{cccc|c} a_{11}^{(1)} & 0 & \dots & 0 & b_1^{(m)} \\ 0 & a_{22}^{(2)} & \dots & 0 & b_2^{(m)} \\ \dots & \dots & \dots & \dots & \dots \\ 0 & 0 & \dots & a_{mm}^{(m)} & b_m^{(m)} \end{array} \right).$$

Similarly, for the lower triangular form of reduced matrix A and forward substitution we can determine the analogous formulas that produce system (2.29).

The method requires about $m^3/2$ ([95]) multiplications and a similar number of additions, which is about 50% more expensive than the standard Gaussian elimination. But it does not require any backward or forward substitution at all.

To illustrate the method we consider the following numeric example:

$$\begin{pmatrix} 2 & 2 & 1 \\ 2 & -1 & 2 \\ 1 & -1 & 2 \end{pmatrix} \cdot \begin{pmatrix} x_1 \\ x_2 \\ x_3 \end{pmatrix} = \begin{pmatrix} 9 \\ 6 \\ 5 \end{pmatrix}.$$

or in augmented representation:

$$\left(\begin{array}{ccc|c} 2 & 2 & 1 & 9 \\ 2 & -1 & 2 & 6 \\ 1 & -1 & 2 & 5 \end{array} \right). \tag{2.30}$$

We begin by using ero's to transform the first column of (2.30) into form $e_1 = (1, 0, 0)^T$.

Step 1 Multiply row 1 of (2.30) by $1/2$. This ero yields

$$\left(\begin{array}{ccc|c} 1 & 1 & 1/2 & 9/2 \\ 2 & -1 & 2 & 6 \\ 1 & -1 & 2 & 5 \end{array}\right); \tag{2.31}$$

Step 2 In (2.31) replace row 2 with (row 2)-2(row 1). The result of this ero is:

$$\left(\begin{array}{ccc|c} 1 & 1 & 1/2 & 9/2 \\ 0 & -3 & 1 & -3 \\ 1 & -1 & 2 & 5 \end{array}\right); \tag{2.32}$$

Step 3 In (2.32) replace row 3 with (row 3)-(row 1). The result of this ero is:

$$\left(\begin{array}{ccc|c} 1 & 1 & 1/2 & 9/2 \\ 0 & -3 & 1 & -3 \\ 0 & -2 & 3/2 & 1/2 \end{array}\right). \tag{2.33}$$

The first column of (2.30) has now been transformed into unit form $e_1 = (1,0,0)^T$.

We now transform the second column of system (2.33) into unit form $e_1 = (0,1,0)^T$.

Step 4 In (2.33) multiply row 2 by $-1/3$. This ero yields

$$\left(\begin{array}{ccc|c} 1 & 1 & 1/2 & 9/2 \\ 0 & 1 & -1/3 & 1 \\ 0 & -2 & 3/2 & 1/2 \end{array}\right); \tag{2.34}$$

Step 5 In (2.34) replace row 1 with (row 1)-(row 2). The result of this ero is:

$$\left(\begin{array}{ccc|c} 1 & 0 & 5/6 & 7/2 \\ 0 & 1 & -1/3 & 1 \\ 0 & -2 & 3/2 & 1/2 \end{array}\right); \tag{2.35}$$

Step 6 In (2.35) replace row 3 with (row 3)+2(row 2). The result of this ero is:

$$\left(\begin{array}{ccc|c} 1 & 0 & 5/6 & 7/2 \\ 0 & 1 & -1/3 & 1 \\ 0 & 0 & 5/6 & 5/2 \end{array}\right). \tag{2.36}$$

The second column of (2.30) has now been transformed into unit form$e_1 = (0,1,0)^T$. Observe that ero's of steps 4, 5 and 6 did not change the first column of our matrix.

To complete the Gauss-Jordan elimination method, we have to transform the third column into unit form $e_1 = (0,0,1)^T$. The following steps accomplish this goal.

Step 7 In (2.36) multiply row 3 by $6/5$. This ero yields

$$\left(\begin{array}{ccc|c} 1 & 0 & 5/6 & 7/2 \\ 0 & 1 & -1/3 & 1 \\ 0 & 0 & 1 & 3 \end{array}\right); \tag{2.37}$$

Step 8 In (2.37) replace row 1 with (row 1)-5/6(row 3). The result of this ero is:

$$\left(\begin{array}{ccc|c} 1 & 0 & 0 & 1 \\ 0 & 1 & -1/3 & 1 \\ 0 & 0 & 1 & 3 \end{array}\right); \tag{2.38}$$

Step 9 In (2.38) replace row 2 with (row 2)+1/3(row 3). The result of this ero is:

$$\left(\begin{array}{ccc|c} 1 & 0 & 0 & 1 \\ 0 & 1 & 0 & 2 \\ 0 & 0 & 1 & 3 \end{array}\right). \tag{2.39}$$

The last represents the diagonal system

$$\begin{pmatrix} 1 & 0 & 0 \\ 0 & 1 & 0 \\ 0 & 0 & 1 \end{pmatrix} \cdot \begin{pmatrix} x_1 \\ x_2 \\ x_3 \end{pmatrix} = \begin{pmatrix} 1 \\ 2 \\ 3 \end{pmatrix}$$

that has the following unique solution

$$x_1 = 1, \quad x_2 = 2, \quad x_3 = 3.$$

Observe that in this example we have not used elementary row operations of type 3 (interchanging rows) since all pivots we have used were nonzero.

Consider the following system in the augmented form

$$\left(\begin{array}{ccc|c} 0 & 2 & 1 & 9 \\ 2 & -1 & 2 & 6 \\ 1 & -1 & 2 & 5 \end{array}\right).$$

It is obvious that having zero entry in a_{11} we cannot use elementary row operations of type 1 (multiplying rows) to produce $a_{11} = 1$. If, however, we interchange row 1 with row 2 (ero of type 3), we obtain

$$\left(\begin{array}{rrr|r} 2 & -1 & 2 & 6 \\ 0 & 2 & 1 & 9 \\ 1 & -1 & 2 & 5 \end{array}\right)$$

and we may proceed as usual with the Gauss-Jordan method.

Gauss-Jordan Algorithm

```
{Reducing to Diagonal Form}
For k := 1 To m Do                                {Loop over all columns}
   Begin
   If a_kk = 0 Then Stop;                          {Stop if pivot is zero}
   {Loop for rows above pivot entry}
   For i := 1 To k − 1 Do
     Begin
      μ := a_ik/a_kk;                              {Compute multiplier}
      b_i := b_i − μb_k;                           {Update RHS entry}
      For j := k To m Do a_ij := a_ij − μa_kj;     {Update row i}
     End
   {Loop for rows below pivot entry}
   For i := k + 1 To m Do
      Begin
      μ := a_ik/a_kk;                              {Compute multiplier}
      b_i := b_i − μb_k;                           {Update RHS entry}
      For j := k To m Do a_ij := a_ij − μa_kj;     {Update row i}
      End
   End
{Calculating solution}
For j := 1 To m Do
   Begin
   If a_jj = 0 Then Stop;                          {Stop if matrix is singular}
   x_j := b_j/a_jj;
   End
```

Figure 2.5. Algorithm – Gauss-Jordan elimination.

One of the possible ways to implement this method is presented in the algorithm shown in Figure 2.5.

9. Multiple RHS's and Inverses

In the previous sections we considered the case of system of linear equations with a single right-hand side vector b, but all we have considered for Gaussian elimination and Gauss-Jordan method may be easily extended to the case with multiple right-hand side vectors $b_l = (b_{1l}, b_{2l}, \ldots, b_{ml})^T$, $l = 1, 2, \ldots, v$. In the case of backward substitution (upper triangular form) in elimination phase of Gauss method, formula (2.17) must be replaced with the following one:

$$b_{il}^{(k+1)} = b_{il}^{(k)} - (a_{ik}^{(k)}/a_{kk}^{(k)})\, b_{kl}^{(k)}, \quad i > k, \; l = 1, 2, \ldots, v. \tag{2.40}$$

Further, the formulas (2.18)-(2.19) used during backward substitution have to be expanded to the following form:

$$x_{ml} = b_{ml}^{(m)}/a_{mm}^{(m)}, \quad l = 1, 2, \ldots, v; \tag{2.41}$$

and

$$x_{jl} = (b_{jl}^{(j)} - \sum_{i=j+1}^{m} a_{ji}^{(j)} x_{il})/a_{jj}^{(j)}, \quad j = m-1, m-2, \ldots, 1; \quad l = 1, 2, \ldots, v; \tag{2.42}$$

correspondingly.

An important special case of multiple right-hand side vectors b_l is the one where the inverse A^{-1} matrix is required, since it may be obtained as a solution of the following set of systems:

$$Ax = e_i, \quad i = 1, 2, \ldots, m,$$

where $e_i = (0, 0, \ldots, 0, 1, 0, \ldots, 0)^T$ is an ith column of the unit matrix I_m.

If a sequence of systems $Ax = b$ with the same matrix A but different right-hand side vectors b is to be solved, it is may be worth calculating the inverse matrix A^{-1} and use it to form the products $x = A^{-1}b$ for different right-hand side vectors b.

For inverting a matrix, Gauss-Jordan elimination is almost as computationally efficient as any other direct methods. For solving sets of linear equations, this method produces not only the solution for the given system of linear equations (with one or more right-hand-side vectors b) but may also be used to obtain the inverse of the original matrix A. This fact may be considered as a great advantage of the method, if the inverse matrix A^{-1} is *desired.* On the other hand, if the inverse matrix A^{-1} is *not* desired this property of the method may be considered as a principle weakness. For these reasons, Gauss-Jordan elimination should usually not be the method of the first choice, for solving systems of linear equations. The decomposition methods we consider later in section 2 are better in this sense. The Gauss-Jordan elimination method is approximately three times as slow ([147]) as the best alternative techniques for solving systems of linear equations $Ax = b$.

10. Discussion Questions and Exercises

2.1 Using Gaussian elimination with backward substitution calculate inverse matrix A^{-1} for matrix

$$A = \begin{pmatrix} 2 & 0 & -1 \\ 3 & 1 & 2 \\ -1 & 0 & 1 \end{pmatrix}$$

considered in section 5.

2.2 Solve the following system using Gaussian elimination with reduction to upper triangular form and backward substitution

$$A\,|b = \left(\begin{array}{ccc|c} 1 & 5 & 4 & 25 \\ 4 & 1 & 5 & 15 \\ 3 & 2 & 1 & 30 \end{array}\right).$$

Then reduce this system to lower triangular form, perform forward substitution and recalculate solution. Is the solution obtained the same?

2.3 Find products AB and BA of the following two matrices A and B:

$$A = \begin{pmatrix} 1 & 0 & 0 & 0 \\ 2 & 1 & 0 & 0 \\ 3 & 0 & 1 & 0 \\ 4 & 0 & 0 & 1 \end{pmatrix}, \quad B = \begin{pmatrix} 1 & 0 & 0 & 0 \\ 0 & 1 & 0 & 0 \\ 0 & 5 & 1 & 0 \\ 4 & 6 & 1 & 1 \end{pmatrix}.$$

2.4 For system given in exercise 2.2 perform Gauss-Jordan elimination.

2.5 Check if the given matrix

$$A = \begin{pmatrix} 1 & 2 & 3 \\ 2 & 0 & 0 \\ 4 & 0 & 3 \end{pmatrix}$$

is orthogonal. What properties must be possessed by the columns of an orthogonal matrix?

2.6 Find A^{-1} (if exists) for the following matrices

$$\begin{pmatrix} 1 & 3 \\ 2 & 5 \end{pmatrix}, \quad \begin{pmatrix} 1 & 0 & 1 \\ 4 & 1 & 2 \\ 3 & 1 & -1 \end{pmatrix}, \quad \begin{pmatrix} 1 & 0 & 1 \\ 1 & 1 & 1 \\ 2 & 1 & 1 \end{pmatrix}, \quad \begin{pmatrix} 1 & 2 & 1 \\ 1 & 2 & 0 \\ 2 & 4 & 1 \end{pmatrix}.$$

2.7 Suppose that matrices A and B both have inverses. Find the inverse of the product matrix AB.

2.8 Check matrices A and B given on page 12 if they are orthogonal matrices.

Chapter 3

INTRODUCTION TO LFP

This is an introduction to the theory of linear-fractional programming. Here we define the common problem of LFP, give its main economic interpretation, formulate and prove the main lemmas and theorems of LFP and describe some important real-world applications where the use of LFP may prove to be quite useful.

Problems of LFP arise when there appears a necessity to optimize the efficiency of some activity: profit gained by company per unit of expenditure of labor, cost of production per unit of produced goods, nutritiousness of ration per unit of cost, etc. Nowadays because of a deficit of natural resources the use of specific criteria becomes more and more topical. So an application of LFP to solving real-world problems connected with optimizing efficiency could be as useful as in the case of LP.

1. What is a Linear-Fractional Problem ?

In 1960, Hungarian mathematician Béla Martos ([131], [132]) formulated and considered a so-called *hyperbolic programming* problem, which in the English language special literature is referred as a *linear-fractional programming* problem. In a typical case the **common problem** of LFP may be formulated as follows:

Given objective function

$$Q(x) = \frac{P(x)}{D(x)} = \frac{\sum_{j=1}^{n} p_j x_j + p_0}{\sum_{j=1}^{n} d_j x_j + d_0}, \tag{3.1}$$

which must be maximized (or minimized) subject to

$$\left.\begin{array}{ll} \sum_{j=1}^{n} a_{ij} x_j \leq b_i, & i = 1, 2, \dots, m_1, \\ \sum_{j=1}^{n} a_{ij} x_j \geq b_i, & i = m_1 + 1, m_1 + 2, \dots, m_2, \\ \sum_{j=1}^{n} a_{ij} x_j = b_i, & i = m_2 + 1, m_2 + 2, \dots, m, \end{array}\right\} \tag{3.2}$$

$$x_j \geq 0, \quad j = 1, 2, \dots, n_1, \tag{3.3}$$

where $m_1 \leq m_2 \leq m, \quad n_1 \leq n.$ Here and in what follows we suppose that $D(x) \neq 0, \quad \forall x = (x_1, x_2, \dots, x_n) \in S,$ where S denotes a **feasible set** or **set of feasible solutions** defined by constraints (3.2)-(3.3).

Because denominator $D(x) \neq 0 \quad \forall x \in S,$ without loss of generality we can assume that

$$D(x) > 0, \quad \forall x \in S. \tag{3.4}$$

In the case of $D(x) < 0$ we can multiply numerator $P(x)$ and denominator $D(x)$ of objective function $Q(x)$ with (-1).

Here and in what follows throughout the book we deal with just such linear-fractional programming problems that satisfy condition (3.4). Furthermore, we suppose that all constraints in system (3.2) are linearly independent and so the rank of matrix $A = ||a_{ij}||_{m \times n}$ is equal to m.

So in an LFP problem our aim is to find such a vector x of **decision variables** $x_j, \ j = 1, 2, \dots, n,$ which

1 maximizes (or minimizes) function $Q(x),$ called **objective function**, and at the same time

2 satisfies a set of **main constraints** (3.2) and **sign restrictions** (3.3).

1.1 Main Definitions

Here we introduce the main conceptions that will be used throughout the rest of the book.

DEFINITION 3.1 *If given vector* $x = (x_1, x_2, \ldots, x_n)$ *satisfies constraints (3.2) and (3.3), we will say that vector* x *is a* **feasible solution** *of LFP problem (3.1)-(3.3).*

DEFINITION 3.2 *If given vector* $x = (x_1, x_2, \ldots, x_n)$ *is a feasible solution of maximization (minimization) LFP problem (3.1)-(3.3), and provides maximal (minimal) value for objective function* $Q(x)$ *over the feasible set* S*, we say that vector x is an* **optimal solution** *of maximization (minimization) linear-fractional programming problem (3.1)-(3.3).*

DEFINITION 3.3 *We say that a maximization (minimization) linear-fractional programming problem is* **solvable***, if its feasible set* S *is not empty, that is* $S \neq \emptyset$, *and objective function* $Q(x)$ *has finite upper (lower) bound on* S*.*

DEFINITION 3.4 *If the feasible set is empty, that is* $S = \emptyset$, *we say that the LFP problem is* **infeasible***.*

DEFINITION 3.5 *If objective function* $Q(x)$ *of a maximization (minimization) LFP problem has no upper (lower) finite bound, we say that the problem is* **unbounded***.*

1.2 Relationship with Linear Programming

It is obvious that if all $d_j = 0, j = 1, 2, \ldots n,$ and $d_0 = 1,$ then LFP problem (3.1)-(3.3) becomes an LP problem. This is a reason why we say that an LFP problem (3.1)-(3.3) is a generalization of an LP problem:
Given objective function

$$P(x) = \sum_{j=1}^{n} p_j x_j + p_0, \tag{3.5}$$

which must be maximized (or minimized) subject to

$$\left.\begin{array}{ll} \sum_{j=1}^{n} a_{ij}x_j \leq b_i, & i = 1, 2, \ldots, m_1, \\ \sum_{j=1}^{n} a_{ij}x_j \geq b_i, & i = m_1 + 1, m_1 + 2, \ldots, m_2, \\ \sum_{j=1}^{n} a_{ij}x_j = b_i, & i = m_2 + 1, m_2 + 2, \ldots, m, \end{array}\right\} \tag{3.6}$$

$$x_j \geq 0, \quad j = 1, 2, \ldots, n_1, \tag{3.7}$$

There are also a few special cases when the original LFP problem may be replaced with an appropriate LP problem:

1 If $d_j = 0, \ j = 1, 2, \ldots n, \ d_0 \neq 0,$ then objective function $Q(x)$ becomes a linear one:

$$Q(x) = \sum_{j=1}^{n} \frac{p_j}{d_0} x_j + \frac{p_0}{d_0} = \frac{P(x)}{d_0}.$$

In this case maximization (minimization) of the original objective function $Q(x)$ may be substituted with maximization (minimization) of linear function $P(x)/d_0$ correspondingly on the same feasible set S.

2 If $p_j = 0, \ j = 1, 2, \ldots n,$ then objective function

$$Q(x) = \frac{P(x)}{D(x)} = \frac{p_0}{\sum_{j=1}^{n} d_j x_j + d_0}$$

may be replaced with function $D(x)$. In this case maximization (minimization) of the original objective function $Q(x)$ must be substituted with minimization (maximization) of a new objective function $D(x)$ on the same feasible set S.

3 If vectors $p = (p_1, p_2, \ldots, p_n)$ and $d = (d_1, d_2, \ldots, d_n)$ are linearly dependent, that is there exists such $\mu \neq 0$, that $p = \mu d$, then objective function

$$Q(x) = \frac{P(x)}{D(x)} = \frac{\sum_{j=1}^{n} \mu d_j x_j + p_0}{\sum_{j=1}^{n} d_j x_j + d_0} = \ldots = \mu + \frac{p_0 - \mu d_0}{\sum_{j=1}^{n} d_j x_j + d_0} \tag{3.8}$$

may be replaced with function $D(x)$. Obviously, in this case maximization (minimization) of the original objective function $Q(x)$ must be substituted with

- minimization (maximization) of $D(x)$, if $p_0 - \mu d_0 > 0$,
- maximization (minimization) of $D(x)$, if $p_0 - \mu d_0 < 0$.

We should note here that in the case of $p_0 - \mu d_0 = 0$ as it follows from formula (3.8), we have $Q(x) = \mu$, which means that $Q(x) = const, \ \forall x \in S$. We will not consider such problems because of their pointlessness (see next section).

Here and in what follows throughout the book we exclude from our consideration the following three trivial cases:

1 $P(x) = const, \ \forall x \in S;$

2 $D(x) = const, \ \forall x \in S;$

3 $Q(x) = const, \ \forall x \in S;$

because in these cases the original LFP problem may be reduced to an LP problem (first two cases), or becomes absolutely aimless (case 3).

1.3 Main Forms of the LFP Problem

We have seen that LFP problems may have both equality and inequality constraints. They may also have unknown variables that are required to be nonnegative and variables that are allowed to be unrestricted in sign (*urs variable*). Before the simplex method is discussed we should introduce some special forms of formulating an LFP problem and show how these forms may be converted to one another and to the form that is required by simplex method.

DEFINITION 3.6 *An LFP problem is said to be in* ***standard form*** *if all constraints are equations and all unknown variables are nonnegative, that is*

$$Q(x) = \frac{P(x)}{D(x)} = \frac{\sum_{j=1}^{n} p_j x_j + p_0}{\sum_{j=1}^{n} d_j x_j + d_0} \longrightarrow \max(\min),$$

subject to

$$\sum_{j=1}^{n} a_{ij} x_j = b_i, \quad i = 1, 2, \dots, m,$$

$$x_j \geq 0, \quad j = 1, 2, \dots, n,$$

where $D(x) > 0, \ \forall x \in S$.

DEFINITION 3.7 *An LFP problem is said to be in* ***general form*** *if all constraints are '$\leq$' ('less than') inequalities and all unknown variables are nonnegative, that is*

$$Q(x) = \frac{P(x)}{D(x)} = \frac{\sum_{j=1}^{n} p_j x_j + p_0}{\sum_{j=1}^{n} d_j x_j + d_0} \rightarrow \max(\min),$$

subject to

$$\sum_{j=1}^{n} a_{ij} x_j \leq b_i, \quad i = 1, 2, \dots, m,$$

$$x_j \geq 0, \quad j = 1, 2, \dots, n,$$

where $D(x) > 0, \ \forall x \in S$.

It is obvious that standard and general forms of LFP problems are special cases of a LFP problem formulated in form (3.1)-(3.3). Indeed, if in the common LFP problem (3.1)-(3.3) we put $m_1 = m_2 = 0$ and $n_1 = n$, then we get a standard LFP problem. But if $m_1 = m$ and $n_1 = n$, then we have a general LFP problem.

To convert one form to another we should use the following converting procedures:

1 '$\geq$' ('greater than') $\rightarrow$ '$\leq$' ('less than').
Both sides of the '$\geq$' constraint must be multiplied by (-1).

2 '$\leq$' ('less than') $\rightarrow$ '$=$' ('equal').
Define for '$\leq$' constraint a nonnegative *slack variable* s_i ($s_i \geq 0$ - slack-variable for i-th constraint) and put this variable into the left-side of the constraint, where it will play a role of difference between the left and right sides of the original i-th constraint. Also add the sign restrictions $s_i \geq 0$ to the set of constraints. So

$$\sum_{j=1}^{n} a_{ij} x_j \leq b_i \quad \rightarrow \quad \begin{cases} \sum_{j=1}^{n} a_{ij} x_j + s_i = b_i, \\ s_i \geq 0. \end{cases}$$

3 unrestricted in sign variable x_j $\rightarrow$ restricted in sign nonnegative variable(s). For each urs variable x_j, we begin by defining two new nonnegative variables x'_j and x''_j. Then substitute $x'_j - x''_j$ for x_j in each constraint and in objective function. Also add the sign restrictions $x'_j \geq 0$ and $x''_j \geq 0$ to the set of constraints.

Because all three forms of an LFP problem (the most common form (3.1)-(3.3), standard and general) may be easily converted to one another, instead of an LFP problem in form (3.1)-(3.3) sometimes we will consider its equivalent LFP problem in standard or in general form. Obviously, such substitution does not lead to any loss of generality, but allows us to simplify our consideration.

Let us introduce the following notations:

$$A_j = (a_{1j}, a_{2j}, \dots, a_{mj})^T, \; j = 1, 2, \dots, n;$$

$$b = (b_1, b_2, \dots, m)^T, \quad A = (A_1, A_2, \dots, A_n),$$

$$x = (x_1, x_2, \dots, x_n)^T, \quad p = (p_1, p_2, \dots, p_n)^T, \quad d = (d_1, d_2, \dots, d_n)^T.$$

Using this notation we can re-formulate an LFP problem in a matrix form:

Standard problem

$$Q(x) = \frac{p^T x + p_0}{d^T x + d_0} \longrightarrow \max,$$

subject to

$$\sum_{j=1}^{n} A_j x_j = b,$$

$$x \geq 0,$$

where $D(x) = d^T x + d_0 > 0, \;\; \forall x \in S,$

General problem:

$$Q(x) = \frac{p^T x + p_0}{d^T x + d_0} \longrightarrow \max,$$

subject to

$$Ax \leq b,$$

$$x \geq 0,$$

where $D(x) = d^T x + d_0 > 0, \; \forall x \in S.$

We should note here that in accordance with the theory of mathematical programming

$$\min_{x \in S} F(x) \equiv \max_{x \in S}(-F(x)), \tag{3.9}$$

which means that to convert a minimization LFP problem to a maximization one, we should multiply its objective function by (−1). So there is no reason to consider both cases (i.e. maximization and minimization) separately. This is why in our further discussions we will focus our consideration only on maximization LFP problems.

2. The Graphical Method

We now go on to discuss how any LFP problem with only two variables can be solved graphically.

Consider the following LFP problem with two unknown variables:

$$Q(x) = \frac{P(x)}{D(x)} = \frac{p_1x_1 + p_2x_2 + p_0}{d_1x_1 + d_2x_2 + d_0} \longrightarrow \max(\min) \tag{3.10}$$

subject to

$$a_{i1}x_1 + a_{i2}x_2 \leq b_i, \quad i = 1, 2, \ldots, m; \tag{3.11}$$

$$x_1 \geq 0, \quad x_2 \geq 0\ . \tag{3.12}$$

2.1 The Single Optimal Vertex

Let us suppose that constraints (3.11) and (3.12) define feasible set S shown by shading in Figure 3.1. Let $Q(x) = K,$ where K is an arbitrary constant.

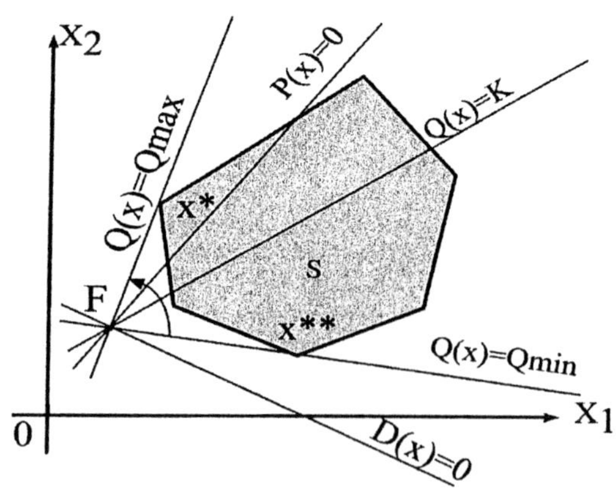

Figure 3.1. Two-variable LFP problem – Single optimal vertex.

For any real value K, equation

$$Q(x) = K,$$

or

$$(p_1 - Kd_1)x_1 + (p_2 - Kd_2)x_2 + (p_0 - Kd_0) = 0$$

represents all the points on a straight line in the two-dimensional plane x_1Ox_2. If this so-called *level-line* (or *isoline*) intersects the set of feasible solutions S, the points of intersection are the feasible solutions that give the value K to the objective function $Q(x)$. Changing the value of K translates the entire line to another line that intersects the previous line in *focus point* (point F in Figure 3.1) with coordinates defined as solution of system

$$\left.\begin{array}{rcl} p_1x_1 + p_2x_2 & = & -p_0, \\ d_1x_1 + d_2x_2 & = & -d_0, \end{array}\right\} \tag{3.13}$$

In other words, in the focus point F straight lines with equations $P(x) = 0$ and $D(x) = 0$ intersect one another.

If lines $P(x) = 0$ and $D(x) = 0$ are not parallel with one another, then the determinant of system (3.13) is not equal to zero and the system has a unique solution (coordinates of focus point F). In the other case, if lines $P(x) = 0$ and $D(x) = 0$ are parallel with one another, the determinant of system (3.13) is equal to zero and the system has no solution. It means that there is no focus point and all level-lines are also parallel with one another. In accordance with case 3 (see page 44) the given LFP problem (3.10)-(3.12) degenerates to an LP one. Hence, to maximize objective function $Q(x)$ we should minimize or maximize its denominator $D(x)$ depending on the sign of expression $p_0 - \mu d_0$ (see formula (3.8)).

Let us return to the case when level-lines are not parallel with one another. Pick an arbitrary value of K and draw the line $Q(x) = K$ (see Figure 3.1). Let us rewrite equality $Q(x) = K$ as follows

$$x_2 = -\frac{p_1 - Kd_1}{p_2 - Kd_2}x_1 - \frac{p_0 - Kd_0}{p_2 - Kd_2}.$$

In such a case the slope

$$k = -\frac{p_1 - Kd_1}{p_2 - Kd_2}$$

of level-line $Q(x) = K$ depends on value K of objective function $Q(x)$, and is a monotonic function on K because

$$\frac{dk}{dK} = \frac{d_1p_2 - d_2p_1}{(p_2 - Kd_2)^2}.$$

Further, the sign of $\frac{dk}{dK}$ does not depend on the value of K, so we can write

$$\text{sign}\,\{\frac{dk}{dK}\} = \text{sign}\,\{d_1p_2 - d_2p_1\} = const.$$

It means that when rotating level-line around its focus point F in positive direction (i.e. counterclockwise), the value of objective function $Q(x)$ increases or decreases depending on the sign of expression $(d_1p_2 - d_2p_1)$.

Obviously, Figure 3.1 represents the case when rotating level-line in positive direction leads to growth of value $Q(x)$. When rotating level-line around its focus point F the line $Q(x) = K$ intersects feasible set S in two vertices (extreme points) x^* and x^{**}. In the point x^* objective function $Q(x)$ takes its maximal value over set S and in the point x^{**} it takes its minimal value.

2.2 Multiple Optimal Solutions

It may occur that when rotating level-line on its focus point F the level-line $Q(x) = K$ captures some edge of set S (see edge e in Figure 3.2). In this case

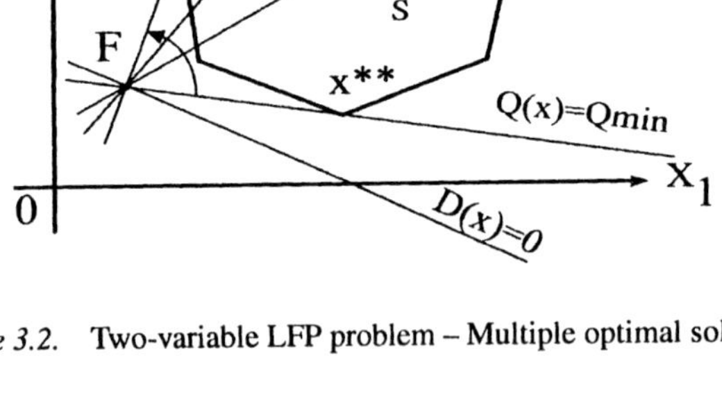

Figure 3.2. Two-variable LFP problem – Multiple optimal solutions.

the problem has an infinite number of optimal solutions (all points x of edge e) that may be represented as a linear combination of two vertex points x^* and x^{***}:

$$x = \lambda x^* + (1 - \lambda)x^{***}, \quad 0 \le \lambda \le 1.$$

2.3 Mixed cases

If feasible set S is unbounded and an appropriate unbounded edge concurs with extreme level-line (see Figure 3.3), then the problem has an infinite number of optimal solutions - one of them in vertex x^* and others are non-vertex points of unbounded edge. We should note here that among these non-vertex points there is one infinite point too. This is why we say in this case that the problem has 'mixed' solutions, i.e. finite optimal solution(s) and asymptotic one(s).

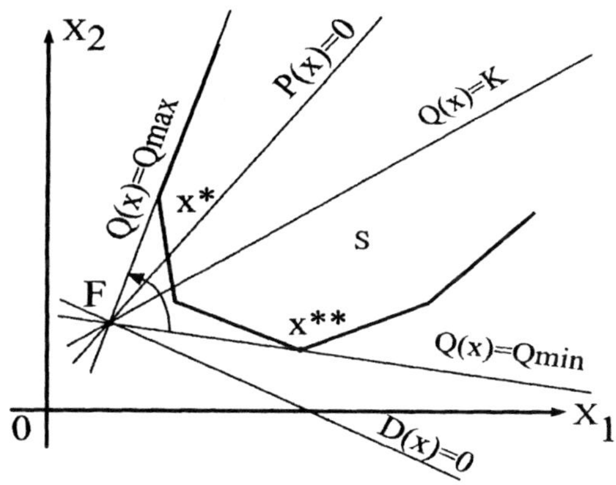

Figure 3.3. Two-variable LFP problem – Mixed case.

2.4 Asymptotic cases

Let use suppose that constraints (3.11) and (3.12) define an unbounded feasible set S shown in Figure 3.4. It may occur that when rotating level-line, after an extreme vertex (see vertex x^* in Figure 3.4) we can rotate the level-line a bit more in the same direction because the intersection between level-line and feasible set S is still not empty (see line $Q(x) = K$ in Figure 3.4). In this case we can rotate level-line until it becomes parallel with the appropriate unbounded edge (see edge e in Figure 3.4). If such a case occurs we should compute the value of objective function $Q(x)$ in infinite point x of the unbounded edge e, i.e. the following limit:

$$\lim_{\substack{x \to \infty \\ x \in e}} Q(x) .$$

Depending on the value of this limit, the maximal (minimal) value of objective function $Q(x)$ may be finite or infinite.

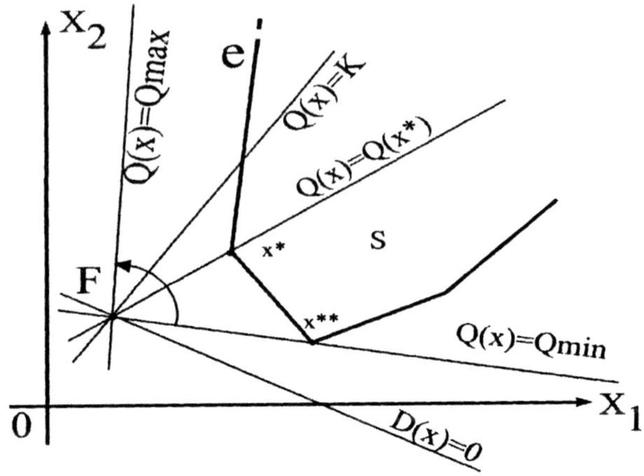

Figure 3.4. Two-variable LFP problem – Asymptotic case.

To illustrate this method we consider the following numeric example with a bounded feasible set

$$Q(x) = \frac{6\,x_1 + 3\,x_2 + 6}{5\,x_1 + 2\,x_2 + 5} \longrightarrow \max(\min)$$

subject to

$$\begin{array}{rcrcr} 4\,x_1 & - & 2\,x_2 & \leq & 20\,, \\ 3\,x_1 & + & 5\,x_2 & \leq & 25\,, \end{array}$$

$$x_1 \geq 0, \quad x_2 \geq 0.$$

First, we have to construct a feasible set. The convex set S of all feasible solutions for this problem is shown as the shaded region in Figure 3.5. Then, to determine coordinates of the focus point F we solve system

$$\begin{array}{rcrcr} 6\,x_1 & + & 3\,x_2 & = & -6\,, \\ 5\,x_1 & + & 2\,x_2 & = & -5\,, \end{array}$$

which gives us $F = (-1, 0)$. Level-lines being rotated around focus point F give the following extremal points

$$A = (0, 5), \quad B = (5, 0), \quad \text{and} \quad C = (0, 0)\,,$$

with objective values

$$Q(A) = 21/15, \quad Q(B) = 18/15, \quad \text{and} \quad Q(C) = 18/15\,,$$

respectively. So, the objective function $Q(x)$ reaches its maximal value in the point $A = (0, 5)$, while the minimization problem has multiple optimal

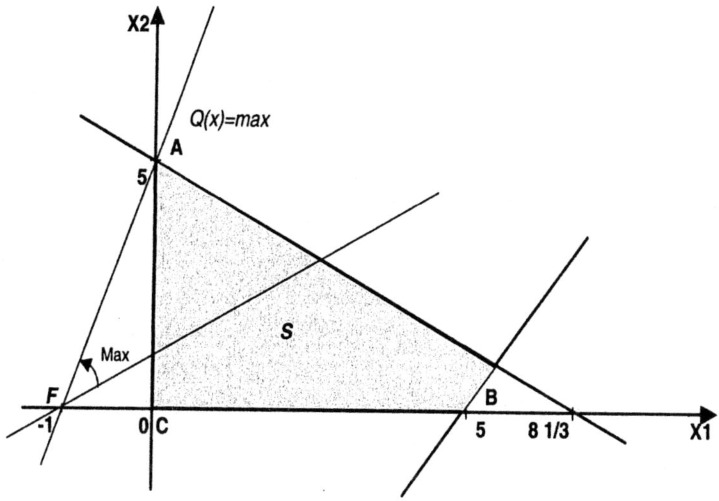

Figure 3.5. Graphical example – Bounded feasible set.

solutions: two extremal points $B = (5, 0)$ and $C = (0, 0)$, and all points x representable as a linear combination of B and C.

The following LFP problem illustrates an asymptotic case:

$$Q(x) = \frac{1\,x_1 - 2\,x_2 + 1}{1\,x_1 + 1\,x_2 + 4} \longrightarrow \max(\min)$$

subject to

$$\begin{array}{rcrcl} 1\,x_1 & + & 1\,x_2 & \geq & 2\,, \\ 1\,x_1 & - & 2\,x_2 & \leq & 4\,, \end{array}$$

$$x_1 \geq 0, \quad x_2 \geq 0.$$

Set S of all feasible solutions for this problem is shown in Figure 3.6. Solving system

$$\begin{array}{rcrcl} 1\,x_1 & - & 2\,x_2 & = & -1\,, \\ 1\,x_1 & + & 1\,x_2 & = & -4\,; \end{array}$$

we obtain a focus point with coordinates $F = (-3, -1)$. Then, rotating level-lines around focus point F in both directions (i.e. clockwise and counter-clockwise) we realize that the maximization problem has an optimal solution on the point $(4, 0)$ where $Q(x) = 5/8$, and the minimization problem has an

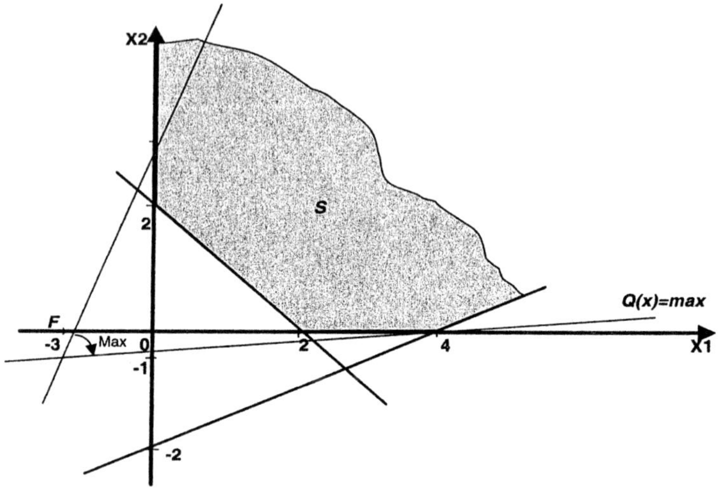

Figure 3.6. Graphical example – Unbounded feasible set.

asymptotic optimal solution in infinite point $(0, \infty)$ on the axes Ox_2

$$\min_{x \in S} Q(x) = \lim_{x_2 \to \infty} Q(0, x_2) = -2.$$

In the case of $n > 2$ 'level-lines' of objective function $Q(x)$ define a bundle of level-surfaces $P(x) - K \cdot D(x) = 0$ that are rotating on their $(n-2)$-dimensional focus axes, where surfaces $P(x) = 0$ and $D(x) = 0$ intersect one another.

Closing this section, we note that this geometrical interpretation of an LFP problem may be used for any number n of unknown variables and any number m of main constraints, if the original system of constraints may be reduced to 2 independent variables, i.e. $n - m = 2$.

3. Charnes & Cooper's Transformation

In 1962 A. Charnes and W.W. Cooper [38] showed that any linear-fractional programming problem with a bounded set of feasible solutions may be converted to a linear programming problem.

Consider the common LFP problem (3.1)-(3.3). Let us introduce the following new variables:

$$t_j = \frac{x_j}{D(x)}, \quad j = 1, 2, \ldots, n, \qquad t_0 = \frac{1}{D(x)},$$

where

$$D(x) = \sum_{j=1}^{n} d_j x_j + d_0. \tag{3.14}$$

Using these new variables $t_j,\ j = 0, 1, \ldots, n,$ we can rewrite the original objective function $Q(x)$ in the following form:

$$L(t) = \sum_{j=0}^{n} p_j t_j \longrightarrow \max(\text{ or } \min)\ . \tag{3.15}$$

Since we suppose that $D(x) > 0\ \forall x \in S,$ we can multiply all constraints of (3.2) and (3.3) by $1/D(x)$, so we obtain the following constraints:

$$\left.\begin{array}{ll} -b_i t_0 + \sum_{j=1}^{n} a_{ij} t_j \leq 0, & i = 1, 2, \ldots, m_1, \\ -b_i t_0 + \sum_{j=1}^{n} a_{ij} t_j \geq 0, & i = m_1 + 1, m_1 + 2, \ldots, m_2, \\ -b_i t_0 + \sum_{j=1}^{n} a_{ij} t_j = 0, & i = m_2 + 1, m_2 + 2, \ldots, m, \end{array}\right\} \tag{3.16}$$

$$t_j \geq 0, \quad j = 0, 1, 2, \ldots, n_1, \tag{3.17}$$

The connection between the original variables x_j and the new variables t_j will be completed if we multiply equality (3.14) by the same value $1/D(x)$, and then append the new constraint to the new problem:

$$\sum_{j=0}^{n} d_j t_j = 1. \tag{3.18}$$

Here and in what follows the new problem (3.15)-(3.18) will be referred to as a *linear analogue* of an LFP problem.

Since feasible set S is bounded, function $D(x)$ is linear and $D(x) > 0,\ \forall x \in S,$ the following statement may be formulated and proved:

LEMMA 3.1 *If vector* $t = (t_0, t_1, \ldots, t_n)^T$ *is a feasible solution of problem (3.15)-(3.18), then* $t_0 > 0$.

Proof. Let us suppose that vectors

$$x' = (x'_1, x'_2, \ldots, x'_n)^T, \quad \text{and} \quad t' = (t'_0, t'_1, t'_2, \ldots, t'_n)^T$$

are feasible solutions of the original LFP problem (3.1)-(3.3) and problem (3.15)-(3.18), respectively. Assume that

$$t'_0 = 0, \quad \text{i.e.} \quad t' = (0, t'_1, t'_2, \dots, t'_n)^T.$$

Since vectors x' and t' are feasible solutions of their problems, from (3.2)-(3.3) and (3.16)-(3.17) respectively, follows that

$$\left.\begin{array}{ll} \sum_{j=1}^{n} a_{ij} x'_j \leq b_i, & i = 1, 2, \dots, m_1, \\ \sum_{j=1}^{n} a_{ij} x'_j \geq b_i, & i = m_1 + 1, m_1 + 2, \dots, m_2, \\ \sum_{j=1}^{n} a_{ij} x'_j = b_i, & i = m_2 + 1, m_2 + 2, \dots, m, \end{array}\right\} \tag{3.19}$$

$$x'_j \geq 0, \quad j = 1, 2, \dots, n_1, \tag{3.20}$$

and

$$\left.\begin{array}{ll} \sum_{j=1}^{n} a_{ij} t'_j \leq 0, & i = 1, 2, \dots, m_1, \\ \sum_{j=1}^{n} a_{ij} t'_j \geq 0, & i = m_1 + 1, m_1 + 2, \dots, m_2, \\ \sum_{j=1}^{n} a_{ij} t'_j = 0, & i = m_2 + 1, m_2 + 2, \dots, m, \end{array}\right\} \tag{3.21}$$

$$t'_j \geq 0, \quad j = 1, 2, \dots, n_1. \tag{3.22}$$

Let us multiply each i-th constraint of system (3.21) by arbitrary positive λ and then add it to an appropriate i-th constraint of system (3.19). The same λ we will use to multiply each j-th restriction (3.22) and then to add it to the appropriate j-th constraint of (3.20), $j = 1, 2, \dots, n_1$. We have:

$$\begin{array}{ll} \sum_{j=1}^{n} a_{ij}(x'_j + \lambda t'_j) \leq b_i, & i = 1, 2, \dots, m_1, \\ \sum_{j=1}^{n} a_{ij}(x'_j + \lambda t'_j) \geq b_i, & i = m_1 + 1, m_1 + 2, \dots, m_2, \\ \sum_{j=1}^{n} a_{ij}(x'_j + \lambda t'_j) = b_i, & i = m_2 + 1, m_2 + 2, \dots, m, \end{array}$$

$$(x'_j + \lambda t'_j) \geq 0, \quad j = 1, 2, \dots, n_1.$$

It means that vector $x' + \lambda t'$ is a feasible solution of the original LFP problem for any positive λ. But λ may be as large as required, and hence it follows that feasible set S is unbounded. The latter contradicts our assumption that S is a bounded set. $\diamond$

This transformation (usually referred to as *Charnes&Cooper transformation*) of variables establishes a "*one* $\leftrightarrow$ *one*" connection between the original LFP problem (3.1)-(3.3) and its linear analogue (3.15)-(3.18):

THEOREM 3.1 *If vector* $t^* = (t_0^*, t_1^*, \dots, t_n^*)^T$ *is an optimal solution of problem (3.15)-(3.18), then vector* $x^* = (x_1^*, x_2^*, \dots, x_n^*)^T$ *is an optimal solution of original LFP problem (3.1)-(3.3), where*

$$x_j^* = \frac{t_j^*}{t_0^*}, \quad j = 1, 2, \dots, n. \tag{3.23}$$

Proof. We prove this statement only for the case of maximization problems. In the case of minimization the proof may be implemented in an analogous way.

Since vector t^* is the optimal solution of maximization linear analogue (3.15)-(3.18), it follows that

$$L(t^*) \geq L(t), \quad \forall t \in T, \tag{3.24}$$

where T denotes a feasible set of linear analogue (3.15)-(3.18). Let us suppose that vector x^* is not an optimal solution of the maximization LFP problem (3.1)-(3.3). Hence, there exists some another vector $x' \in S$, such that $Q(x') \geq Q(x^*)$. But at the same time

$$\begin{aligned} Q(x^*) &= \frac{\sum_{j=1}^{n} p_j x_j^* + p_0}{\sum_{j=1}^{n} d_j x_j^* + d_0} \overset{(3.23)}{=} \frac{\sum_{j=1}^{n} p_j \frac{t_j^*}{t_0^*} + p_0}{\sum_{j=1}^{n} d_j \frac{t_j^*}{t_0^*} + d_0} = \\ &= \frac{\sum_{j=1}^{n} p_j t_j^* + p_0 t_0^*}{\sum_{j=1}^{n} d_j t_j^* + d_0 t_0^*} \overset{(3.18)}{=} \frac{\sum_{j=1}^{n} p_j t_j^* + p_0 t_0^*}{1} \overset{(3.15)}{=} L(t^*). \end{aligned}$$

It means that

$$Q(x') \geq L(t^*). \tag{3.25}$$

Since vector x' is a feasible solution of the original LFP problem (3.1)-(3.3), it is easy to show that vector

$$t' = (t'_0, t'_1, \dots, t'_n)^T, \text{ where } t'_0 = \frac{1}{D(x')}, \quad t'_j = \frac{x'_j}{D(x')}, \quad j = 1, 2, \dots, n,$$

is a feasible solution of linear analogue (3.15)-(3.18) and

$$L(t') \geq L(t^*).$$

But the latter contradicts our assumption that vectort^* is an optimal solution of the maximization problem (3.15)-(3.18). It means that vectorx^* is an optimal solution of the maximization LFP problem (3.1)-(3.3). ◇

Consider the following numeric example

$$Q(x) = \frac{8\,x_1 + 9\,x_2 + 4\,x_3 + 4}{2\,x_1 + 3\,x_2 + 2\,x_3 + 7} \longrightarrow \max$$

subject to

$$\begin{array}{rcrcrcr} 1\,x_1 & + & 1\,x_2 & + & 2\,x_3 & \leq & 3\,, \\ 2\,x_1 & + & 1\,x_2 & + & 4\,x_3 & \leq & 4\,, \\ 5\,x_1 & + & 3\,x_2 & + & 1\,x_3 & \leq & 15\,, \end{array}$$

$$x_j \geq 0, \quad j = 1, 2, 3.$$

Solving this LFP problem we obtain

$$x^* = (1, 2, 0)^T, \quad P(x^*) = 30, \quad D(x^*) = 15, \quad Q(x^*) = 2\,.$$

In accordance with (3.15)-(3.18) we construct the following linear analogue of our LFP problem

$$L(t) = 4\,t_0 + 8\,t_1 + 9\,t_2 + 4\,t_3 \longrightarrow \max$$

subject to

$$\begin{array}{rcrcrcrcr} 7\,t_0 & + & 2\,t_1 & + & 3\,t_2 & + & 2\,t_3 & = & 1\,, \\ -3\,t_0 & + & 1\,t_1 & + & 1\,t_2 & + & 2\,t_3 & \leq & 0\,, \\ -4\,t_0 & + & 2\,t_1 & + & 1\,t_2 & + & 4\,t_3 & \leq & 0\,, \\ -15\,t_0 & + & 5\,t_1 & + & 3\,t_2 & + & 1\,t_3 & \leq & 0\,, \end{array}$$

$$t_j \geq 0, \quad j = 0, 1, 2, 3.$$

If we solve this linear programming problem we have

$$t^* = (\frac{1}{15}, \frac{1}{15}, \frac{2}{15}, 0)^T, \quad L(t^*) = 2\,.$$

So,

$$t_0^* = \frac{1}{15}\,,\ t_1^* = \frac{1}{15}\,,\ t_2^* = \frac{2}{15}\,,\ t_3^* = \frac{0}{15}\,.$$

We should note here that in the case of an unbounded feasible set S it may occur that in the optimal solution of the linear analogue $t_0^* = 0$. It means that the optimal solution of the original LFP problem is asymptotic and the optimal solution x^* contains variables with an infinite value. For more information on this topic see [68].

The connection between the optimal solutions of the original LFP problem and its linear analogue formulated in Theorem 3.1 seems to be very useful and at least from the point of view of theory allows to substitute the original LFP problem with its linear analogue and in this way to use LP theory and methods. However, in practice this approach based on the Charnes and Cooper transformation may not always be utilized. The problems arise when we should transform an LFP problem with some special structure of constraints, for example *transportation problem*, or *assignment problem* (see Chapter 9) or any other problem with a fixed structure of constraints, and would like to apply appropriate special methods and algorithms. Indeed, if in the original LFP problem we have n unknown variables and m main conditions, then in its linear analogue we obtain $n+1$ variables and $m+1$ constraints. Moreover, in the right-hand side of system (3.16) we have no vector b. Instead of the original vector b we have a vector of zeros. As we will see later in Chapter 5, the latter means that we cannot apply the main results of duality theory formulated for LP problems.

All these changes in the structure of constraints means that the use of special methods and algorithms in this case becomes very difficult or absolutely impossible.

This is why, in spite of the existence of the Charnes and Cooper transformation, we will focus on a direct approach to the investigation of an LFP problem and as we have seen the use of such a direct approach is necessary and unavoidable.

4. Dinkelbach's Algorithm

One of the most popular and general strategies for fractional programming (not necessary linear) is the parametric approach described by W.Dinkelbach in [54]. In the case of linear-fractional programming this method reduces the solution of a problem to the solution of a sequence of linear programming problems.

Consider the common LFP problem (3.1)-(3.3) and function

$$F(\lambda) = \max_{x \in S}\{P(x) - \lambda D(x)\}, \quad \lambda \in R,$$

where S denotes the feasible set of (3.1)-(3.3).

The following theorem plays the role of the theoretical foundation of Dinkelbach's method.

THEOREM 3.2 *Vector x^* is an optimal solution of the LFP problem (3.1)-(3.3) if and only if*

$$F(\lambda^*) = \max_{x \in S}\{P(x) - \lambda^* D(x)\} \ = \ 0 \tag{3.26}$$

where

$$\lambda^* = \frac{P(x^*)}{D(x^*)}. \tag{3.27}$$

Proof. If vector x^* is an optimal solution of problem (3.1)-(3.3) then

$$\lambda^* = \frac{P(x^*)}{D(x^*)} \geq \frac{P(x)}{D(x)}, \ \forall x \in S.$$

The latter means that

$$P(x) - \lambda^* D(x) \leq 0, \ \forall x \in S.$$

Taking into account equality (3.27) we obtain

$$\max_{x \in S}\{P(x) - \lambda^* D(x)\} \ = \ 0.$$

Conversely, if vector x^* is an optimal solution of problem (3.26) then

$$P(x) - \lambda^* D(x) \leq P(x^*) - \lambda^* D(x^*) \ = \ 0, \ \forall x \in S.$$

This means that vector x^* is an optimal solution of LFP problem (3.1)-(3.3). ◇

This theorem also gives a procedure for calculating the optimal solution of linear-fractional programming problem (3.1)-(3.3). Indeed, since $D(x) > 0, \ \forall x \in S$, we have

$$\frac{\partial F(\lambda)}{\partial \lambda} \ = \ -D(x) \ < \ 0.$$

The latter means that $F(\lambda)$ is strictly decreasing in λ. So, the algorithm consists of the steps shown in Figure 3.7.

To illustrate this algorithm we consider the following numeric example

$$Q(x) \ = \ \frac{P(x)}{D(x)} \ = \ \frac{x_1 + x_2 + 5}{3x_1 + 2x_2 + 15} \longrightarrow \max$$

Dinkelbach's Algorithm

Step 0. Take $x^{(0)} \in S$, compute $\lambda^{(1)} := \dfrac{P(x^{(0)})}{D(x^{(0)})}$, and let $k := 1$;

Step 1. Determine $x^{(k)} := \arg\max_{x \in S}\{P(x) - \lambda^{(k)} D(x)\}$;

Step 2. If $F(\lambda^{(k)}) = 0$ then $x^* = x^{(k)}$ is an optimal solution; Stop;

Step 3. Let $\lambda^{(k+1)} := \dfrac{P(x^{(k)})}{D(x^{(k)})}$; let $k := k + 1$; go to Step 1;

Figure 3.7. Algorithm – Dinkelbach's Algorithm.

subject to

$$\begin{array}{rcrcr} 3x_1 & + & x_2 & \leq & 6\,, \\ 3x_1 & + & 4x_2 & \leq & 12\,, \end{array} \tag{3.28}$$
$$x_1 \geq 0, \;\; x_2 \geq 0$$

Step 0: Since vector $x = (0,0)^T$ satisfies all constraints of the problem, we may take it as a starting point $x^{(0)} \in S$. So, for $x^{(0)} = (0,0)^T$ we obtain

$$\lambda^{(1)} := \frac{P(x^{(0)})}{D(x^{(0)})} = \frac{5}{15} = \frac{1}{3},$$

Step 1: Now, we have to solve the following linear programming problem

$$P(x) - \lambda^{(1)} D(x) = P(x) - \frac{1}{3} D(x) = \frac{1}{3}\, x_2 \longrightarrow \max$$

subject to constraints (3.28).

Solving this problem we obtain

$$x^{(1)} = (0,3)^T, \quad F(\lambda^{(1)}) = 1\,.$$

Step 2: Since $F(\lambda^{(1)}) \neq 0$ we have to perform
Step 3: We have to calculate

$$\lambda^{(2)} := \frac{P(x^{(1)})}{D(x^{(1)})} = \frac{1 \times 3 + 5}{2 \times 3 + 15} = \frac{8}{21},$$

then to put $k := k + 1 = 2$ and repeat
Step 1: Solve the following LP problem:

$$P(x) \;-\; \lambda^{(2)} D(x) \;=$$

$$= (1 - \frac{8}{21} \times 3)x_1 + (1 - \frac{8}{21} \times 2)x_2 + (5 - \frac{8}{21} \times 15) =$$
$$= -\frac{1}{7}x_1 + \frac{5}{21}x_2 - \frac{5}{7} \longrightarrow \max$$

subject to constraints (3.28).

The optimal solution for this problem is

$$x^{(2)} = (0,3)^T \quad \text{with} \quad F(\lambda^{(2)}) = 0 \,.$$

Step 2: Since $F(\lambda^{(2)}) = 0$ vector $x^* = x^{(2)}$ is the optimal solution; Stop;

In accordance with the algorithm, the optimal solution of our LFP problem is $x^* = (0,3)^T$ with optimal objective value $Q(x^*) = 8/21$.

5. LFP models

The applications of linear programming to various branches of human activity, and especially to economics, are well known. The applications of linear-fractional programming are less known, and, until now, less numerous. Of course, the linearity of a problem makes it easier to deal with, and hence leads to its greater popularity. However, not all real-life problems may be adequately described in the frames of linear models. Linear-fractional programming is a branch of nonlinear programming that was introduced only in the early 60's but since the first publications devoted to LFP problems, this branch has attracted the attention of more and more researchers and specialists because there is a broad field of real-world problems, where the use of LFP is more suitable.

In this section of the book we set out to consider several problems that may be formulated in the form of LFP problems.

5.1 Main Economic Interpretation

Let a certain company manufacture n different products. Further, let p_j be the profit gained by the company from a unit of the j-th product, p_0 be some constant profit gained whose magnitude is independent of the output volume. The manufacturing of one unit of product j costs d_j and there is some constant expenditure d_0 whose value does not depend on the production activity of the company and must be paid for in any case, even if the company does not manufacture anything.

Let b_i be the volume of some scarce resource i available to the company and a_{ij} be the expenditure quota of the i-th resource for manufacturing a unit of j-th kind of the product. The company must decide how many units of each product j should be produced if the efficiency calculated as the ratio (total profit)/(total cost) must be maximized.

This problem leads us to define decision variables x_j - the unknown output volume of some j-th product, $j = 1, 2, \ldots, n$. The company's total profit (including constant profit p_0) may be expressed as

$$P(x) = \sum_{j=1}^{n} p_j x_j + p_0,$$

while the total cost of production activity (including constant expenditure d_0) is

$$D(x) = \sum_{j=1}^{n} d_j x_j + d_0.$$

So the company's objective function may be written as

$$Q(x) = \frac{P(x)}{D(x)} = \frac{\sum_{j=1}^{n} p_j x_j + p_0}{\sum_{j=1}^{n} d_j x_j + d_0} \longrightarrow \max$$

The company's main constraints are the following:

$$\sum_{j=1}^{n} a_{ij} x_j \leq b_i, \quad i = 1, 2, \ldots, m.$$

Since unknown variables express the amount of production to be produced, of course, we also require

$$x_j \geq 0, \quad \text{for all} \quad j = 1, 2, \ldots, n.$$

This problem is formulated in general form as an LFP problem with n unknown nonnegative variables and m main constraints.

5.2 A Maritime Transportation Problem

Let us suppose that in port A we have to load a ship of limited carrying capacity C with n types of goods and transport these goods to port B. Our aim is to determine how much of each type of goods must be loaded such that the profit gained per unit of transportation cost be maximal. Let U_j be the maximum available quantity of j-th good, and p_j and d_j be the profit gained per unit of this good and cost of its transportation respectively, $j = 1, 2, \ldots, n$. If w_j denotes the weight of unit of j-th good, and x_j is an unknown variable, which expresses the quantity of j-th good be loaded, the mathematical model of such a problem may be formulated as follows:

$$\frac{\sum_{j=1}^{n} p_j x_j}{\sum_{j=1}^{n} d_j x_j} \longrightarrow \max$$

subject to

$$\sum_{j=1}^{n} w_j x_j \leq C$$

$$0 \leq x_j \leq U_j, \quad j = 1, 2, \ldots, n.$$

The problem formulated in this way is an LFP problem with one main constraint and n unknown nonnegative bounded variables.

5.3 Product Planning

Suppose that a refrigerator manufacturer is able to produce five types of refrigerator: Lehel 220, Lehel 120, Star 200, Star 160 and Star 250. The manufacturer has an order from dealers to produce 150, 70 and 290 units of Star 200, Star 160 and Star 250 respectively and240 units without type detailing (that is, they can be of any type). The manufacturer wishes to formulate a production plan that maximizes its profit gained per unit of cost. All necessary resources excluding Freon 12 and TL 16 aren't scarce. The maximal available quantities of Freon 12 and TL 16 are 125 and 80 liters respectively.

Manufacturing has the following requirements and known data:

	L 200	L 120	S 200	S 160	S 250
TL 16 $(liter/unit)$	0.2	0.13	–	–	–
F 12 $(liter/unit)$	–	–	0.22	0.21	0.26
$Price$ $\$/unit$	420.0	365.0	395.0	355.0	450.0
$Cost$ $\$/unit$	320.0	290.0	300.0	280.0	340.0

The manufacturer wishes to satisfy given orders and to get maximum profit gained per unit of total cost of production.

Let $x_j, j = 1, 2, 3, 4, 5,$ denotes the unknown quantities of Lehel 200, Lehel 120, Star 200, Star 160 and Star 250 respectively to be produced. The total profit gained by the manufacturer may be expressed in the following form:

$$\begin{aligned} P(x) &= (420 - 320)x_1 + (365 - 290)x_2 + (395 - 300)x_3 + \\ &+ (355 - 280)x_4 + (450 - 340)x_5. \end{aligned}$$

Obviously, total cost is

$$D(x) = 320x_1 + 290x_2 + 300x_3 + 280x_4 + 340x_5.$$

In this case the objective function will be the following:

$$Q(x) = \frac{P(x)}{D(x)} = \frac{420x_1 + 365x_2 + 395x_3 + 355x_4 + 450x_5}{320x_1 + 290x_2 + 300x_3 + 280x_4 + 340x_5} \longrightarrow \max$$

The main constraints of the problem will be:

$$\begin{array}{llllllll}
\text{Freon 12} & : & & & +0.22x_3 & +0.21x_4 & +0.26x_5 & \leq 125 \\
\text{TL 16} & : & 0.20x_1 & +0.13x_2 & & & & \leq 80 \\
\text{Star 200} & : & & & +1.00x_3 & & & \geq 150 \\
\text{Star 160} & : & & & & +1.00x_4 & & \geq 70 \\
\text{Star 250} & : & & & & & +1.00x_5 & \geq 290 \\
\text{Totally} & : & 1.00x_1 & +1.00x_2 & +1.00x_3 & +1.00x_4 & +1.00x_5 & = 750
\end{array}$$

Obviously, all unknown variables must be nonnegative:

$$x_j \geq 0, j = 1, 2, 3, 4, 5.$$

In this LFP problem we have the objective function to be maximized,6 main constraints and 5 unknown variables. Note that it would be more realistic to restrict variables x_j to integer values. Indeed, if we solve this problem we obtain the following optimal solution:

$$x_1^* = 232.69, \quad x_2^* = 0.00, \quad x_3^* = 150.00, \quad x_4^* = 70.00, \quad x_5^* = 297.31$$

which means that, for example the quantity of refrigerators Lehel 220 and Star 250 to be produced is 232.69 and 297.31 units respectively. Obviously, such an optimal solution cannot be applied in real life.

We will return to the so-called *integer linear-fractional programming* problems later in Chapter 8.

5.4 A Financial Problem

Suppose that the financial advisor of a university's endowment fund must invest up to $\$100,000$ in two types of securities: bond *7Stars*, paying a dividend of 7%, and stock *MaxMay*, paying a dividend of 9%. The adviser has been advised that no more than $\$30,000$ can be invested in stock *MaxMay*, while the amount invested in bond *7Stars* must be at least twice the amount invested in stock *MaxMay*. Independent of the amount to be invested, the service of the broker company which serves the adviser costs $\$100$.

How much should be invested in each security to maximize the efficiency of investment?

Let x and y denote the amounts invested in bond *7Stars* and stock *MaxMay*, respectively. We must then have

$$x + y \leq 100000;$$
$$x \geq 2y;$$
$$y \leq 30000.$$

Of course, we also require

$$x \geq 0, \quad \text{and} \quad y \geq 0.$$

The return to the university is

$$R(x, y) = 0.07x + 0.09y$$

while the total amount of investment will be

$$D(x, y) = x + y + 100.$$

Thus, our mathematical model is as follows

$$Q(x, y) = \frac{R(x, y)}{D(x, y)} = \frac{0.07x + 0.09y}{x + y + 100} \longrightarrow \max$$

subject to

$$x + y \leq 100000;$$
$$x - 2y \geq 0;$$
$$y \leq 30000.$$
$$x \geq 0, \quad y \geq 0.$$

5.5 A Transportation Problem

We formulate here a special problem of LFP that represents a very important special class of LFP problems considered in Chapter 9.

A company has three electric power plants that supply the power needs of four cities[1]. Each power plant can supply the following numbers of kilowatt-hours (kwh) of electricity:

	Plant 1	*Plant* 2	*Plant* 3
Supply (*Million kwh*)	35	50	40

[1]This model is based on an example considered in [188].

The peak power demands in these cities are as follows (in kwh):

	$City\ 1$	$City\ 2$	$City\ 3$	$City\ 4$
$Demand\ (Million\ kwh)$	45	20	30	30

The costs of sending 1 million kwh of electricity from plant to city depend on the distance the electricity must be transported (see Table 3.1) and the profit of the company gained per 1 million kwh electricity supplied is presented in Table 3.2. We formulate now an LFP problem to maximize the profit gained

	$City\ 1$	$City\ 2$	$City\ 3$	$City\ 4$
$Plant\ 1$	\$8	\$6	\$10	\$9
$Plant\ 2$	\$9	\$12	\$13	\$7
$Plant\ 3$	\$14	\$9	\$16	\$5

Table 3.1. Transportation problem – Shipping costs.

	$City\ 1$	$City\ 2$	$City\ 3$	$City\ 4$
$Plant\ 1$	\$5	\$4	\$4	\$3
$Plant\ 2$	\$6	\$2	\$3	\$4
$Plant\ 3$	\$10	\$5	\$6	\$2

Table 3.2. Transportation problem – Profit of company.

by the company per 1 unit of shipping cost.

Define a variable for each possible path of electricity: x_{ij} - unknown quantity of kwh of electricity sent from i-th plant to j-th city, $i = 1,2,3;\ j = 1,2,3,4.$ In terms of these variables, the total profit and total cost of the company respectively may be written as:

$$\begin{aligned} P(x) &= 5x_{11} + 4x_{12} + 4x_{13} + 3x_{14} + \\ &+ 6x_{21} + 2x_{22} + 3x_{23} + 4x_{24} + \\ &+ 10x_{31} + 5x_{32} + 6x_{33} + 2x_{34}, \end{aligned}$$

$$\begin{aligned} D(x) &= 8x_{11} + 6x_{12} + 10x_{13} + 9x_{14} + \\ &+ 9x_{21} + 12x_{22} + 13x_{23} + 7x_{24} + \\ &+ 14x_{31} + 9x_{32} + 16x_{33} + 5x_{34}. \end{aligned}$$

So we can formulate an objective function as:

$$Q(x) = \frac{P(x)}{D(x)} \longrightarrow \max$$

The company's plan should satisfy two types of constraints. First, the total power supplied by each plant cannot exceed the plant's capacity. So we have the following *supply constraints*:

$$\begin{array}{ccccccccc} x_{11} & + & x_{12} & + & x_{13} & + & x_{14} & \leq & 35\ , \\ x_{21} & + & x_{22} & + & x_{23} & + & x_{24} & \leq & 50\ , \\ x_{31} & + & x_{32} & + & x_{33} & + & x_{34} & \leq & 40\ . \end{array}$$

Second, each city must receive sufficient power to meet its demand. A constraint that ensures that a city receives its demand is a *demand constraint*. The company's plan must satisfy the following demand constraints:

$$\begin{array}{ccccccc} x_{11} & + & x_{21} & + & x_{31} & \geq & 45\ , \\ x_{12} & + & x_{22} & + & x_{32} & \geq & 20\ , \\ x_{13} & + & x_{23} & + & x_{33} & \geq & 30\ , \\ x_{14} & + & x_{24} & + & x_{34} & \geq & 30\ . \end{array}$$

Since all the unknown variables x_{ij} must be nonnegative, we add the sign restrictions

$$x_{ij} \geq 0, \quad i = 1, 2, 3; \ j = 1, 2, 3, 4.$$

Combining the objective function, supply constraints, demand constraints, and sign restrictions yields the LFP problem with 12 nonnegative variables and 7 main constraints.

5.6 A Blending Problem

A metal processor wishes to produce at least 15 kilograms of a new alloy *NA* of lead and tin, containing at least 60% of lead and at least 35% of tin. This new product may be sold for $200 per kilogram. There are four different alloys $A1$, $A2$, $A3$, and $A4$ available in amount of 12, 15, 16, and 10 kilograms, respectively. These alloys have the percentage compositions and prices per kilogram shown in the table bellow

	$A1$	$A2$	$A3$	$A4$
Lead	40%	60%	80%	70%
Tin	60%	40%	20%	30%
Costs	\$240	\$180	\$160	\$210

How should the processor blend alloys $A1$, $A2$ $A3$, and $A4$ to maximize efficiency of the business? In other words, the processor would like to know how many of each alloy must be blended so that the income/cost ratio would be maximal?

First of all, we define variables x_1, x_2, x_3, and x_4 which express the amount of each alloy to be blended. It is obvious, that the total cost of the blend is

$$D(x) = 240x_1 + 180x_2 + 160x_3 + 210x_4,$$

while the total income expected from the blend produced and sold is

$$P(x) = 200(x_1 + x_2 + x_3 + x_4) = 200x_1 + 200x_2 + 200x_3 + 200x_4.$$

The explicit conditions of the problem may be expressed as a following system of inequalities

$$\text{for lead:} \quad \frac{0.4x_1 + 0.6x_2 + 0.8x_3 + 0.7x_4}{x_1 + x_2 + x_3 + x_4} \geq 0.60\ ,$$

$$\text{for tin:} \quad \frac{0.6x_1 + 0.4x_2 + 0.2x_3 + 0.3x_4}{x_1 + x_2 + x_3 + x_4} \geq 0.35\ ;$$

which gives us the following system of linear inequalities

$$-0.20x_1 + 0.00x_2 + 0.20x_3 + 0.10x_4 \geq 0\ ,$$

$$0.25x_1 + 0.05x_2 - 0.15x_3 - 0.05x_4 \geq 0\ .$$

Since the available value of each alloy is limited, we have the following restrictions

$$x_1 \leq 12, \quad x_2 \leq 15, \quad x_3 \leq 16, \quad x_4 \leq 10.$$

Finally, we have to add to the system the following condition

$$x_1 + x_2 + x_3 + x_4 \geq 15$$

since the processor wishes to produce at least 15 kilograms of new alloy. Of course, we also require

$$x_j \geq 0, \quad j = 1, 2, 3, 4.$$

Combining objective function $Q(x) = P(x)/D(x)$ with restrictions leads to the following LFP problem with four bounded variables

$$Q(x) = P(x)/D(x) = \frac{200x_1 + 200x_2 + 200x_3 + 200x_4}{240x_1 + 180x_2 + 160x_3 + 210x_4} \longrightarrow \max$$

subject to

$$\begin{array}{rcrcrcrcr} -0.20x_1 & + & & + & 0.20x_3 & + & 0.10x_4 & \geq & 0\,, \\ 0.25x_1 & + & 0.05x_2 & - & 0.15x_3 & - & 0.05x_4 & \geq & 0\,, \\ x_1 & & & & & & & \leq & 12\,, \\ & & x_2 & & & & & \leq & 15\,, \\ & & & & x_3 & & & \leq & 16\,, \\ & & & & & & x_4 & \leq & 10\,, \end{array}$$

$$x_1 \geq 0, \quad x_2 \geq 0, \quad x_3 \geq 0, \quad x_4 \geq 0.$$

5.7 A Location Problem

An example of the recent interest in linear-fractional location models is given by the practical situation of locating and sizing offshore oil platforms. Initially, the problem was modelled as a *Multi-capacitated facility location problem* [89] where the investment costs must be minimized. Later, in [154] it was suggested that for this practical situation it would be more preferable to minimize the *cost/production* ratio and the problem was re-formulated with a fractional objective function.

One of the best known and most widely used discrete location models is the so-called *un-capacitated facility location* problem. The problem may be described as follows: there is a discrete set of possible locations for given facilities, and a set of consumers with known demands for production to be produced. The aim of optimization is to find such a location for facilities which satisfies all given constraints for demand, and maximizes the profit or the efficiency calculated as the profit/cost ratio (sometimes in the special literature referred to as a *profitability index*). Facilities are assumed to have unlimited capacity (un-capacitated facility), i.e. any facility can satisfy the demand of all consumers. In the case if each facility can only supply demand up to a given limit, the problem is called the *capacitated facility location* problem.

In its most general form, the un-capacitated facility location problem in LFP form may be formulated as follows [18]. Let $I = \{1, 2, \dots, m\}$ denote the set of consumers and $J = \{1, 2, \dots, n\}$ the set of sites where the given facilities may be located. Let also f_j denote the fixed cost of opening facility in site j, and c_{ij} the profit associated with satisfying the demand of consumer i from facility j. Usually, c_{ij} is a function of the production costs at site j, the demand

and selling price of consumer i, and the transportation costs between consumer i and site j. Obviously, without loss of generality we can assume that the fixed costs f_j are nonnegative. Introducing variables

$$y_j = \begin{cases} 1, & \text{if facility } j \text{ is open,} \\ 0, & \text{otherwise ;} \end{cases} \qquad j = 1, 2, \ldots, n,$$

and

$$x_{ij} \geq 0, \quad i = 1, 2, \ldots, m, \quad j = 1, 2, \ldots, n,$$

where x_{ij} is an unknown fraction of the demand of consumer i served by facility j, we can formulate the un-capacitated facility location problems in the following form

$$Q(x) = \frac{P(x)}{D(x)} = \frac{\sum_{i=1}^{m}\sum_{j=1}^{n} c_{ij}x_{ij}}{\sum_{j=1}^{n} f_j y_j} \longrightarrow \max,$$

subject to

$$\sum_{i=1}^{m}\sum_{j=1}^{n} c_{ij}x_{ij} - \sum_{j=1}^{n} f_j y_j \geq P_{min}, \tag{3.29}$$

$$\sum_{j=1}^{n} x_{ij} = 1, \quad i = 1, 2, \ldots, m,$$

$$x_{ij} \leq y_j, \quad i = 1, 2, \ldots, m, \quad j = 1, 2, \ldots, n,$$

$$x_{ij} \geq 0, \quad i = 1, 2, \ldots, m, \quad j = 1, 2, \ldots, n,$$

$$y_j = 0 \text{ or } 1, \quad j = 1, 2, \ldots, n,$$

where it is assumed that $f_j \geq 0,\ j = 1, 2, \ldots, n$, and $P_{min} > 0$. Additional constraint (3.29) here guarantees a minimum profit P_{min}. Note that the given LFP problem contains the discrete unknown variables y_j, and hence, belongs to the class of *integer LFP problems* (see Chapter 8). For more detailed information on location models see [18], [90], [136].

Such an enumeration of possible real-world applications of LFP may be as long as in the case of LP. Here we just note that LFP problems are particularly useful in the solution of economic problems, where various activities use certain scarce resources in various proportions while the aim is to find such a plan which optimizes a usually profit-per-cost like ratio subject to the constrained imposed on the limited resources.

Applications of LFP can also be found, for instance in information theory, stochastic programming, numerical analysis, game theory and maintenance. A rich collection of references on applications of linear fractional programming like cutting stock problems, shipping schedules, optimal policy for a Markov chain, macroeconomic planning model, etc. can be found in a book by B.D.Craven [48]. Another review of various applications for LFP is given by S.Schaible [163]. Special multi-objective models with ratio-type objective functions used in decision analysis may be found in [129]. Some other interesting applications of LFP may be found in the following papers: models of location analysis [18], financial planning [75], military applications [103], linear-fractional programming problems with multiple objective functions [116], and numerous other examples in [173].

6. Discussion Questions and Exercises

In the following exercises set up a linear-fractional programming model of the situation described. Determine if the model is in standard form. If it is not, state what must be changed to put the model into standard form.

3.1 (Blending problem) A new plastic material is being prepared by using two available products: PRS and SRA. Each kilogram of PRS contains 30 grams of substance CRA and 40 grams of substance MAL, while each kilogram of SPA contains 40 grams of CRA and 20 grams of MAL. The final blend may be sold for \$3.50 per kilogram and must contain at least 130 grams of CRA and at most 80 grams of MAL. Each kilogram of PRS costs\$3.00 and each kilogram of SRA costs \$2.50. How many kilograms of PRS and SRA should be used to maximize the ration income/cost, if we have only 2 kilograms of PRS and 3 kilograms of SRA?

3.2 (Agricultural problem) A farmer owns a farm which produces corn, soybeans, and oats. There are 25 acres of land available for cultivation. Each crop which is planted has certain requirements for labor and capital. These data along with the net profit figures are given in the accompanying table

$Crops$	$Labor(hrs)$	$Capital$	$Net\,profit$
$Corn(peracre)$	6	\$36	\$40
$Soybeans(peracre)$	6	\$24	\$30
$Oats(peracre)$	2	\$18	\$20

The farmer has available \$800 for capital and knows that there 280 hours available for working these crops. How much of each crop should be planted to maximize efficiency (net profit)/cost, if the farmer has to pay a constant land tax of \$500 independent of the crops planted?

3.3 (Investment problem) The administrator of a $\$250,000$ trust fund set up by Mr. Inco Gnito will have to adhere to certain guidelines. The total amount of $\$250,000$ need not be fully invested at any one time. The money may be invested in three different types of securities: a utilities stock paying a9% dividend, an electronics stock paying a4% dividend, and a bond paying a5% interest. Suppose that the amount invested in the stocks cannot be more than half the total amount invested. Moreover, the amount invested in the utilities stock cannot exceed $\$40,000$. At the same time, the amount invested in the bond must be at least $\$70,000$. What investment policy should be pursued to maximize efficiency of investments (total income)/(total investment)?

In the following exercises sketch the feasible set S defined by the given constraints, then find all vertices (extreme points) of$S$, define where the focus point of the objective function is, and finally, for the given objective function find the optimal solution(s).

3.4

$$Q(x) = \frac{x_1 + x_2 + 5}{3x_1 + 2x_2 + 15} \longrightarrow \max$$

subject to

$$\begin{array}{rcrcr} 3x_1 & + & x_2 & \leq & 6\,, \\ 3x_1 & + & 4x_2 & \leq & 12\,, \end{array}$$

$$x_1 \geq 0, \;\; x_2 \geq 0$$

3.5

$$Q(x) = \frac{3x_1 + x_2 - 5}{7x_1 + 2x_2 + 15} \longrightarrow \max$$

subject to

$$\begin{array}{rcrcr} -3x_1 & + & x_2 & \geq & 6\,, \\ 3x_1 & + & 5x_2 & \leq & 15\,, \end{array}$$

$$x_1 \geq 0, \;\; x_2 \geq 0$$

3.6

$$Q(x) = \frac{5x_1 - 3x_2 + 2}{4x_1 + 1x_2 + 5} \longrightarrow \min$$

subject to

$$\begin{array}{rcrcl} x_1 & + & 2x_2 & \leq & 4\,, \\ x_1 & + & 3x_2 & \geq & 6\,, \end{array}$$

$$x_1 \geq 0, \quad x_2 \geq 0$$

3.7

$$Q(x) = \frac{5x_1 - 3x_2 + 2}{4x_1 + 1x_2 - 2} \longrightarrow \max$$

subject to

$$\begin{array}{rcrcl} x_1 & + & 2x_2 & \geq & 4\,, \\ x_1 & + & 3x_2 & \leq & 6\,, \end{array}$$

$$x_1 \geq 0, \quad x_2 \geq 0$$

For the LFP problems given in exercises 3.4-3.7 formulate their linear analogue problems using the Charnes-Cooper transformation.

Chapter 4

THE SIMPLEX METHOD

In 1947, George Dantzig [51] developed an efficient method, the simplex algorithm, for solving linear programming problems. Since the development of the simplex method, LP has been used to solve optimization problems anywhere where there appears a necessity of optimizing some absolute criteria. It might be, for example, cost of trucking, profit gained by some company, number of full-time employees, cost of nutrition rations, etc.

Later, in 1960, Béla Martos [131], [132] upgraded the simplex method for solving LFP problems formulated in the following*standard* form:

$$Q(x) = \frac{P(x)}{D(x)} = \frac{\sum_{j=1}^{n} p_j x_j + p_0}{\sum_{j=1}^{n} d_j x_j + d_0} \rightarrow \max, \tag{4.1}$$

subject to

$$\sum_{j=1}^{n} a_{ij} x_j = b_i, \quad i = 1, 2, \dots, m, \tag{4.2}$$

$$x_j \geq 0, \quad j = 1, 2, \dots, n, \tag{4.3}$$

where $D(x) > 0$ for all $x = (x_1, x_2, \cdots, x_n)^T$, which satisfy constraints (4.2)-(4.3). We assume that feasible set$S$ is a regular set, i.e. is non-empty and bounded.

1. Main Definitions and Theorems

In this section we formulate the main definitions and theorems for the standard LFP problem formulated in the form of (4.1)-(4.3). Some important facts connected with convex sets and the monotonicity of linear-fractional function are discussed too.

DEFINITION 4.1 *We say that linear-fractional programming problem* ***issolvable****, if*

- *feasible set S is not empty, i.e. there exists at least one such vector x that satisfies constraints (4.2)-(4.3) and*
- *objective function $Q(x)$ has a finite upper bound over set S.*

In other cases an LFP problem is said to be ***unsolvable****.*

Consider the following system of linear equations:

$$\sum_{j=1}^{n} A_j x_j = b,$$

where

$$A_j = \begin{pmatrix} a_{1j} \\ a_{2j} \\ \vdots \\ a_{mj} \end{pmatrix}, \; j = 1, 2, \dots, n, \quad b = \begin{pmatrix} b_1 \\ b_2 \\ \vdots \\ b_m \end{pmatrix}, \quad \text{and } m \leq n.$$

DEFINITION 4.2 *We say that system $B = \{A_{s_1}, A_{s_2}, \dots, A_{s_m}\}$ of vectors A_j is a* ***basis****, if vectors $A_{s_1}, A_{s_2}, \dots, A_{s_m}$ are linearly independent.*

Let us suppose that given system $B = \{A_{s_1}, A_{s_2}, \dots, A_{s_m}\}$ is a basis. Let J_B be a set of indices j corresponding to vectors A_j of basis B, i.e. $J_B = \{s_1, s_2, \dots, s_m\}$. If $J = \{1, 2, \dots, n\}$, then set $J_N = J \backslash J_B$ denotes the indices of those vectors Aj, which are not in basis B.

DEFINITION 4.3 *The given vector $x = (x_1, x_2, \dots, x_n)^T$ is a* ***basic solution*** *(or* ***basic vector****) to system $Ax = b$, if vector x satisfies system*

$$\sum_{j \in J_B} A_j x_j = b \text{ and } x_j = 0, \quad \forall j \in J_N.$$

Those variables x_j whose indices are in the set J_B are said to be *basic variables* or *BV's*. If variable x_j is such that $j \in J_N$, we will say that this variable is a *nonbasic variable* or *NBV*.

DEFINITION 4.4 *A point x in a convex set S is called an* ***extreme point*** *of S if x cannot be expressed as a convex combination of any other two distinct points of S.*

Equivalently, we can say that x is an extreme point if x is not an in-between (inner) point of any line segment of S. This fact may be expressed more precisely in the form of the following definition.

DEFINITION 4.5 *A point x in a convex set S is an* ***extreme point*** *of S if there do not exist distinct points x' and x'' in S and number λ, where $0 < \lambda < 1$, such that $x = \lambda x' + (1 - \lambda)x''$.*

Other suggestive names for extreme point are *corner point* and *vertex.*

DEFINITION 4.6 *For any convex set corresponding to the system $Ax = b$, with m constraints, two basic solutions are said to be* ***adjacent*** *(or* ***neighbouring****) if their sets of basic variables have $(m - 1)$ basic variables in common.*

Extreme points play a very important role in solving optimization problems related to convex polyhedrons.

As a preamble to the important Theorem 4.1, let us recall that a *polyhedral, convex set* is an intersection set of a finite number of closed half-spaces of R^n, while a *hyperplane* in R^n is the set of points satisfying an equation of the form

$$a_1x_1 + a_2x_2 + \ldots + a_nx_n = b.$$

THEOREM 4.1 *A point x of set corresponding to system $Ax = b$, is its extreme point if and only if it is its basic solution.*

In other words, at least one basic solution corresponds to any extreme point.

DEFINITION 4.7 *We will say that basic solution x is* ***degenerate,*** *if at least one of its basic variables is equal to zero, i.e. $\exists j : j \in J_B$, such that $x_j = 0$. In the case if $x_j \neq 0$, $\forall j \in J_B$, basic solution x is said to be* ***non-degenerate.***

The conception of a (non-)degenerate basic feasible solution has a very important role in the simplex method because in the case of degeneracy an extreme point may have more than one basis and hence, more than one basic solution.

DEFINITION 4.8 *Basic solution $x = (x_1, x_2, \dots, x_n)^T$ of system $Ax = b$ is said to be a **basic feasible solution (BFS)** of LFP problem (4.1)-(4.3)if all elements $x_j,\ j = 1, 2, \dots, n,$ of vector x satisfy nonnegativity constraints (4.3).*

DEFINITION 4.9 *Standard LFP problem (4.1)-(4.3) is said to be **normal** (or **canonical**) if all elements $b_i,\ i = 1, 2, \dots, m,$ of right-hand side (RHS) vector $b = (b_1, b_2, \dots, b_m)^T$ are nonnegative, i.e. $b_i \geq 0,\ i = 1, 2, , , \dots, m$.*

The applicability of the simplex method to an LFP problem is based on the following theorem:

THEOREM 4.2 **(Monotonicity)** *Objective function $Q(x)$ is monotonic on any segment of a straight line in feasible set S.*

Proof. We begin the proof by choosing two arbitrary points x' and x'' from feasible set S, i.e. $x' \in S$ and $x'' \in S$. Let us consider objective function $Q(x)$ on segment $x'x''$, in other words, let $x = \lambda x' + (1 - \lambda)x''$, where $0 \leq \lambda \leq 1$. It is clear that

$$Q(\lambda) = \frac{P(\lambda x' + (1-\lambda)x'')}{D(\lambda x' + (1-\lambda)x'')} = \dots = \frac{\lambda P(x') + (1-\lambda)P(x'')}{\lambda D(x') + (1-\lambda)D(x'')}$$

and

$$\frac{dQ(\lambda)}{d\lambda} = \frac{P(x')D(x'') - P(x'')D(x')}{(\lambda D(x') + (1-\lambda)D(x''))^2}.$$

The latter means that on the line segment $x'x''$ objective function $Q(x)$

- is increasing, if $P(x')D(x'') - P(x'')D(x') > 0$,
- is decreasing, if $P(x')D(x'') - P(x'')D(x') < 0$,
- is constant, if $P(x')D(x'') - P(x'')D(x') = 0$.

Thus the theorem is proved. ◇

Since feasible set S is a convex set, from Theorem 4.2 it follows that

THEOREM 4.3 *If feasible set S in linear-fractional programming problem (4.1)-(4.3) is bounded, then objective function $Q(x)$ attains its maximal value over S in an extreme point of S.*

REMARK 4.1 *Theorem 4.3 is true for LFP problems in which the feasible set S is bounded. It may not be true for a problem with an unbounded feasible set (see Section 2.3 and 2.4 of Chapter 3).*

2. Criteria of Optimality

Let us suppose that standard LFP problem (4.1)-(4.3) is normal (canonical), i.e. $b_i \geq 0, \quad i = 1, 2, \ldots, m$. We suppose also that vector $x = (x_1, x_2, \ldots, x_n)^T$ is a non-degenerate basic feasible solution of this problem with basis $B = (A_{s_1}, A_{s_2}, \ldots, A_{s_m})$. It means that

$$x_j \begin{cases} > 0, & j \in J_B = \{s_1, s_2, \ldots, s_m\}, \\ = 0, & j \in J_N = J \setminus J_B, \end{cases}$$

where $J = \{1, 2, , \ldots, n\}$. In accordance with our assumption, we obtain the following

$$\sum_{j=1}^{n} A_j x_j = \sum_{j \in J_B} A_j x_j + \sum_{j \in J_N} A_j x_j = \sum_{j \in J_B} A_j x_j + 0 = \sum_{i=1}^{m} A_{s_i} x_{s_i}.$$

Since vector x is a feasible solution, we have

$$\sum_{i=1}^{m} A_{s_i} x_{s_i} = b. \tag{4.4}$$

In accordance with the theory of the simplex method, let us choose some non-basic vector A_j (i.e. $j \in J_N$) and bring it into the basis. Let θ denote the value of a new basic variable x_j in the new basis, and $x_j(\theta)$ be new values of other basic variables. Then from (4.4) we get the following:

$$\sum_{i=1}^{m} A_{s_i} x_{s_i}(\theta) + A_j \theta = b. \tag{4.5}$$

Since vectors $A_{s_1}, A_{s_2}, \ldots, A_{s_m}$ of basis B are linearly independent, we can represent vector A_j as their linear combination:

$$A_j = \sum_{i=1}^{m} A_{s_i} x_{ij}. \tag{4.6}$$

Replacing vector A_j in formula (4.5) with its representation (4.6) we obtain that

$$\sum_{i=1}^{m} A_{s_i} x_{s_i}(\theta) + \theta \sum_{i=1}^{m} A_{s_i} x_{ij} = b. \tag{4.7}$$

The right-hand sides of expressions (4.4) and (4.7) are identical, so

$$\sum_{i=1}^{m} A_{s_i} x_{s_i}(\theta) + \theta \sum_{i=1}^{m} A_{s_i} x_{ij} = \sum_{i=1}^{m} A_{s_i} x_{s_i}$$

or

$$\sum_{i=1}^{m} A_{s_i} x_{s_i}(\theta) = \sum_{i=1}^{m} A_{s_i}(x_{s_i} - \theta x_{ij}) .$$

Since vectors $A_{s_1}, A_{s_2}, \ldots, A_{s_m}$ are linearly independent, the latest means that

$$x_{s_i}(\theta) = x_{s_i} - \theta x_{ij}, \quad i = 1, 2, \ldots, m. \tag{4.8}$$

Formula (4.8) being used for calculating the new basic vector $x(\theta)$ guarantees that main constraints (4.2) of LFP problem (4.1)-(4.3) will be satisfied. However, there is no guarantee that all components $x_j(\theta), \ j = 1, 2, \ldots, n,$ of the new basic vector $x(\theta)$ will be nonnegative, and hence, vector $x(\theta)$ will be a basic feasible solution of LFP problem (4.1)-(4.3). This is why we have to select such θ that

$$x_{s_i}(\theta) \geq 0, \quad i = 1, 2, \ldots, m,$$

or, in accordance with (4.8)

$$x_{s_i} - \theta x_{ij} \geq 0, \quad i = 1, 2, \ldots, m.$$

It is obvious that the latter may be rewritten as follows:

$$\theta \leq \frac{x_{s_i}}{x_{ij}}, \quad \text{for those index } i \text{ that } x_{ij} > 0,$$

$$\theta \geq \frac{x_{s_i}}{x_{ij}}, \quad \text{for those index } i \text{ that } x_{ij} < 0,$$

or in more compact form:

$$\max_{x_{ij}<0} \frac{x_{s_i}}{x_{ij}} \leq \theta \leq \min_{x_{ij}>0} \frac{x_{s_i}}{x_{ij}} \tag{4.9}$$

Since θ is the new value of the new basic variable x_j, we may choose only nonnegative θ, so instead of (4.9) we have to use the following range

$$0 \leq \theta \leq \min_{x_{ij}>0} \frac{x_{s_i}}{x_{ij}}. \tag{4.10}$$

Moreover, we cannot choose θ to be zero because in this case remain only $(m-1)$ vectors in the basis and hence, in accordance with Definition 4.2, it is not a basis. For the same reason, we cannot select a value for θ from within of range (4.10) (as an inner point), because in this case the new system of vectors will consist of $(m+1)$ vectors A_j. This is why we have to select

$$\theta = \min_{x_{ij}>0} \frac{x_{s_i}}{x_{ij}}. \tag{4.11}$$

This formula (4.11) is called the*minimum ratio test.* Note that when performing this minimum ratio test such a case may occur when for a given vectorA_j there is no such index i that $x_{ij} > 0$ and hence, the upper bound for range (4.9) does not exist. Here we do not discuss this situation but will return to this case later in Section 3. Another 'bad' case, when a minimum ratio test results in more than one index i, is called *tie*, and is discussed in detail in Section 8.2.

Once we have cleared the rule for choosing the value ofθ, let us suppose that

$$\min_{x_{ij}>0} \frac{x_{s_i}}{x_{ij}} = \frac{x_r}{x_{rj}}.$$

It means that in the new basis $x_r(\theta) = 0,\ \ x_j = \theta$ and vector A_j will replace in the basis vector A_r. So instead of basis

$$B = \{A_{s_1}, A_{s_2}, \ldots, A_r, \ldots, A_{s_m}\}$$

we obtain a new basis

$$B' = \{A_{s_1}, A_{s_2}, \ldots, A_j, \ldots, A_{s_m}\}.$$

Now we have to calculate the new value of objective function$Q(x)$ for the new basic feasible solution $x(\theta)$:

$$Q(x(\theta)) = \frac{P(x(\theta))}{D(x(\theta))} = \frac{\sum_{j=1}^{n} p_j x_j(\theta) + p_0}{\sum_{j=1}^{n} d_j x_j(\theta) + d_0} = \frac{\sum_{i=1}^{m} p_{s_i} x_{s_i}(\theta) + p_j\theta + p_0}{\sum_{i=1}^{m} d_{s_i} x_{s_i}(\theta) + d_j\theta + d_0} =$$

$$= \frac{\sum_{i=1}^{m} p_{s_i}(x_{s_i} - \theta x_{ij}) + p_j\theta + p_0}{\sum_{i=1}^{m} d_{s_i}(x_{s_i} - \theta x_{ij}) + d_j\theta + d_0} =$$

$$= \frac{\sum_{i=1}^{m} p_{s_i} x_{s_i} + p_0 - \theta(\sum_{i=1}^{m} p_{s_i} x_{ij} - p_j)}{\sum_{i=1}^{m} d_{s_i} x_{s_i} + d_0 - \theta(\sum_{i=1}^{m} d_{s_i} x_{ij} - d_j)} = \frac{P(x) - \theta\Delta'_j}{D(x) - \theta\Delta''_j},$$

where

$$\Delta'_j = \sum_{i=1}^{m} p_{s_i} x_{ij} - p_j, \quad \Delta''_j = \sum_{i=1}^{m} d_{s_i} x_{ij} - d_j.$$

Once we have calculated $Q(x(\theta))$, we can estimate the change in the value of objective function $Q(x)$:

$$Q(x(\theta)) - Q(x) = \frac{P(x) - \theta\Delta'_j}{D(x) - \theta\Delta''_j} - \frac{P(x)}{D(x)} = \ldots = \frac{-\theta\Delta_j(x)}{D(x(\theta))}, \qquad (4.12)$$

where

$$\Delta_j(x) = \Delta'_j - Q(x)\Delta''_j = \begin{vmatrix} \Delta'_j & Q(x) \\ \Delta''_j & 1 \end{vmatrix}.$$

Formula (4.12) has a very important role in the simplex method because it allows us to check if we have made a right choice bringing vector A_j into the basis or not. Indeed, since $\theta > 0$ and $D(x(\theta)) > 0$ $(D(x) > 0,\ \forall x \in S)$, when replacing basic vector A_r with nonbasic vector A_j (and hence, changing point x to point $x(\theta)$), the value of objective function $Q(x)$ increases or decreases depending on the sign of determinant $\Delta_j(x)$. If $\Delta_j(x) < 0$, then the value of function $Q(x)$ increases, if $\Delta_j(x) > 0$, then $Q(x)$ decreases. In the case if $\Delta_j(x) = 0$, then the value of $Q(x)$ remains without any change.

In this way we have shown that the following takes place

THEOREM 4.4 (CRITERIA OF OPTIMALITY) *A basic feasible solution x is a basic optimal solution of linear-fractional programming problem (4.1)-(4.3) if and only if $\Delta_j(x) \geq 0,\ j = 1, 2, \ldots, n$.*

Obviously, if $d_j = 0,\ j = 1, 2, \ldots, n$ and $d_0 = 0$, then $\Delta_j(x) = \Delta'_j,\ j = 1, 2, \ldots, n$ and from Theorem 4.4 we obtain criteria of optimality for the simplex method in LP.

Before closing this section we should remark that in linear programming

$$\Delta'_j = \sum_{i=1}^{m} p_{s_i} x_{ij} - p_j, \quad j = 1, 2, \ldots, n,$$

usually are referred to as *reduced costs* or *relative costs*. If p_j denotes the direct cost related to a unit of jth product to be produced, and the aim of the objective function of an LP problem is minimization of the total cost, then

$$z_j = \sum_{i=1}^{m} p_{s_i} x_{ij}, \quad j = 1, 2, \ldots, n,$$

express the so-called *indirect costs*. So,

$$\Delta'_j = z_j - p_j,$$

is the difference between the indirect cost z_j and the direct cost p_j, and indicates how much the optimal value of objective function $P(x)$ would change per unit change in the optimal value of x_j.

Observe that in LFP, $\Delta_j(x)$ cannot be interpreted in this manner. Even so, for the sake of similarity with LP sometimes we will refer to Δ'_j, Δ''_j, and $\Delta_j(x)$ as *reduced cost of numerator, reduced cost of denominator* and *reduced cost of LFP*, respectively.

3. General Scheme of the Simplex Method

Here we describe how the simplex method can be used to solve an LFP problem in which its objective function must be maximized. The solution of a minimization LFP problem may be obtained in the same way if we substitute the original minimization problem with its appropriate maximization equivalent (see formula (3.9), page 48).

The simplex method proceeds as follows:

1 Convert the LFP problem to standard form (see Section 1.3).

2 Find an initial basic feasible solution, if possible. This may be very easy if all constraints in the original LFP problem are "$\leq$" constraints with nonnegative right-hand sides. Then the slack variables s_i may be used as the basic variable for i-th row. If no BFS is readily apparent, we use the techniques discussed in Section 6.1 and Section 6.2 to find a basic feasible solution.

3 If all nonbasic variables $x_j, \ \forall j \in J_N,$ have nonnegative determinants $\Delta_j(x) \geq 0, \ \forall j \in J_N,$ the current basic feasible solution is optimal. If there exists at least one index j such that $\Delta_j(x) < 0, \ j \in J_N,$, choose the appropriate variable to bring it into the basis. We call this variable the *entering variable* and the corresponding vector A_j the *entering vector.*

4 Bring chosen entering variable into the basis, recalculate reduced costs of LFP $\Delta_j(x)$ and then go to step 3.

Let us focus on step 3 of this procedure. Suppose that vector x is a nondegenerate basic feasible solution of a standard LFP problem (4.1)-(4.3) with basis $B = (A_{s_1}, A_{s_2}, \dots, A_{s_m})$. As in the previous sections, let J_B denote a set of indices j which correspond to basic vectors A_j, that is $J_B = \{s_1, s_2, \dots, s_m\}$. Let $J = \{1, 2, \dots, n\}$, and $J_N = J \setminus J_B$ be a set of indices of nonbasic vectors.

An analysis of vector x for the purpose of its optimality begins by calculating the following values (in the given order):

1 reduced costs $\Delta'_j,\ \Delta''_j,\ j = 1, 2, \ldots, n,$

2 objective function $Q(x)$ in the point x and then

3 determinants $\Delta_j(x) = \begin{vmatrix} \Delta'_j & Q(x) \\ \Delta''_j & 1 \end{vmatrix},\ j = 1, 2, \ldots, n.$

When checking calculated reduced costs of LFP $\Delta_j(x)$, the following 3 cases may occur:

1 All nonbasic determinants $\Delta_j(x)$ are nonnegative, that is

$$\Delta_j(x) \geq 0,\ \forall j \in J_N;$$

2 There does exist at least one nonbasic index j_0, such that $\Delta_{j_0}(x)$ has a negative value, and all m appropriate coefficients x_{ij_0} are non-positive, that is

$$J_0 = \{j :\ j \in J_N;\ \Delta_j(x) < 0\} \neq \emptyset;$$

and

$$J_0^- = \{j :\ j \in J_0;\ x_{ij} \leq 0,\ \forall i = 1, 2, \ldots, m\} \neq \emptyset;$$

3 There does exist at least one nonbasic index j_0, such that $\Delta_{j_0}(x)$ has a negative value, and for all such indices j_0 at least one coefficient x_{ij_0} is positive, that is

$$J_0 = \{j :\ j \in J_N;\ \Delta_j(x) < 0\} \neq \emptyset;$$

and

$$J_0^- = \{j :\ j \in J_0;\ x_{ij} \leq 0,\ \forall i = 1, 2, \ldots, m\} = \emptyset;$$

In case 1, in accordance with criteria of optimality (see Theorem 4.4), vector x is an optimal basic solution of LFP problem (4.1)-(4.3). The method must be terminated here because the problem has been solved.

REMARK 4.2 *If among nonbasic determinants $\Delta_j(x)$ there is at least one zero value $\Delta_j(x)$, it means that the LFP problem has alternative optimal solutions.*

In case 2, feasible set S of LFP problem (4.1)-(4.3) is unbounded (we excluded this case from our consideration, see page 75). Indeed, in this case we can find such an index j_0 that $\Delta_{j_0}(x) < 0$ and all m coefficients x_{ij_0} are non-positive, that is $x_{ij_0} \leq 0,\ i = 1, 2, \ldots, m$. In accordance with formula (4.10) it means that θ has no finite upper bound and its value may be arbitrarily large. In this case, as it follows from formula (4.8), new vector $x(\theta)$ remains a feasible solution of LFP problem (4.1)-(4.3) for any $x(\theta)$, and may contain arbitrarily large components $x_j(\theta)$. The latter means that feasible set S in this case is unbounded. Here the simplex method must be terminated because for the given LFP problem the simplex method is not applicable.

REMARK 4.3 *Case 2 does not mean that a given LFP problem is unsolvable in principle because of the unboundedness of objective function $Q(x)$ from above. Since $Q(x)$ has fraction form, the limit*

$$\lim_{\substack{x \to \infty \\ x \in S}} Q(x)$$

may have a finite value too.

We have to note here that several attempts were made to expand the simplex method to the case of unbounded LFP, see for example [27], [94].

In case 3, there does exist such a new basic feasible solution $x(\theta)$ that

$$Q(x(\theta)) > Q(x).$$

Indeed, in accordance with our assumptions in this we can find at least one such nonbasic index j that $J_0 \neq \emptyset$ and $J_0^- = \emptyset$. Hence, from the range (4.10) it follows that the value of θ is bounded from above, and its maximal possible value is defined by formula (4.11). Since $x(\theta)$ is a feasible solution of LFP problem (4.1)-(4.3), and $D(x) > 0,\ \forall x \in S,$ we are sure that $D(x(\theta)) > 0$. From the latter it follows that under the conditions of the current case ($J_0 \neq \emptyset$) we can choose such an index $j_0 \in J_0$ that (see formula (4.12))

$$Q(x(\theta)) - Q(x) = \frac{-\theta \Delta_{j_0}(x)}{D(x(\theta))} > 0.$$

It means that bringing vector A_{j_0} into the new basis we can construct such a new basic feasible solution $x(\theta)$ for LFP problem (4.1)-(4.3) which is better than the current basic feasible solution x, that is $Q(x(\theta)) > Q(x)$. Thus, we have proceeded from one BFS to a better adjacent BFS. The procedure used to get from one BFS to another (and perhaps, better) one is called an *iteration* of the simplex method.

Since set S of feasible solutions x is bounded, and we can choose only such new basic feasible solutions $x(\theta)$ that are better than the current BFS x, the simplex method guarantees that after a finite number of such iterations we get case 1 or case 2.

REMARK 4.4 *In this section we assumed that the current basic feasible solution x is a non-degenerate vector, i.e. contains exactly m positive basic* *variables (see Definition 4.7). This assumption guarantees in formula (4.11) that $\theta > 0$ and hence, that the value of objective function $Q(x)$ does increase. In the case of the degenerate vector x, the value of $\theta = 0$ and hence, the value of objective function $Q(x)$ does not change. In this situation there may occur so-called* ***cycling***[1]*, which may be avoided by using special techniques described in Section 9.*

It may occur that one (or more) nonbasic determinant $\Delta_j(x)$ calculated for optimal basic feasible solution x, has zero value. It means that corresponding nonbasic vector A_j may be entered into the new basis but it does not lead to any change in the value of objective function $Q(x)$ (see formula (4.12)). So we can obtain a new basic feasible solution $x(\theta)$ with the same optimal value for objective function $Q(x)$, that is $Q(x) = Q(x(\theta))$. Obviously, vector $x(\theta)$ is a so-called *alternative* basic optimal solution of LFP problem (4.1)-(4.3). Since every basic feasible solution x corresponds to some vertex of polyhedron S, all points x' that may be represented as a linear combination of these two optimal basic solutions x and $x(\theta)$

$$x' = \lambda x + (1 - \lambda)x(\theta), \quad \text{where } 0 \leq \lambda \leq 1,$$

are also optimal solutions for LFP problem (4.1)-(4.3). In this situation, an LFP problem has two basic optimal solutions x and $x(\theta)$, and an infinite number of nonbasic optimal solutions x'.

4. Simplex Tableau

When applying the simplex method to solve an LFP problem we must examine the current basic feasible solution for its optimality and attempt to arrive at a basic feasible solution where the optimum value (i.e. maximum or minimum value) of the objective function $Q(x)$ is reached. Thus, it is clear that it would be useful to organize all necessary data in some tableau.

Such a *simplex tableau* is presented in Table 4.1. In this tableau the first two rows contain coefficients of numerator $P(x)$ and denominator $D(x)$ of objective function $Q(x)$. The third row contains only headers B - for basis, P_B, D_B and x_B - for basic components of numerator $P(x)$, denominator $D(x)$, and basic feasible solution x, respectively. Then follow m rows containing: identifier for basic vector A_{s_i}, appropriate basic components of numerator $P(x)$, denominator $D(x)$, and basic feasible solution x, respectively, and, finally, n coefficients x_{ij} for linear representation (4.6) of vectors A_j, $j = 1, 2, \dots, n$ in basic vectors $A_{s_1}, A_{s_2}, \dots, A_{s_m}$. Coefficients Δ'_j, Δ''_j and determinants

[1]A sequence of iterations that goes through the same simplex tableaus and repeats itself indefinitely

				p_1	p_2	$\cdots$	p_n
				d_1	d_2	$\cdots$	d_n
B	P_B	D_B	x_B	A_1	A_2	$\cdots$	A_n
A_{s_1}	p_{s_1}	d_{s_1}	x_{s_1}	x_{11}	x_{12}	$\cdots$	x_{1n}
A_{s_2}	p_{s_2}	d_{s_2}	x_{s_2}	x_{21}	x_{22}	$\cdots$	x_{2n}
$\vdots$	$\vdots$	$\vdots$	$\vdots$	$\vdots$	$\vdots$	$\vdots$	$\vdots$
A_{s_m}	p_{s_m}	d_{s_m}	x_{s_m}	x_{m1}	x_{m2}	$\cdots$	x_{mn}
$P(x)$				Δ'_1	Δ'_2	$\cdots$	Δ'_n
$D(x)$				Δ''_1	Δ''_2	$\cdots$	Δ''_n
$Q(x)$				$\Delta_1(x)$	$\Delta_2(x)$	$\cdots$	$\Delta_n(x)$

Table 4.1. Simplex tableau for an LFP problem.

$\Delta_j(x)$ may be stored in the last three rows. The current values of$P(x),\ D(x)$ and $Q(x)$ are in the left lower corner of the tableau.

5. Connection Between Iterations

In this section we deal with the operation of interchanging vectors in the simplex tableau. This operation is usually called a*pivot transformation.* When replacing a basic vector in simplex tableau with some nonbasic vector we have to recalculate the simplex tableau. Our aim now is to discuss how it may be performed.

5.1 Theoretical Background

Consider the following two systems of linearly independent column vectors:

$$V_1 = \{P_i\}_{i \in I} \text{ and } V_2 = \{P_i\}_{i \in I'}, \text{ where } \{I\}_{i \neq r} \equiv \{I'\}_{i \neq k},$$

that is these two systems differ from each other only in one vector in the position r and position k respectively.

Let W denote an arbitrary column vector of the same size as vectors P_i. Because systems V_1 and V_2 consist of linearly independent vectors, we can represent vector W as a linear combination of vectors P_i:
for system V_1

$$W_j = \sum_{i \in I} P_i q_{ij}. \tag{4.13}$$

and, for system V_2

$$W_j = \sum_{i \in I'} P_i q'_{ij}. \tag{4.14}$$

Let us suppose that coefficients x_{ik} of the representation of column vector P_k in system V_1 are given, so

$$P_k = \sum_{i \in I} P_i x_{ik}. \tag{4.15}$$

We should note here that $x_{rk} \neq 0$, because systems V_1 and V_2 are linearly independent.

Indeed, if we assume that $x_{rk} = 0$, then we can rewrite (4.15) in the following form:

$$\begin{aligned} P_k &= \sum_{\substack{i \in I \\ i \neq r}} P_i x_{ik} + P_r x_{rk} = \\ &= \sum_{\substack{i \in I \\ i \neq r}} P_i x_{ik} + P_r 0 = \sum_{\substack{i \in I' \\ i \neq k}} P_i x_{ik}. \end{aligned} \tag{4.16}$$

The latter means that

$$\sum_{\substack{i \in I' \\ i \neq k}} P_i x_{ik} - P_k = 0,$$

i.e. system V_2 is linearly dependent. This contradiction assures us that $x_{rk} \neq 0$.

Using (4.16), we have

$$P_r = \frac{1}{x_{rk}} (P_k - \sum_{\substack{i \in I \\ i \neq r}} P_i x_{ik}), \tag{4.17}$$

so from (4.13) we obtain the following:

$$\begin{aligned} W_j &= \sum_{i \in I} P_i \, q_{ij} = \sum_{\substack{i \in I \\ i \neq r}} P_i \, q_{ij} + P_r \, q_{rj} \overset{(4.17)}{=} \\ &= \sum_{\substack{i \in I \\ i \neq r}} P_i \, q_{ij} + \frac{q_{rj}}{x_{rk}} (P_k - \sum_{\substack{i \in I \\ i \neq r}} P_i \, x_{ik}) = \\ &= \sum_{\substack{i \in I \\ i \neq r}} P_i \, q_{ij} + \frac{q_{rj}}{x_{rk}} P_k - \frac{q_{rj}}{x_{rk}} \sum_{\substack{i \in I \\ i \neq r}} P_i \, x_{ik} = \end{aligned}$$

$$= \sum_{\substack{i \in I \\ i \neq r}} P_i (q_{ij} - \frac{q_{rj}\, x_{ik}}{x_{rk}}) + \frac{q_{rj}}{x_{rk}} P_k =$$

$$= \sum_{\substack{i \in I' \\ i \neq k}} P_i\, (q_{ij} - \frac{q_{rj}\, x_{ik}}{x_{rk}}) + \frac{q_{rj}}{x_{rk}} P_k,$$

i.e.

$$W_j = \sum_{\substack{i \in I' \\ i \neq k}} P_i (q_{ij} - \frac{q_{rj}\, x_{ik}}{x_{rk}}) + \frac{q_{rj}}{x_{rk}} P_k.$$

The latter is a representation of vector W_j in the linearly independent system V_2. At the same time, for vector W_j we have representation (4.14) in the same system V_2 of linearly independent vectors P_i. But any vector W_j may be represented in a linearly independent system in only one unique way. It means that

$$\sum_{\substack{i \in I' \\ i \neq k}} P_i (q_{ij} - \frac{q_{rj}\, x_{ik}}{x_{rk}}) + \frac{q_{rj}}{x_{rk}} P_k = \sum_{i \in I'} P_i q'_{ij}$$

and hence, we have:

$$q'_{ij} = \begin{cases} q_{ij} - \dfrac{q_{rj}\, x_{ik}}{x_{rk}}, & i \in I', \ i \neq r, \\ \dfrac{q_{rj}}{x_{rk}}, & i = r. \end{cases} \tag{4.18}$$

Formula (4.18) establishes the connection between two linearly independent systems of vectors, which differ from each other in only one vector in the following way: if we have a linear representation of vector P_k and some vector W_j in system V_1, then after replacing vector P_r in system V_1 with vector P_k, we obtain some new linearly independent system V_2 and some new representation of vector W_j in system V_2. Formula (4.18) allows us to compute the representation of vector W_j in system V_2 on the basis of its known representation in system V_1.

Using formula (4.18) we can easily perform iterations of the simplex method.

5.2 Pivot Transformation

The *pivot transformation* is indicated by the diagram presented in Table 4.2.

The calculations indicated in the diagram are as follows:

x_{ij}	$\cdots$	x_{ik}
$\vdots$	$\cdots$	$\vdots$
x_{rj}	$\cdots$	x_{rk}

$\longrightarrow$

$x_{ij} - \dfrac{x_{rj}\, x_{ik}}{x_{rk}}$	$\cdots$	0
$\vdots$	$\cdots$	$\vdots$
$\dfrac{x_{rj}}{x_{rk}}$	$\cdots$	1

Table 4.2. Pivot transformation in a simplex tableau.

1 All elements of pivot row r must be divided by pivot element x_{rk} (note that $x_{rk} \neq 0$). Thus pivot element x_{rk} goes to 1, and all other remaining entries x_{rj} of the pivot row go to x_{rj}/x_{rk}, $j = 1, 2, \dots, n$.

2 All entries x_{ij} of all remaining non-pivot rows go to

$$x_{ij} - (x_{rj}\, x_{ik})/x_{rk}.$$

Note that here elements x_{rj} and x_{ik} are the two entries that "form a rectangle" with entry x_{ij} and pivot element x_{rk}.

3 All remaining elements of pivot column must be recalculated on the basis of the same formula $x'_{ij} = x_{ij} - (x_{rj}x_{ik})/x_{rk}$, where $j = k$. Thus, we have that $x'_{ik} = x_{ik} - (x_{rk}x_{ik})/x_{rk} = 0$. So all entries of pivot column go to 0, excluding pivot element x_{rk}, which goes to 1.

Finally, we must remember to interchange the marginal labels of the pivot row and pivot column. All other marginal labels remain unchanged.

6. Initialization of the Simplex Method

When discussing theoretical backgrounds of the simplex method we assumed that there is a basic feasible solution x.

Sometimes it may occur that we can easily find such vectors A_j which form a linearly independent system, and may be used as initial basis B. One of such 'easy' special cases is described below.

Let us suppose that the main constraints of an LFP problem to be solved contain only " $\leq$ " relations, i.e.

$$\sum_{j=1}^{n} A_j x_j \leq b, \quad x_j \geq 0, \ \ j = 1, 2, \dots, n,$$

and all elements $b_i, \ i = 1, 2, \dots, m,$ of the right-hand side vector $b = (b_1, b_2, \dots, b_m)^T$ are nonnegative. In this case, when converting the problem to standard form we augment m nonnegative *artificial variables*

$$x_{n+1}, \ x_{n+2}, \ \dots, \ x_{n+m},$$

to the original LFP problem (see Section 1.3, page 46). These new variables correspond to the unit column-vectors

$$A_{n+1}, \ A_{n+2}, \ \dots, \ A_{n+m},$$

respectively, which form a unit matrix of order m, where

$$A_{n+i} = (\underbrace{\overbrace{0, 0, \dots, 0, 1}^{i}, 0, 0, \dots, 0}_{m})^T, \ \ i = 1, 2, \dots, m.$$

Now the augmented LFP problem has an initial BFS solution that is obtained directly:

$$x = (\underbrace{0, 0, \dots, 0}_{n}, b_1, b_2, \dots, b_m)^T.$$

REMARK 4.5 *In the objective function of an augmented LFP problem these nonnegative artificial variables* $x_{n+1}, x_{n+2}, \dots, x_{n+m}$ *have zero value coefficients* p_j *and* d_j, $j = n+1, n+2, \dots, n+m$, *so this augmented LFP problem may be presented in the form of (4.19)-(4.21).*

$$Q(x) = \frac{\sum_{j=1}^{n} p_j x_j + \sum_{j=n+1}^{n+m} 0 x_j + p_0}{\sum_{j=1}^{n} d_j x_j + \sum_{j=n+1}^{n+m} 0 x_j + d_0} \longrightarrow \max(\text{ or } \min) \tag{4.19}$$

subject to

$$\left.\begin{array}{llllllll}
a_{11}x_1 + & \dots & + a_{1n}x_n & + x_{n+1} & & & & = b_1 \\
a_{21}x_1 + & \dots & + a_{2n}x_n & & + x_{n+2} & & & = b_2 \\
\vdots & & \vdots & & & & & \vdots \\
a_{m1}x_1 + & \dots & + a_{mn}x_n & & & & + x_{n+m} & = b_m
\end{array}\right\} \tag{4.20}$$

$$x_j \geq 0, \quad j = 1, 2, \ldots, n+m. \tag{4.21}$$

The initial simplex tableau for LFP problem (4.19)-(4.21) will be as follows in Table 4.3.

				p_1	$\cdots$	p_n	0	$\cdots$	0
				d_1	$\cdots$	d_n	0	$\cdots$	0
B	P_B	D_B	x_B	A_1	$\cdots$	A_n	A_{n+1}	$\cdots$	A_{n+m}
A_{n+1}	0	0	b_1	a_{11}	$\cdots$	a_{1n}	1	$\cdots$	0
A_{n+2}	0	0	b_2	a_{21}	$\cdots$	a_{2n}	0	$\cdots$	0
$\vdots$	$\vdots$	$\vdots$	$\vdots$	$\vdots$	$\vdots$	$\vdots$	$\vdots$	$\vdots$	$\vdots$
A_{n+m}	0	0	b_m	a_{m1}	$\cdots$	a_{mn}	0	$\cdots$	1
	$P(x)$			Δ'_1	$\cdots$	Δ'_n	0	$\cdots$	0
	$D(x)$			Δ''_1	$\cdots$	Δ''_n	0	$\cdots$	0
	$Q(x)$			$\Delta_1(x)$	$\cdots$	$\Delta_n(x)$	0	$\cdots$	0

Table 4.3. Initial simplex tableau for an LFP problem.

REMARK 4.6 *Note that instead of coefficients x_{ij} (see Table 4.1), in Table 4.3 we use coefficients a_{ij} of the original matrix A because the basic vectors A_{n+i}, $i = 1, 2, \ldots, m$, are unit column-vectors and hence*

$$A_j = \sum_{i=1}^{m} A_{n+i} a_{ij}, \; j = 1, 2, \ldots, n.$$

Unfortunately, in LFP problems to find an *initial* or *starting* BFS may not be a trivial problem. Below in this section, we discuss the *Big M method*, a version of the simplex method that first finds a BFS by adding artificial variables to the problem, and then we consider the *Two-phase simplex method*, which may be used as an alternative to the Big M method. These two methods allow us to find an initial basic feasible solution of an LFP problem to be solved. Other modern techniques of the initialization simplex method and special implementation issues may be found in [30].

6.1 The Big M Method

When this method is used the original standard normalized LFP problem (4.1)-(4.3) to be solved must be replaced with the so-called$M$-*problem* [40]:

$$\tilde{Q}(x) = \frac{P(x) - M\sum_{i=1}^{m} x_{n+i}}{D(x)} \longrightarrow \max \tag{4.22}$$

$$\sum_{j=1}^{n} a_{ij}x_j + x_{n+i} = b_i, \quad i = 1, 2, \ldots, m, \tag{4.23}$$

$$x_j \geq 0, \quad j = 1, 2, \ldots, n+m, \tag{4.24}$$

where M denotes an arbitrarily large positive number, and

$$x_{n+1},\ x_{n+2},\ \ldots,\ x_{n+m}$$

are artificial variables.

It is not necessary to give a specific value to$M$, but it is treated as a parameter that is very much larger than any number with which it is compared. This$M$-problem has an initial basic feasible solution

$$x = (\underbrace{0, 0, \ldots, 0}_{n}, b_1, b_2, \ldots, b_m),$$

and hence, the simplex method can be applied to this problem directly to solve it.

In the initial simplex tableau the coefficients Δ'_j, Δ''_j and determinants $\Delta_j(x)$ are of the form

$$\begin{aligned}
\Delta'_j &= \sum_{i=1}^{m}(-M)a_{ij} - p_j &&= -M\sum_{i=1}^{m} a_{ij} - p_j, \\
\Delta''_j &= \sum_{i=1}^{m} 0a_{ij} - d_j &&= -d_j, \\
\Delta_j(x) &= \Delta'_j - Q(x)\Delta''_j,
\end{aligned}$$

where $j = 1, 2, \ldots, n,$ and $Q(x) = (p_0 - M\sum_{i=1}^{m} x_{n+i})/d_0.$

When applied to this M-problem, the simplex method might terminate in several ways. The corresponding cases are considered below.

Let us suppose that vector

$$\tilde{x} = (\tilde{x}_1, \tilde{x}_2, \ldots, \tilde{x}_n, \tilde{x}_{n+1}, \ldots, \tilde{x}_{n+m})^T$$

is an optimal basic solution of M-problem (4.22)-(4.24).

Relations between original LFP problem (4.1)-(4.3) and its$M$-problem, and between their optimal solutions are established by the following statements.

THEOREM 4.5 *If vector $\tilde{x}$ is an optimal basic solution of M-problem (4.22)-(4.24) and*

$$\tilde{x}_{n+i} = 0, \quad i = 1, 2, \ldots, m,$$

i.e.

$$\tilde{x} = (\tilde{x}_1, \tilde{x}_2, \ldots, \tilde{x}_n, \underbrace{0, 0, \ldots, 0}_{m}),$$

then vector $x^* = (\tilde{x}_1, \tilde{x}_2, \ldots, \tilde{x}_n)^T$ is an optimal basic solution of original standard normalized LFP problem (4.1)-(4.3).

Proof. First of all let us observe that if vector $\tilde{x}$ is an optimal solution of M-problem (4.22)-(4.24) and $\tilde{x}_{n+i} = 0, \ i = 1, 2, \ldots, m,$ then vector $x^* = (\tilde{x}_1, \tilde{x}_2, \ldots, \tilde{x}_n)^T$ satisfies constraints (4.2)-(4.3) and hence, is a feasible solution of original LFP problem (4.1)-(4.3).
Now to prove this theorem we should show that vectorx^* is an optimal solution of problem (4.1)-(4.3). Let us suppose that it is not true, i.e. vectorx^* is not an optimal solution of problem (4.1)-(4.3) and there exists some vector $x' = (x'_1, x'_2, , x'_n)^T$, such that

$$x' \in S \text{ and } Q(x') > Q(x^*).$$

Since vector x' is a feasible solution of the original problem (4.1)-(4.3), i.e. satisfies constraints (4.2)-(4.3), it is obvious that vector

$$\tilde{x'} = (x'_1, x'_2, \ldots, x'_n, \underbrace{0, 0, \ldots, 0}_{m})^T$$

satisfies constraints (4.23)-(4.24) and so is a feasible solution of M-problem (4.22)-(4.24). In this case, we have

$$\tilde{Q}(\tilde{x'}) = Q(x') > Q(x^*) = \tilde{Q}(\tilde{x}).$$

The latter means that vector $\tilde{x}$ is not an optimal solution of M-problem (4.22)-(4.24). The contradiction obtained assures us that our assumption thatx^* is not an optimal solution of problem (4.1)-(4.3) is not correct and thus proves the theorem. $\Diamond$

THEOREM 4.6 *If vector $\tilde{x}$ is an optimal solution of M-problem (4.22)-(4.24) and among elements $\tilde{x}_{n+1}, \tilde{x}_{n+2}, \dots, \tilde{x}_{n+m}$ there is at least one with a positive value, i.e. $\exists\, i_0 : \tilde{x}_{n+i_0} > 0,\ 1 \le i_0 \le m,$ then the original LFP problem (4.1)-(4.3) is unsolvable because its feasible set S is empty, i.e. $S = \emptyset$.*

Proof. To prove this theorem we suppose that $S \neq \emptyset$ and vector $x' = (x'_1, x'_2, \dots, x'_n)^T$ is a feasible solution of the original LFP problem (4.1)-(4.3). Obviously, in such a case vector

$$\tilde{x'} = (x'_1, x'_2, \dots, x'_n, \underbrace{0, 0, \dots, 0}_{m})^T$$

is a feasible solution of M-problem (4.22)-(4.24). Furthermore, since M is a very big positive number and $\tilde{x}_{n+i_0} > 0$, we have that

$$\tilde{Q}(\tilde{x'}) = \frac{P(\tilde{x'})}{D(\tilde{x'})} > \frac{P(\tilde{x}) - M\tilde{x}_{n+i_0}}{D(\tilde{x})} = \tilde{Q}(\tilde{x}).$$

The latter contradicts our assumption that vector $\tilde{x}$ is an optimal solution of M-problem (4.22)-(4.24), and hence proves this theorem. $\diamondsuit$

REMARK 4.7 *Feasible set S_M of M-problem (4.22)-(4.24) is not an empty set because it contains at least one vector*

$$x = (\underbrace{0, 0, \dots, 0}_{n}, b_1, b_2, \dots, b_m)$$

which satisfies constraints (4.23) and (4.24). This is why the case when M-problem is unsolvable because its feasible set S_M is empty is excluded and may not occur.

REMARK 4.8 *Since M is a very big positive number, objective function $\tilde{Q}(x)$ of M-problem (4.22)-(4.24) is bounded from above on feasible set S_M.*

REMARK 4.9 *In the M-problem instead of objective function in form (4.22) we may use*

$$\tilde{Q}(x) = \frac{P(x)}{D(x) + M\sum_{i=1}^{m} x_{n+i}} \longrightarrow \max$$

or

$$\tilde{Q}(x) = \frac{P(x) - M\sum_{i=1}^{m} x_{n+i}}{D(x) + M\sum_{i=1}^{m} x_{n+i}} \longrightarrow \max.$$

Therefore, if the original LFP problem (4.1)-(4.3) has an optimal solution, the Big M method will find it after the application of the simplex method once. If LFP problem (4.1)-(4.3) is infeasible, i.e. its feasible set$S$ is empty, or objective function $Q(x)$ is unbounded on its feasible set S, it will be also determined by the Big M method.

The initial simplex tableau for the Big$M$ method, i.e for M-problem (4.22)-(4.24) is presented in Table 4.4.

				p_1	$\dots$	p_n	$-M$	$\dots$	$-M$
				d_1	$\dots$	d_n	0	$\dots$	0
B	P_B	D_B	x_B	A_1	$\dots$	A_n	A_{n+1}	$\dots$	A_{n+m}
A_{n+1}	$-M$	0	b_1	a_{11}	$\dots$	a_{1n}	1	$\dots$	0
A_{n+2}	$-M$	0	b_2	a_{21}	$\dots$	a_{2n}	0	$\dots$	0
$\vdots$	$\vdots$	$\vdots$	$\vdots$	$\vdots$		$\vdots$	$\vdots$		$\vdots$
A_{n+m}	$-M$	0	b_m	a_{m1}	$\dots$	a_{mn}	0	$\dots$	1
	$P(x)$			Δ'_1	$\dots$	Δ'_n	0	$\dots$	0
	$D(x)$			Δ''_1	$\dots$	Δ''_n	0	$\dots$	0
	$Q(x)$			$\Delta_1(x)$	$\dots$	$\Delta_n(x)$	0	$\dots$	0

Table 4.4. The Big M-method – Initial simplex tableau.

To illustrate this method we consider the following maximization LFP problem

$$Q(x) = \frac{P(x)}{D(x)} = \frac{3x_1 + 3x_2 + 4x_3 + 6}{4x_1 + 5x_2 + 3x_3 + 8} \longrightarrow \max \tag{4.25}$$

subject to

$$\begin{array}{rcrcrcr} 1x_1 & + & 3x_2 & + & 2x_3 & = & 24\,, \\ 2x_1 & + & 1x_2 & + & 3x_3 & = & 18\,, \\ 1x_1 & + & 2x_2 & + & 2x_3 & \leq & 16\,, \end{array} \tag{4.26}$$
$$x_1 \geq 0,\ \ x_2 \geq 0,\ \ x_3 \geq 0$$

First of all, we have to convert the given problem to the standard form. So, we enter slack variable x_4 into the third constraint. We have

$$Q(x) = \frac{P(x)}{D(x)} = \frac{3x_1 + 3x_2 + 4x_3 + 0x_4 + 6}{4x_1 + 5x_2 + 3x_3 + 0x_4 + 8} \longrightarrow \max \tag{4.27}$$

subject to

$$\begin{array}{rcrcrcrcl} 1x_1 & + & 3x_2 & + & 2x_3 & & & = & 24\,, \\ 2x_1 & + & 1x_2 & + & 3x_3 & & & = & 18\,, \\ 1x_1 & + & 2x_2 & + & 2x_3 & + & 1x_4 & = & 16\,, \end{array} \tag{4.28}$$

$$x_j \geq 0,\ j = 1,2,3,4.$$

Since all main constraints are in the form of equality ("=") and all right-hand side $b_i,\ i = 1,2,3$, are non-negative, the problem is in canonical form. Observe that this problem has only one unit vector $A_4 = (0,0,1)^T$. This is why to construct a complete unit submatrix we have to enter two artificial variables x_5 and x_6. So, the M-problem will be as follows:

$$Q(x) = \frac{P(x)}{D(x)} = \frac{3x_1 + 3x_2 + 4x_3 + 0x_4 - Mx_5 - Mx_6 + 6}{4x_1 + 5x_2 + 3x_3 + 0x_4 + 0x_5 + 0x_6 + 8} \longrightarrow \max$$

subject to

$$\begin{array}{rcrcrcrcrcrcl} 1x_1 & + & 3x_2 & + & 2x_3 & & & + & 1x_5 & & & = & 24\,, \\ 2x_1 & + & 1x_2 & + & 3x_3 & & & & & + & 1x_6 & = & 18\,, \\ 1x_1 & + & 2x_2 & + & 2x_3 & + & 1x_4 & & & & & = & 16\,, \end{array}$$

$$x_j \geq 0,\ j = 1,2,3,4,5,6.$$

Now we can initiate the simplex method with initial basis

$$B = (A_5, A_6, A_4) = \begin{vmatrix} 1 & 0 & 0 \\ 0 & 1 & 0 \\ 0 & 0 & 1 \end{vmatrix}$$

and initial BFS $x = (0,0,0,16,24,18)^T$. The initial tableau for the M-problem is shown in Table 4.5, where

$$\begin{aligned} P(x) &= 3 \times 0 + 3 \times 0 + 4 \times 0 + 0 \times 16 + 24(-M) + 18(-M) + 6 = \\ &= -42M + 6, \\ D(x) &= 4 \times 0 + 5 \times 0 + 3 \times 0 + 0 \times 16 + 0 \times 24 + 0 \times 18 + 8 = 8, \\ Q(x) &= \frac{-42M + 6}{8}, \end{aligned}$$

$$\begin{aligned} \Delta'_1 &= (-M) \times 1 + (-M) \times 2 + 0 \times 1 - 3 = -3M - 3, \\ \Delta''_1 &= 0 \times 1 + 0 \times 2 + 0 \times 1 - 4 = -4, \\ \Delta_1(x) &= \Delta'_1 - Q(x)\Delta''_1 = (-3M - 3) - \frac{-42M + 6}{8}(-4) = -24M, \end{aligned}$$

$$\begin{aligned} \Delta'_2 &= (-M) \times 3 + (-M) \times 1 + 0 \times 2 - 3 = -4M - 3, \\ \Delta''_2 &= 0 \times 3 + 0 \times 1 + 0 \times 2 - 5 = -5, \end{aligned}$$

					3	3	4	0	$-M$	$-M$	
					4	5	3	0	0	0	
B	P_B	D_B	x_B		A_1	A_2	A_3	A_4	A_5	A_6	
A_5	$-M$	0	24		1	3	2	0	1	0	
A_6	$-M$	0	18		2	1	3	0	0	1	$\Rightarrow$
A_4	0	0	16		1	2	2	1	0	0	
$P(x) = 6 - 42M$					$-3M-3$	$-4M-3$	$-5M-4$	0	0	0	
$D(x) = 8$					-4	-5	-3	0	0	0	
$Q(x) = \frac{6-42M}{8}$					$-24M$	$\frac{3-121M}{4}$	$\frac{-83M-7}{4}$	0	0	0	
							$\Uparrow$				

Table 4.5. The Big M-method example – Initial simplex tableau.

$$\begin{aligned}
\Delta_2(x) &= \Delta'_2 - Q(x)\Delta''_2 = (-4M-3) - \frac{-42M+6}{8}(-5) = \\
&= -\frac{121}{4}M + \frac{3}{4}, \\
\Delta'_3 &= (-M)\times 2 + (-M)\times 3 + 0\times 2 - 4 = -5M - 4, \\
\Delta''_3 &= 0\times 2 + 0\times 3 + 0\times 2 - 3 = -3, \\
\Delta_3(x) &= \Delta'_3 - Q(x)\Delta''_3 = (-5M-4) - \frac{-42M+6}{8}(-3) = \\
&= -\frac{83}{4}M - \frac{7}{4}.
\end{aligned}$$

Since the aim of this problem is maximization and the bottom row of the initial tableau contains negative non-basic $\Delta_1(x)$, $\Delta_2(x)$ and $\Delta_3(x)$, it means that the current BFS is not optimal. In this case, we have to choose a non-basic vector A_j with negative reduced cost $\Delta_j(x)$ and enter it into the basis. Let it be vector A_3. Now, we determine the leaving vector: since

$$\theta = \min\{24/2,\ 18/3,\ 16/2\} = 6,$$

we obtain that leaving vector is A_6.

After performing the pivot transformation and recalculating $P(x),\ D(x),\ Q(x)$ and all reduced costs $\Delta'_j,\ \Delta''_j,\ \Delta_j(x)$, we obtain the simplex tableau shown in Table 4.6. As it follows from Theorem 4.4, since the bottom row of this tableau contains negative non-basic reduced costs $\Delta_1(x)$ and $\Delta_2(x)$, it means that the

				3	3	4	0	$-M$	$-M$	
				4	5	3	0	0	0	
B	P_B	D_B	x_B	A_1	A_2	A_3	A_4	A_5	A_6	
A_5	$-M$	0	12	$-1/3$	$7/3$	0	0	1	$-2/3$	
A_3	4	3	6	$2/3$	$1/3$	1	0	0	$1/3$	
A_4	0	0	4	$-1/3$	$4/3$	0	1	0	$-2/3$	$\Rightarrow$
$P(x) = -12M + 30$				$\frac{M-1}{3}$	$\frac{-7M-5}{3}$	0	0	0	$\frac{5M+4}{3}$	
$D(x) = 26$				-2	-4	0	0	0	1	
$Q(x) = \frac{30-12M}{26}$				$\frac{77-23M}{39}$	$\frac{115-163M}{39}$	0	0	0	$\frac{83M+7}{39}$	
					$\Uparrow$					

Table 4.6. The Big M-method example – After first iteration.

				3	3	4	0	$-M$	$-M$	
				4	5	3	0	0	0	
B	P_B	D_B	x_B	A_1	A_2	A_3	A_4	A_5	A_6	
A_5	$-M$	0	5	$1/4$	0	0	$-7/4$	1	$1/2$	
A_3	4	3	5	$3/4$	0	1	$-1/4$	0	$1/2$	$\Rightarrow$
A_2	3	5	3	$-1/4$	1	0	$3/4$	0	$-1/2$	
$P(x) = -5M + 35$				$\frac{-M-3}{4}$	0	0	$\frac{7M+5}{4}$	0	$\frac{M+1}{2}$	
$D(x) = 38$				-3	0	0	3	0	-1	
$Q(x) = \frac{35-5M}{38}$				$\frac{153-49M}{76}$	0	0	$\frac{163M-115}{76}$	0	$\frac{7M+27}{19}$	
				$\Uparrow$						

Table 4.7. The Big M-method example – After second iteration.

current BFS $x = (0, 0, 6, 4, 12, 0)^T$ is not optimal and we have to continue the process. It results in the simplex tableau with basis $B = (A_5, A_3, A_2)$ and non-optimal BFS $x = (0, 3, 5, 0, 5, 0)^T$ shown in Table 4.7.

The final simplex tableau corresponding to the next iteration is shown in Table 4.8. Observe that optimal basis $B = (A_5, A_1, A_2)$ contains vector-

				3	3	4	0	$-M$	$-M$
				4	5	3	0	0	0
B	P_B	D_B	x_B	A_1	A_2	A_3	A_4	A_5	A_6
A_5	$-M$	0	$10/3$	0	0	$-1/3$	$-5/3$	1	$1/3$
A_1	3	4	$20/3$	1	0	$4/3$	$-1/3$	0	$2/3$
A_2	3	5	$14/3$	0	1	$1/3$	$2/3$	0	$-1/3$
$P(x) = \frac{120-10M}{3}$				0	0	$\frac{M+3}{3}$	$\frac{5M+3}{3}$	0	$\frac{2M+3}{3}$
$D(x) = 58$				0	0	4	2	0	1
$Q(x) = \frac{120-10M}{174}$				0	0	$\frac{49M-153}{87}$	$\frac{155M-33}{87}$	0	$\frac{21M+9}{29}$

Table 4.8. The Big M-method example – Final tableau.

column A_5 corresponding to artificial variable x_5, which was introduced in accordance with the rules of the Big M Method. In this case, as it follows from Theorem 4.6, the original LFP problem is unsolvable since its feasible set is empty.

6.2 The Two-Phase Simplex Method

When an initial basic feasible solution is not readily available, the two-phase simplex method may be used as an alternative to the Big M-method. In *Phase I* of the Two-phase simplex method we add artificial variables to the main constraints of standard normalized LFP problem (4.1)-(4.3), as it was made in the Big M-method, and then find an initial BFS for the original LFP problem (4.1)-(4.3) by solving the following minimization linear programming problem [173]

$$Z(x) = \sum_{i=1}^{m} x_{n+i} \longrightarrow \min \tag{4.29}$$

subject to

$$\sum_{j=1}^{n} a_{ij}x_j + x_{n+i} = b_i, \quad i = 1, 2, \dots, m, \tag{4.30}$$

$$x_j \geq 0, \quad j = 1, 2, \dots, n + m. \tag{4.31}$$

Let us consider this *Phase I problem* (4.29)-(4.31). Since vector

$$x = (\underbrace{0, 0, \dots, 0}_{n}, b_1, b_2, \dots, b_m)^T$$

satisfies constraints (4.30)-(4.31), and objective function (4.29) is bounded from below, it is obvious that problem (4.29)-(4.31) is solvable.

Let us suppose that vector $x' = (x'_1, x'_2, \dots, x'_n, x'_{n+1}, \dots, x'_{n+m})^T$ is an optimal basic solution of Phase I problem (4.29)-(4.31). In this case, the following two cases are possible:

1 $Z(x') = 0$, i.e. $x'_{n+i} = 0$, $\forall i = 1, 2, \dots, m$. In this case vector $x'' = (x'_1, x'_2, \dots, x'_n)^T$ is a basic feasible solution of the original LFP problem (4.1)-(4.3). So in *Phase II* we have to solve original LFP problem (4.1)-(4.3) using vector x' as initial BFS. The optimal solution obtained in Phase II is the optimal solution of the original LFP problem (4.1)-(4.3).

2 $Z(x') > 0$, i.e. $\exists\, i_0 : x'_{n+i_0} > 0$, $1 \le i_0 \le m$. In this case the original LFP problem (4.1)-(4.3) is unsolvable because its feasible set is empty, i.e. $S = \emptyset$ (the proof may be carried out analogously to the proof of Theorem 4.6).

Here we omitted the proofs for these two cases because intuitively it is clear that the ideas used in the Big M-method and in the Two-phase simplex method are almost the same. Indeed, in both methods our aim is to minimize the sum of artificial variables. If this sum is equal to zero, we obtain an optimal solution of the original LFP problem in the case of the Big M-method, or initial BFS in the case of the Two-phase simplex method. If the sum of artificial variables is greater than zero, it means in both methods that the feasible set of the original LFP problem is empty and hence the problem is unsolvable.

To illustrate how this method works, we consider the maximization LFP problem (4.25)-(4.26) from the previous section (see page 96). After converting the original LFP problem to the standard form, we obtain problem (4.27)-(4.28) which contains slack variable x_4 with corresponding unit vector $A_4 = (0, 0, 1)^T$. In accordance with the Two-Phase Simplex Method rules we enter two artificial variables x_5 and x_6 to the constraints (4.28) and formulate in Phase I the following linear programming minimization problem

$$Z(x) = 1x_5 + 1x_6 \longrightarrow \min \tag{4.32}$$

subject to

$$\begin{array}{rcrcrcrcrcrcl} 1x_1 & + & 3x_2 & + & 2x_3 & & & + & 1x_5 & & & = & 24\,, \\ 2x_1 & + & 1x_2 & + & 3x_3 & & & & & + & 1x_6 & = & 18\,, \\ 1x_1 & + & 2x_2 & + & 2x_3 & + & 1x_4 & & & & & = & 16\,, \end{array} \qquad (4.33)$$

$$x_j \geq 0,\ j = 1,2,3,4,5,6.$$

So, we can initiate the LP simplex method with initial basis

$$B = (A_5, A_6, A_4) = \begin{vmatrix} 1 & 0 & 0 \\ 0 & 1 & 0 \\ 0 & 0 & 1 \end{vmatrix}$$

and initial BFS $x = (0,0,0,16,24,18)^T$. The initial tableau for the this LP

			0	0	0	0	1	1	
B	P_B	x_B	A_1	A_2	A_3	A_4	A_5	A_6	
A_5	1	24	1	3	2	0	1	0	
A_6	1	18	2	1	3	0	0	1	$\Rightarrow$
A_4	0	16	1	2	2	1	0	0	
$Z(x) = 42$			3	4	5	0	0	0	
			$\Uparrow$						

Table 4.9. The Two-Phase Simplex Method example – Initial simplex tableau.

problem is shown in Table 4.9, where

$$\begin{aligned} Z(x) &= 0 \times 0 + 0 \times 0 + 0 \times 0 + 0 \times 16 + 24 \times 1 + 18 \times 1 = 42, \\ \Delta_1 &= 1 \times 1 + 1 \times 2 + 0 \times 1 - 0 = 3, \\ \Delta_2 &= 1 \times 3 + 1 \times 1 + 0 \times 2 - 0 = 4, \\ \Delta_3 &= 1 \times 2 + 1 \times 3 + 0 \times 2 - 0 = 5, \\ \Delta_4 &= 1 \times 0 + 1 \times 0 + 0 \times 0 - 0 = 0, \\ \Delta_5 &= 1 \times 1 + 1 \times 0 + 0 \times 0 - 1 = 0, \\ \Delta_6 &= 1 \times 0 + 1 \times 1 + 0 \times 0 - 1 = 0. \end{aligned}$$

Notice that the aim in the Phase I problem is minimization, and the bottom row in the initial tableau contains positive non-basicΔ_1, Δ_2 and Δ_3. The latter means that the current BFS is not optimal. In this case, we have to choose a

			0	0	0	0	1	1	
B	P_B	x_B	A_1	A_2	A_3	A_4	A_5	A_6	
A_5	1	15	0	5/2	1/2	0	1	−1/2	
A_1	0	9	1	1/2	3/2	0	0	1/2	
A_4	0	7	0	3/2	1/2	1	0	−1/2	⇒
$Z(x) = 15$			0	$\frac{5}{2}$	$\frac{1}{2}$	0	0	$-\frac{3}{2}$	
				⇑					

Table 4.10. The Two-Phase Simplex Method example – After first iteration.

non-basic vector A_j with positive reduced cost Δ_j and enter it into the basis. Let it be vector A_1. Now, we determine the leaving vector: since

$$\theta = \min\{24/1,\ 18/2,\ 16/1\} = 9,$$

we obtain that the leaving vector is A_6. After performing pivot transformation and recalculating objective function $Z(x)$ and all reduced costs Δ_j we obtain the simplex tableau shown in Table 4.10.

Since the bottom row of this tableau contains positive non-basic reduced costs Δ_2 and Δ_3, it means that the current BFS $x = (9, 0, 0, 7, 15, 0)^T$ is not optimal and we have to continue the process. It results in the simplex tableau with basis $B = (A_5, A_1, A_2)$ and optimal BFS $x = (20/3, 14/3, 0, 0, 10/3, 0)^T$ shown in Table 4.11. Observe that optimal basis $B = (A_5, A_1, A_2)$ contains

			0	0	0	0	1	1
B	P_B	x_B	A_1	A_2	A_3	A_4	A_5	A_6
A_5	1	10/3	0	0	−1/3	−5/3	1	1/3
A_1	0	20/3	1	0	4/3	−1/3	0	2/3
A_2	0	14/3	0	1	1/3	2/3	0	−1/3
$Z(x) = 10/3$			0	0	−1/3	−5/3	0	$-\frac{2}{3}$

Table 4.11. The Two-Phase Simplex Method example – Final tableau.

vector-column A_5 corresponding to artificial variable x_5, which was introduced in accordance with the rules of the Two-Phase Simplex Method. It means that the original LFP problem is unsolvable since its feasible set is empty.

7. Compact Form of the Simplex Tableau

When considering the theoretical backgrounds of The Big$M$ Method in Section 6.1, and The Two-Phase Simplex Method in Section 6.2, we used simplex tableaus with $n+m$ constraint columns $A_j = (a_{1j}, a_{2j}, \dots, a_{mj})^T$, which were associated with n original variables $x_j,\ j = 1,2,\dots,n$, and m slack and/or artificial variables $x_{n+i},\ i = 1,2,\dots,m$. Recall that these slack and/or artificial variables were added to the constraints of the original LFP problem because we had to convert the original LFP problem to the canonical form and then to form a unit submatrix, which was used later in the simplex method as an initial basis. While performing the simplex method we interchange basic and non-basic columns in the simplex tableau and recalculate coefficients $x_{ij},\ i = 1,2,\dots,m,\ j = 1,2,\dots,n+m,$ of the linear combinations

$$\sum_{i=1}^{m} A_{s_i} x_{ij} = A_j, \quad j = 1,2,\dots,n+m,$$

in the current basis $B = (A_{s_1}, A_{s_2}, \dots, A_{s_m})$ using transformation formulas (4.18). These coefficients x_{ij} for non-basic indices j allow us to recalculate new reduced costs for numerator (Δ'_j), denominator (Δ''_j) and objective function ($\Delta_j(x)$), and then to check if the current basis is optimal, while all other coefficients x_{ij}, i.e. those with basic index j, form a unit submatrix of order m, which is stored in the simplex tableaus from iteration to iteration. It is obvious that when implementing the simplex method in a computer code there is no sense to store a unit submatrix in the computer memory and then to recalculate it multiple times from iteration to iteration. This is why when using the simplex method (not only in a computer code!) we may use a so-called *compact simplex tableau* presented in Table 4.12.

The corresponding pivot transformation is indicated by the diagram shown in Table 4.13. The calculations indicated in the diagram are as follows:

1 The pivot x_{rk} is replaced by its reciprocal. Thus, x_{rk} goes to $1/x_{rk}$, note that $x_{rk} \neq 0$.

2 All other elements of pivot row r must be divided by pivot element x_{rk}. Thus, pivot elements x_{rj} of the pivot row goes to $x_{rj}/x_{rk},\ j = 1,2,\dots,n$, $j \neq k$.

3 The remaining entries in the pivot column are divided by the pivot x_{rk} and then the sign is changed. So, x_{ik} goes to $-x_{ik}/x_{rk}$ for all $i = 1,2,\dots,m,\ i \neq r$.

		p_1	p_2	$\cdots$	p_n
		d_1	d_2	$\cdots$	d_n
B	x_B	A_1	A_2	$\cdots$	A_n
A_{n+1}	b_1	a_{11}	a_{12}	$\cdots$	a_{1n}
A_{n+2}	b_2	a_{21}	a_{22}	$\cdots$	a_{2n}
$\vdots$	$\vdots$	$\vdots$	$\vdots$	$\vdots$	$\vdots$
A_{n+m}	b_m	a_{m1}	a_{m2}	$\cdots$	a_{mn}
$P(x)$		Δ'_1	Δ'_2	$\cdots$	Δ'_n
$D(x)$		Δ''_1	Δ''_2	$\cdots$	Δ''_n
$Q(x)$		$\Delta_1(x)$	$\Delta_2(x)$	$\cdots$	$\Delta_n(x)$

Table 4.12. Compact simplex tableau.

4 All entries x_{ij} of all remaining non-pivot rows go to

$$x_{ij} - (x_{rj}\, x_{ik})/x_{rk}.$$

Note that here elements x_{rj} and x_{ik} are the two entries that "form a rectangle" with entry x_{ij} and pivot element x_{rk}.

x_{ij}	$\cdots$	x_{ik}		$x_{ij} - \dfrac{x_{rj}\, x_{ik}}{x_{rk}}$	$\cdots$	$-\dfrac{x_{ik}}{x_{rk}}$
$\vdots$	$\cdots$	$\vdots$	$\longrightarrow$	$\vdots$	$\cdots$	$\vdots$
x_{rj}	$\cdots$	x_{rk}		$\dfrac{x_{rj}}{x_{rk}}$	$\cdots$	$\dfrac{1}{x_{rk}}$

Table 4.13. Pivot transformation in the compact simplex tableau.

Using standard algebraic notation we express this transformation rule as follows:

$$x'_{ij} = \begin{cases} \dfrac{1}{x_{rk}}, & j = k,\ i = r; \\ -\dfrac{x_{ij}}{x_{rk}}, & j = k,\ i = 1,2,\dots,m,\ i \neq r; \\ \dfrac{x_{ij}}{x_{rk}}, & j = 1,2,\dots,n,\ j \neq k,\ i = r; \\ x_{ij} - \dfrac{x_{rj}\,x_{ik}}{x_{rk}}, & i = 1,2,\dots,m,\ i \neq r, \\ & j = 1,2,\dots,n,\ j \neq k. \end{cases} \qquad (4.34)$$

Unlike the 'wide' simplex tableau used in previous sections, where all columns in the tableau are in a fixed order, when using a compact simplex tableau we really 'interchange' basic and non-basic vectors moving them from rows to columns and vice versa. Let us suppose that we have an LFP problem with current basis $B = (A_{n+1}, A_{n+2}, \dots, A_{n+m})$ and we have to interchange basic vector A_{n+r} and non-basic vector A_k. In this case, we move non-basic vector A_k from column k into row r, meanwhile basic vector A_{n+r} leaves its position in row r and occupies column k. This interchange is reflected in the tableaus presented in Table 4.14 (before interchange) and Table 4.15 (after interchange), where coefficients $x_{ij},\ \ i = 1,2,\dots,m,\ \ j = 1,2,\dots,n,$ in Table 4.14 are

		p_1 d_1	$\cdots$ $\cdots$	p_k d_k	$\cdots$ $\cdots$	p_n d_n	
B	x_B	A_1	$\cdots$	A_k	$\cdots$	A_n	
A_{n+1}	b_1	x_{11}	$\cdots$	x_{1k}	$\cdots$	x_{1n}	
$\vdots$	$\vdots$	$\vdots$	$\cdots$	$\vdots$	$\vdots$	$\vdots$	
A_{n+r}	b_r	x_{r1}	$\cdots$	x_{rk}	$\cdots$	x_{rn}	$\Rightarrow$
$\vdots$	$\vdots$	$\vdots$	$\cdots$	$\vdots$	$\vdots$	$\vdots$	
A_{n+m}	b_m	x_{m1}	$\cdots$	x_{mk}	$\cdots$	x_{mn}	
$P(x)$ $D(x)$ $Q(x)$		Δ'_1 Δ''_1 $\Delta_1(x)$	$\cdots$ $\cdots$ $\cdots$	Δ'_k Δ''_k $\Delta_k(x)$	$\cdots$ $\cdots$ $\cdots$	Δ'_n Δ''_n $\Delta_n(x)$	
				$\Uparrow$			

Table 4.14. Compact simplex tableau – Before interchange.

		p_1	$\cdots$	p_{n+r}	$\cdots$	p_n
		d_1	$\cdots$	d_{n+r}	$\cdots$	d_n
B	x_B	A_1	$\cdots$	A_{n+r}	$\cdots$	A_n
A_{n+1}	b'_1	x'_{11}	$\cdots$	x'_{1k}	$\cdots$	x'_{1n}
$\vdots$	$\vdots$	$\vdots$	$\cdots$	$\vdots$	$\vdots$	$\vdots$
A_k	b'_k	x'_{r1}	$\cdots$	x'_{rk}	$\cdots$	x'_{rn}
$\vdots$	$\vdots$	$\vdots$	$\cdots$	$\vdots$	$\vdots$	$\vdots$
A_{n+m}	b'_m	x'_{m1}	$\cdots$	x'_{mk}	$\cdots$	x'_{mn}
$P(x)$		Δ'_1	$\cdots$	Δ'_{n+r}	$\cdots$	Δ'_n
$D(x)$		Δ''_1	$\cdots$	Δ''_{n+r}	$\cdots$	Δ''_n
$Q(x)$		$\Delta_1(x)$	$\cdots$	$\Delta_{n+r}(x)$	$\cdots$	$\Delta_n(x)$

Table 4.15. Compact simplex tableau – After interchange.

determined from system

$$\sum_{i=1}^{m} A_{n+i}\, x_{ij} \;=\; A_j, \;\; j = 1, 2, \ldots, n,$$

and entries x'_{ij}, $i = 1, 2, \ldots, m$, $j = 1, 2, \ldots, n$, must be determined in accordance with transformation rules shown in (4.34).

Consider the following numerical example:

$$Q(x) \;=\; \frac{P(x)}{D(x)} \;=\; \frac{1x_1 + 3x_2 + 2.5x_3 + 6}{2x_1 + 3x_2 + 2x_3 + 12} \longrightarrow \max \tag{4.35}$$

subject to

$$\begin{array}{rcrcrcl} 1x_1 & + & 2x_2 & + & 2.5x_3 & \leq & 40\,, \\ 2x_1 & + & 2x_2 & + & 2x_3 & \leq & 60\,, \end{array} \tag{4.36}$$
$$x_1 \geq 0, \;\; x_2 \geq 0, \;\; x_3 \geq 0.$$

First of all, we have to convert system (4.36) to canonical form. So, adding two slack variables x_4 and x_5 to system (4.36) we obtain the following canonical LFP problem

$$Q(x) \;=\; \frac{P(x)}{D(x)} \;=\; \frac{1x_1 + 3x_2 + 2.5x_3 + 0x_4 + 0x_5 + 6}{2x_1 + 3x_2 + 2x_3 + 0x_4 + 0x_5 + 12} \longrightarrow \max \tag{4.37}$$

subject to

$$\begin{array}{rcrcrcrcrcl} 1x_1 & + & 2x_2 & + & 2.5x_3 & + & 1x_4 & + & & = & 40\ , \\ 2x_1 & + & 2x_2 & + & 2x_3 & + & & + & 1x_5 & = & 60\ , \end{array} \qquad (4.38)$$

$$x_j \geq 0, \ \ j = 1, 2, 3, 4, 5,$$

which has the initial compact simplex tableau shown in Table 4.16. Note we do not store any unit submatrix in the tableau. Since the bottom row in the initial

		1	3	5/2	
		2	3	2	
B	x_B	A_1	A_2	A_3	
A_4	40	1	2	5/2	$\Rightarrow$
A_5	60	2	2	2	
$P(x) = 6$		-1	-3	$-5/2$	
$D(x) = 12$		-2	-3	-2	
$Q(x) = 1/2$		0	$-3/2$	$-3/2$	
			$\Uparrow$		

Table 4.16. Compact tableau example – Initial simplex tableau.

tableau contains negative reduced costs $\Delta_2(x) = -3/2$ and $\Delta_3(x) = -3/2$ it means that we have to choose any non-basic vector A_j with negative reduced cost $\Delta_j(x)$ and enter it into the basis. Let us choose vector A_2. The leaving variable is determined by calculating the θ-ratios. We have

$$\min\{40/2,\ 60/2\} = \min\{20,\ 30\} = 20$$

so that vector A_4 must leave the current basis $B = (A_4, A_5)$. Now, using transformation rules given by Table 4.13 we perform a simplex iteration and obtain the simplex tableau presented in Table 4.17. Since the new basis $B = (A_2, A_5)$ is not optimal, we perform the next iteration and obtain the optimal tableau shown in Table 4.18. So the optimal solution is $x^* = (0, 0, 16)^T$ with optimal value $Q(x^*) = 46/44$.

8. Rules of Entering and Dropping Variables

Now we discuss various rules used to select the variable to be entered into the basis, and rules for choosing the variable to be dropped from the basis.

		1	0	5/2	
		2	0	2	
B	x_B	A_1	A_4	A_3	
A_2	20	1/2	1/2	5/4	$\Rightarrow$
A_5	20	1	-1	$-1/2$	
$P(x) = 66$		1/2	3/2	5/4	
$D(x) = 72$		$-1/2$	3/2	7/4	
$Q(x) = 11/12$		23/24	1/8	$-17/48$	
				$\Uparrow$	

Table 4.17. Compact tableau example – After first iteration.

		1	0	3
		2	0	3
B	x_B	A_1	A_4	A_2
A_3	16	2/5	2/5	4/5
A_5	28	6/5	$-4/5$	2/5
$P(x) = 46$		0	1	-1
$D(x) = 44$		$-6/5$	4/5	$-7/5$
$Q(x) = 23/22$		69/55	9/55	51/110

Table 4.18. Compact tableau example – Final tableau.

8.1 Entering Rules

8.1.1 Steepest Ascent Rule

This rule usually covered in LP texts is the *steepest ascent method* (referred to here as the *old method* or *old rule*). In this rule the entering variable is chosen in such a way that the rate of increase of value in the objective function of the maximization LFP problem per unit change in the value of the entering variable from its present value of zero, is the highest among all eligible variables. By the results in Section 2, this is achieved by choosing variable x_{j_0}, where index j_0 is determined to satisfy

$$\Delta_{j_0}(x) = \max_{j \in J_N^-} |\Delta_j(x)|, \tag{4.39}$$

where set J_N^- denotes indices j of negative nonbasic reduced costs of LFP $\Delta_j(x)$, i.e.

$$J_N^- = \{j : j \in J_N, \ \Delta_j(x) < 0\}.$$

8.1.2 Greatest Increase Rule

The alternative rule to the steepest ascent rule is known as the *highest step method* or *highest step rule* (referred to here as the *new method* or *new rule*). In this rule, the actual change in the value of the objective function that will occur if variable x_j is chosen as the entering variable is computed for each eligible variable x_j at this stage. This is, of course,

$$\frac{-\theta \Delta_j(x)}{D(x(\theta))}$$

(see formula (4.12) on page 82), where $\Delta_j(x)$ is a present determinant connected with variable x_j, θ is the minimum ratio determined by formula (4.11), and $D(x(\theta))$ is a new value of denominator $D(x)$ in the new basic feasible solution if variable x_j is chosen as the entering variable. Then the entering variable is chosen as the eligible variable that corresponds to the greatest increase of the objective function. So, if variable x_{j_0} is chosen in accordance with this rule, it means that

$$\max_{j \in J_N^-}\{\frac{-\theta \Delta_j(x)}{D(x(\theta))}\} = \frac{-\theta \Delta_{j_0}(x)}{D(x(\theta))}$$

The old method is likely to require more iterations than the new, especially in phase 1. However, it has the advantage of being quicker per iteration. If we are going to solve a reasonably large problem repeatedly, it might be worth comparing the two methods for speed of solution. They will both give the same optimal solution for most problems.

However, for very large problems, where rounding errors may compound, the new method should be more accurate as it allows to avoid the selection of small pivots and usually requires fewer iterations.

8.1.3 Lexicographical Rules

These entering variable choice rules require that the variables be arranged in some specific order before the algorithm is initiated. This order can be arbitrary, but once it is selected, it is fixed during the entire algorithm. In each pivot step, these rules choose the entering variable to be that eligible variable that is the first among all eligible variables in this step, in the specific order selected for the variables. So, if the specific order chosen for the variables is the natural order $x_1, x_2, \ldots, x_n$ in increasing order of their indices, this rule is said to be *leftmost of the eligible variables rule.* In accordance with this rule if variable x_{j_0} is chosen to be entered into basis, it means that

$$j_0 = \min_{j \in J_N^-} j.$$

REMARK 4.10 *If an LFP problem has multiple optimal solutions, these pivot rules will generally speaking lead to different optimal solutions.*

8.2 Dropping Rules

When performing a simplex iteration it may occur that for the chosen entering variable x_j the minimum ratio test for θ (4.11) does not provide us a unique index of variable to be dropped from the basis, because minimum ratio (4.11) results in multiple indices. Thus, we need a dropping variable choice rule for those simplex iterations where there are ties in the minimum ratio test. The following are some of the rules that can be used for this choice.

8.2.1 The Topmost Rule

In this rule, the dropping variable is chosen among those eligible, so that the pivot row is the topmost in the tableau, among the possible.

8.2.2 The Lexico-Minimum Rule

This rule selects the dropping variable in simplex iterations uniquely and unambiguously. Once it has been selected, it has to be used in every iteration of algorithm from the beginning. If the minimum ratio test (4.11) identifies the dropping variable uniquely, that will be the dropping variable under this rule too. Whenever there are ties in the minimum ratio test (4.11), this rule selects the dropping variable among those tied by carrying out additional minimum ratio steps using the columns from the inverse of the present basis in place of the updated right-hand side column. Here we just note that if this rule is used, the simplex method is guaranteed to terminate in a finite number of iterations, because it resolves the problem of cycling under degeneracy.

8.2.3 Lexicographical Rules

As in the case of lexicographical rules for entering variable, here also we are required to select a specific ordering of the variables before initiating the simplex method. Once this order is selected, it is fixed during the entire algorithm. In each simplex iteration, these rules select the dropping variable to be that blocking variable that is the first among all the blocking variables in the specific order chosen for the variables. It can be proved that the simplex method executed using such rules terminates after a finite number of iterations (see Section 9).

More detailed and extended discussion of classical and recently developed various entering and dropping rules, and their comparative analysis may be found in [98], [121], [127], [144], [181] [189].

9. Degeneracy and Cycling

In this section we deal with phenomenon of degeneracy and formulate the rule, which allow us to avoid cycling.

In all previous sections we assumed that the considered BFS x is non-degenerate one (see Definition 4.7 in Section 1). However, when using the simplex method, it may occur that the current basic feasible solution$x$ has at least one zero value basic variable. In this situation we may encounter the same BFS more than once. This occurrence is called*cycling*. Fortunately, the simplex method can be modified to ensure that cycling will never occur [31], [52]. In practice, however, degeneracy does not necessarily lead to cycling, so cycling is an extremely rare occurrence [118], [119]. Most of numeric examples which lead to cycling were constructed artificially [155]. However, there are investigations encountered some classes of non-artificial linear programming problems which lead to cycling, for example some special problems of queuing theory [119].

When considering the theoretical backgrounds of the simplex method we assumed that

1 The minimum ratio test (4.11) results in a positive value, i.e.

$$\theta = \min_{x_{ij}>0} \frac{x_{s_i}}{x_{ij}} > 0.$$

2 The dropping variable, or equivalently, the pivot row is uniquely determined, i.e. there will be no ties in determining the minimum ratio.

3 The value of the objective function makes a strict increase after every pivot step.

If the current BFS x is non-degenerate, assumption 1 means that the new basic variable x_j in the new basis will attain strictly positive valueθ, and hence the new basic solution will be non-degenerate too. If we choose such a new basic variable x_j that the corresponding determinant $\Delta_j(x)$ is strictly negative, then in accordance with formula (4.12), the value of the objective function $Q(x)$ makes a strict increase after this pivot step and hence assumption 3 takes place. If such a situation occurs in every pivot step, it guarantees that after a finite number of iterations we obtain the maximal value of the objective function $Q(x)$ over feasible set S (if the problem is solvable). So it is impossible to encounter the same BFS x twice and occurrence of cycling is excluded. For example, suppose we are solving an LFP problem with 10 variables and 5 main constraints, and all basic feasible solutions are non-degenerate. Such an LFP problem has at most

$$C_{10}^{5} = \frac{10!}{5!\,(10-5)!} = 252$$

basic feasible solutions. Since we will never repeat the same BFS, the simplex method guarantees to find an optimal solution after, at most, 252 iterations.

Assumption 2 allows us to avoid a case when the new basic feasible solution is degenerate. Indeed, let us suppose that

$$\theta = \min_{x_{ij}>0} \frac{x_{s_i}}{x_{ij}} = \frac{x_{s_1}}{x_{1j}} = \frac{x_{s_2}}{x_{2j}}.$$

Hence the new basic variable x_j can enter the new basic feasible solution replacing either x_{s_1} or x_{s_2}. If the current basic variable x_{s_1} is the dropping variable, the value of the basic variable x_{s_2} is zero in the BFS obtained after the pivot. So the new basic feasible solution becomes degenerate. Conversely, if the pivot row corresponds to the variable x_{s_2}, the value of the variable x_{s_1} remaining in the basis becomes zero.

Consider the following LFP problem

$$Q(x) = \frac{2x_1 + 4x_2 + 5}{2x_1 + 3x_2 + 10} \longrightarrow \max$$

subject to

$$\begin{array}{rcrcr} 2x_1 & + & 1x_2 & \leq & 6\,, \\ 4x_1 & + & 2x_2 & \leq & 12\,, \end{array}$$

$$x_1 \geq 0, \quad x_2 \geq 0.$$

After entering slack variables, the initial simplex tableau will be as follows:

Initial tableau				2 2	4 3	0 0	0 0	
B	P_B	D_B	x_B	A_1	A_2	A_3	A_4	
A_3	0	0	6	**2**	1	1	0	$\Rightarrow$
A_4	0	0	12	4	2	0	1	
$P(x) = 5$				-2	-4	0	0	
$D(x) = 10$				-2	-3	0	0	
$Q(x) = 1/2$				-1	$-5/2$	0	0	
				$\Uparrow$				

In the bottom of this tableau both original variables x_1 and x_2 have negative reduced costs $\Delta_1(x) = -1$ and $\Delta_2(x) = -5/2$. Hence, either x_1 or x_2 may enter the new basis. Let us choose x_1 and the corresponding vector $A_1 = (2, 4)^T$. Since the ratio test gives

$$\frac{x_3}{x_{11}} = \frac{6}{2} = 3, \quad \text{and} \quad \frac{x_4}{x_{12}} = \frac{12}{4} = 3,$$

it means that if we choose x_1 to enter the basis, degeneracy will occur. We choose x_3 as the leaving variable and perform a simplex iteration. This yields the second tableau below

Tableau 2				2 2	4 3	0 0	0 0	
B	P_B	D_B	x_B	A_1	A_2	A_3	A_4	
A_1	2	2	3	1	$1/2$	$1/2$	0	$\Rightarrow$
A_4	0	0	0	0	0	-2	1	
$P(x) = 11$				0	-3	1	0	
$D(x) = 16$				0	-2	1	0	
$Q(x) = 11/16$				0	$-13/8$	$5/16$	0	
					$\Uparrow$			

Observe that the BFS obtained has a basic variable (namely x_4) which is equal to zero. Since $\Delta_2(x) = -13/8 < 0$ it indicates that the current BFS is not optimal and we have to continue the steps of the simplex method. The only non-basic variable with negative reduced cost $\Delta_j(x)$ is x_2, hence it must be entered into the basis. The minimum ratio test in row 1 gives

$$\theta = \frac{x_1}{x_{12}} = \frac{3}{1/2} = 6,$$

so after performing the corresponding re-calculations we obtain

Tableau 3				2 / 2	4 / 3	0 / 0	0 / 0
B	P_B	D_B	x_B	A_1	A_2	A_3	A_4
A_2	4	3	6	2	1	1	0
A_4	0	0	0	0	0	-2	1
$P(x) = 29$				6	0	4	0
$D(x) = 28$				4	0	3	0
$Q(x) = 29/28$				13/7	0	25/28	0

Since all $\Delta_j(x) \geq 0, \ j = 1,2,3,4,$ it means that we obtain an optimal solution. This is vector $x^* = (0,6)^T$ with $Q(x^*) = 29/28$. As we can see, in this example degeneracy occurred but did not prevent the simplex method finding the optimal solution.

In 1977 R.G.Bland [31] proposed for LP problems a pivoting rule that uses ordered indices of variables and prevents the simplex method cycling. Fortunately, this rule may be applied for LFP problems too.

Bland rule (Least index rule)

- When more than one non-basic variables are candidates for entering the basis (i.e in a maximization problem more than one $\Delta_j(x) < 0, j \in J_N$), then we choose the variable with the smallest index.
- When more than one basic variables are candidates for leaving the basis (i.e degeneracy will occur), then we choose the variable with the smallest index.

Also, R.G.Bland formulated and proved [31] the following

THEOREM 4.7 *When the Least index rule is applied, then the simplex method cannot cycle and hence, terminates after a finite number of steps.*

Closing this section we remark[2] that even if the degeneracy may have relatively frequent occurrence, there are not too many reasons for applying this anti-cycling rule in computer codes because

1 Although degeneracy is frequent, cycling is extremely rare.

2 The precision of computer arithmetic takes care of cycling by itself: round off errors accumulate and eventually get the method out of cycling.

[2]Michael A. Trick's Operations Research Web-site:
http://mat.gsia.cmu.edu/QUANT/NOTES/chap/node5.html

10. Unrestricted-In-Sign Variables

When solving LFP problems with the simplex method, we used the ratio test to determine the row in which the entering variable became a basic variable. Let us recall that the ratio test depended on the fact that any feasible point required all variables to be nonnegative. Thus, if some variables are allowed to be unrestricted in sign (*urs*), the ratio test and therefore the simplex algorithm are no longer valid. In this section, we show how an LFP problem with unrestricted-in-sign variables can be transformed into LFP in which all variables are required to be nonnegative. Note that in some textbooks such unrestricted-in-sign variables may be referred to as *free* variables.

For each urs variable x_j, we begin by defining two new nonnegative variables x'_j and x''_j. Then substitute $x'_j - x''_j$ for x_j in each constraint and in objective function. Also, add the sign restrictions $x'_j \geq 0$ and $x''_j \geq 0$. The effect of this substitution is to express urs x_j as the difference of the two nonnegative variables x'_j and x''_j. Since all variables are now required to be nonnegative, we can proceed with the simplex method. As we will see soon, no basic feasible solution can have both $x'_j > 0$ and $x''_j > 0$. This means that for any basic feasible solution, each urs variable x_j must fall into one of the following three cases:

1 $x'_j > 0$ and $x''_j = 0$.

2 $x'_j = 0$ and $x''_j > 0$.

3 $x'_j = 0$ and $x''_j = 0$.

Consider the following example

$$Q(x) = \frac{x_1 + x_2 + 5}{3x_1 + 2x_2 + 15} \longrightarrow \max$$

subject to

$$\begin{array}{rcrcr} 3x_1 & + & x_2 & = & 6\,, \\ 3x_1 & + & 4x_2 & = & 12\,, \end{array}$$

$$x_1 - \text{urs}, \quad x_2 \geq 0\,.$$

Introducing two new nonnegative variables x'_1 and x''_1, and substituting $x'_1 - x''_1$ for x_1 in each constraint and in objective function, we obtain the following problem

$$Q(x) = \frac{x'_1 - x''_1 + x_2 + 5}{3x'_1 - 3x''_1 + 2x_2 + 15} \longrightarrow \max$$

subject to

$$\begin{array}{rcrcrcr} 3x'_1 & - & 3x''_1 & + & x_2 & = & 6\,, \\ 3x'_1 & - & 3x''_1 & + & 4x_2 & = & 12\,, \end{array}$$

$$x_1' \geq 0, \;\; x_1'' \geq 0, \;\; x_2 \geq 0\,.$$

Since the problem obtained contains only non-negative unknown variables, we can apply the simplex method to solve it.

11. Bounded Variables

When considering real-world applications of LFP, it may occur that one or more unknown variables x_j not only have a non-negativity constraint but are constrained by upper-bound conditions, i.e.

$$0 \leq x_j \leq u_j, \quad \text{for some } \; j \in J = \{1, 2, \dots, n\}. \tag{4.40}$$

Since constraints of this form provide upper bounds on variables, they are usually called *upper-bound constraints.*

Generally speaking, all unknown variables x_j in LFP problems may have also lower-bound constraints. So, in the more common case, instead of (4.40) we have to write

$$l_j \leq x_j \leq u_j, \;\; j = 1, 2, \dots, n.$$

Consider the following LFP problem

$$Q(x) = \frac{P(x)}{D(x)} = \frac{\sum_{j=1}^{n} p_j x_j + p_0}{\sum_{j=1}^{n} d_j x_j + d_0} \longrightarrow \max, \tag{4.41}$$

subject to

$$\sum_{j=1}^{n} a_{ij} x_j = b_i, \quad i = 1, 2, \dots, m, \tag{4.42}$$

$$l_j \leq x_j \leq u_j, \quad j = 1, 2, \dots, n. \tag{4.43}$$

where

$$l_j \leq u_j, \quad j = 1, 2, \dots, n.$$

Let us assume that $D(x) > 0$ for all $x = (x_1, x_2, \cdots, x_n)^T \in S,$ and S denotes a feasible set defined by constraints (4.42)-(4.43). We assume also that feasible set S is non-empty and bounded.

This problem differs from the standard LFP problem (4.1)-(4.3) with only constraints (4.43), which, generally speaking, bound the unknown variables from the sides. It is possible also, that some $l_j = -\infty$ and some $u_j = +\infty$. If for some index j, $l_j = -\infty$ and $u_j = +\infty$, then we obtain an unrestricted-in-sign (or free) variable (see section 10, Chapter 4). If for some index j, $l_j = u_j$, then the corresponding variable x_j is said to be *fixed.* Obviously, if for all $j = 1, 2, \dots, n$, $l_j = 0$ and $u_j = +\infty$, we obtain standard LFP problem (4.1)-(4.3).

It is obvious also that an LFP problem with bounded variables (4.41)-(4.43) may be easily transformed back to the standard form (4.1)-(4.3) and then be solved by the standard simplex method described in previous sections.

Indeed, if we substitute the original variables x_j with $x'_j = x_j - l_j$, $j = 1, 2, \dots, n$, we obtain

$$Q(x') = \frac{P(x')}{D(x')} = \frac{\sum_{j=1}^{n} p_j x'_j + p'_0}{\sum_{j=1}^{n} d_j x'_j + d'_0} \rightarrow \max, \tag{4.44}$$

subject to

$$\sum_{j=1}^{n} a_{ij} x'_j = b'_i, \quad i = 1, 2, \dots, m, \tag{4.45}$$

$$0 \leq x'_j \leq u'_j, \quad j = 1, 2, \dots, n, \tag{4.46}$$

where

$$p'_0 \; = \; \sum_{j=1}^{n} p_j l_j + p_0, \quad d'_0 \; = \; \sum_{j=1}^{n} d_j l_j + d_0,$$

and

$$b'_i \; = \; b_i - \sum_{j=1}^{n} a_{ij} l_j, \quad i = 1, 2, \dots, m;$$

$$u'_j \; = \; u_j - l_j, \quad j = 1, 2, \dots, n.$$

After including upper-bound constraints (4.46) to the main system (4.45), instead of (4.45) and (4.46) we have

$$\left.\begin{array}{rcll} \sum_{j=1}^{n} a_{ij} x'_j & = & b'_i, & i = 1, 2, \dots, m, \\ x'_j & \leq & u'_j, & j = 1, 2, \dots, n, \end{array}\right\} \tag{4.47}$$

$$x'_j \geq 0, \quad j = 1, 2, \dots, n. \tag{4.48}$$

Since system (4.47) is not in canonical form, we also have to introduce n non-negative artificial variables y_j to transform the problem to canonical form. Finally, we have

$$\left.\begin{array}{rcll} \sum_{j=1}^{n} a_{ij} x'_j & = & b'_i\,, & i = 1, 2, \dots, m, \\ x'_j + y_j & = & u'_j\,, & j = 1, 2, \dots, n, \end{array}\right\} \tag{4.49}$$

$$x'_j \geq 0, \quad y_j \geq 0, \quad j = 1, 2, \dots, n. \tag{4.50}$$

As we can see, if the original LFP problem (4.41)-(4.43) is of size $m \times n$, then after the transformation we made, the obtained LFP problem (4.44), (4.49), (4.50) consists of $m + n$ main constraints and $2n$ unknown variables. Obviously, because of the increased size of the problem obtained, this approach is undesirable computationally.

Below we will see that the standard LFP simplex method can be adapted to an LFP problem with bounded variables in such a way that the constraints (4.43) are considered implicitly.

REMARK 4.11 *For example, if* $m = 50$ *and* $n = 100$ *the ordinary simplex method would require a matrix with* $(50 + 100) \times 2 \times 100 = 30,000$ *elements instead of original matrix with* $50 \times 100 = 5,000$ *elements. The reduction in this case would be 83.33%.*

We consider an LFP problem of the form (4.41)-(4.43), where all the variables in the problem are bounded. We suppose that all bounds l_j and u_j are finite and $l_j \leq u_j$, for all $j = 1, 2, \dots, n$.

To adapt the standard simplex method to an LFP problem with upper-bound constraints we have to re-define the basic conceptions of a basic feasible solution.

Let us suppose that a given system of vectors $B = \{A_{s_1}, A_{s_2}, \dots, A_{s_m}\}$ is a basis, i.e. vectors $A_{s_1}, A_{s_2}, \dots, A_{s_m}$, where $A_j = (a_{1j}, a_{2j}, \dots, a_{mj})^T$, are linearly independent.

Let J_B be a set of indices j corresponding to vectors A_j of basis B, i.e. $J_B = \{s_1, s_2, \dots, s_m\}$. If $J = \{1, 2, \dots, n\}$, then set $J_N = J \backslash J_B$ denotes the indices of those vectors Aj, which are not in basis B.

DEFINITION 4.10 *The given vector* $x = (x_1, x_2, \dots, x_n)^T$ *is a* ***basic feasible solution (BFS)*** *of LFP problem (4.41)-(4.43) if vector* x *satisfies system*

$$\sum_{j \in J_B} A_j\, x_j = b$$

and

$$l_j < x_j < u_j, \quad \forall j \in J_B,$$
$$x_j = l_j \ \textit{or} \ x_j = u_j, \quad \forall j \in J_N.$$

Similarly to the ordinary simplex method we have to introduce

$$\left.\begin{array}{rcl} \Delta'_j & = & \sum_{i=1}^{m} p_{s_i} x_{ij} - p_j, \\ \Delta''_j & = & \sum_{i=1}^{m} d_{s_i} x_{ij} - d_j, \\ \Delta_j(x) & = & \Delta'_j - Q(x)\Delta''_j, \end{array}\right\} \quad j = 1, 2, \dots, n,$$

where coefficients x_{ij} are given from the following systems of linear equations

$$A_j = \sum_{i=1}^{m} A_{s_i} x_{ij}, \quad j = 1, 2, \dots, n.$$

Using this notation we can formulate the following statement.

THEOREM 4.8 (CRITERIA OF OPTIMALITY) *A basic feasible solution x is a basic optimal solution of linear-fractional programming problem (4.41)-(4.43) if and only if $\Delta_j(x) \geq 0, \ j = 1, 2, \dots, n$.*

Since the proof of this statement is similar to the one for Theorem 4.4, we omit it.

Suppose that we have some basis B and corresponding to it BFS vector x with the following index partitioning

$$\begin{array}{ll} l_j < x_j < u_j, & \forall j \in J_B = \{s_1, s_2, \dots, s_m\}, \\ x_j = l_j, & \forall j \in J_N^l \subseteq J_N, \\ x_j = u_j, & \forall j \in J_N^u \subseteq J_N, \end{array}$$

where $J_N^l \cup J_N^u = J_N$ and $J_N \cup J_B = J = \{1, 2, \dots, n\}$.

Assume that vector x is not optimal and $\Delta_k(x) < 0, \ k \in J_N$. In accordance with the general scheme of the ordinary simplex method, it means that we have to enter vector A_k into the basis and perform the simplex iteration.

The main difference between the ordinary simplex method and the simplex method with bounded variables is that when updating BFS we have to use the following rule

$$x_\mu(\theta) = \begin{cases} x_{s_i} - \theta x_{ij}, & \text{if } \mu = s_i, \ i = 1, 2, \dots, m; \\ x_\mu + \theta, & \text{if } \mu \in J_N, \ \mu = k; \\ x_\mu, & \text{if } \mu \in J_N, \ \mu \neq k; \end{cases} \tag{4.51}$$

$$\begin{array}{rcccl} l_{s_i} & \le & x_{s_i}(\theta) & \le & u_{s_i}, \quad i = 1, 2, \ldots, m, \\ l_k & \le & x_k(\theta) & \le & u_k, \end{array}$$

or, in accordance with (4.51)

$$\begin{array}{rcccl} l_{s_i} & \le & x_{s_i} - \theta x_{ij} & \le & u_{s_i}, \quad i = 1, 2, \ldots, m \\ l_k & \le & x_k + \theta & \le & u_k. \end{array}$$

It is obvious that the latest may be rewritten as follows:

$$\begin{array}{ll} \theta \le \dfrac{x_{s_i} - l_{s_i}}{x_{ij}}, & \text{for those index } i \text{ that } x_{ij} > 0, \\ \theta \ge \dfrac{x_{s_i} - l_{s_i}}{x_{ij}}, & \text{for those index } i \text{ that } x_{ij} < 0, \\ \theta \ge \dfrac{x_{s_i} - u_{s_i}}{x_{ij}}, & \text{for those index } i \text{ that } x_{ij} > 0, \\ \theta \le \dfrac{x_{s_i} - u_{s_i}}{x_{ij}}, & \text{for those index } i \text{ that } x_{ij} < 0, \\ \theta \le u_k - x_k, & \\ \theta \ge l_k - x_k, & \end{array}$$

or in more compact form:

$$\begin{array}{rcccl} \max\limits_{x_{ij}<0} \dfrac{x_{s_i} - l_{s_i}}{x_{ij}} & \le & \theta & \le & \min\limits_{x_{ij}>0} \dfrac{x_{s_i} - l_{s_i}}{x_{ij}}, \\ \max\limits_{x_{ij}>0} \dfrac{x_{s_i} - u_{s_i}}{x_{ij}} & \le & \theta & \le & \min\limits_{x_{ij}<0} \dfrac{x_{s_i} - u_{s_i}}{x_{ij}}, \\ l_k - x_k & \le & \theta & \le & u_k - x_k, \end{array}$$

From the latter we obtain

$$\theta_{min} \le \theta \le \theta_{max} \,, \tag{4.52}$$

where

$$\theta_{min} = \max\{l_k - x_k, \ \max_{x_{ij}<0} \frac{x_{s_i} - l_{s_i}}{x_{ij}}, \ \max_{x_{ij}>0} \frac{x_{s_i} - u_{s_i}}{x_{ij}}\}$$

and

$$\theta_{max} = \min\{u_k - x_k, \ \min_{x_{ij}>0} \frac{x_{s_i} - l_{s_i}}{x_{ij}}, \ \min_{x_{ij}<0} \frac{x_{s_i} - u_{s_i}}{x_{ij}}\} \,.$$

Note that in the non-degenerate case

$$\theta_{max} > 0, \quad \text{and} \quad \theta_{min} < 0.$$

When choosing θ we have to distinguish the following two cases:

1 Index $k \in J_N^l$, i.e. non-basic variable x_k is equal to its lower bound l_k and when being entered into basis it must be increased. In this case we have to choose

$$\theta = \theta_{max} > 0. \tag{4.53}$$

2 Index $k \in J_N^u$, i.e. non-basic variable x_k is equal to its upper bound u_k and when being entered into the basis it must be decreased. In this case we have to choose

$$\theta = \theta_{min} < 0. \tag{4.54}$$

Concerning the changes necessary to adapt the standard simplex tableau (see Table 4.1, page 87) to the case of bounded variables, the only difference is that we have to store in the tableau the lower and upper bounds of unknown variables (4.43). One of the possible ways to store necessary data when solving an LFP problem with bounded variables is presented in Table 4.19. In the topmost row

			x	x_1	x_2	$\cdots$	x_k	$\cdots$	x_n
			l	l_1	l_2	$\cdots$	l_k	$\cdots$	l_n
			u	u_1	u_2	$\cdots$	u_k	$\cdots$	u_n
			P	p_1	p_2	$\cdots$	p_k	$\cdots$	p_n
			D	d_1	d_2	$\cdots$	d_k	$\cdots$	d_n
B	P_B	D_B	x_B	A_1	A_2	$\cdots$	A_k	$\cdots$	A_n
A_{s_1}	p_{s_1}	d_{s_1}	x_{s_1}	x_{11}	x_{12}	$\cdots$	x_{1k}	$\cdots$	x_{1n}
A_{s_2}	p_{s_2}	d_{s_2}	x_{s_2}	x_{21}	x_{22}	$\cdots$	x_{2k}	$\cdots$	x_{2n}
$\vdots$	$\vdots$	$\vdots$	$\vdots$	$\vdots$	$\vdots$	$\vdots$	$\vdots$	$\cdots$	$\vdots$
A_{s_r}	p_{s_r}	d_{s_r}	x_{s_r}	x_{r1}	x_{r2}	$\cdots$	x_{rk}	$\cdots$	x_{rn}
$\vdots$	$\vdots$	$\vdots$	$\vdots$	$\vdots$	$\vdots$	$\vdots$	$\vdots$	$\cdots$	$\vdots$
A_{s_m}	p_{s_m}	d_{s_m}	x_{s_m}	x_{m1}	x_{m2}	$\cdots$	x_{mk}	$\cdots$	x_{mn}
	$P(x)$		Δ'	Δ'_1	Δ'_2	$\cdots$	Δ'_k	$\cdots$	Δ'_n
	$D(x)$		Δ''	Δ''_1	Δ''_2	$\cdots$	Δ''_k	$\cdots$	Δ''_n
	$Q(x)$		$\Delta(x)$	$\Delta_1(x)$	$\Delta_2(x)$	$\cdots$	$\Delta_k(x)$	$\cdots$	$\Delta_n(x)$

Table 4.19. Simplex tableau for LFP problem with bounded variables.

this tableau contains the current values of all unknown variables $x_1, x_2, \dots, x_n$, while the lower and upper bounds of variables are in the second and third rows, respectively.

To illustrate how this method works, we consider the following LFP problem with bounded variables

$$Q(x) = \frac{P(x)}{D(x)} = \frac{5x_1 + 1x_2 + 10}{4x_1 + 2x_2 + 12} \longrightarrow \max$$

subject to

$$\begin{array}{lllllll} 5x_1 & + & 1x_2 & + & 1x_3 & & = 20\ , \\ 4x_1 & & & - & 1x_3 & + \ 1x_4 & = 14\ , \end{array}$$

$$\begin{array}{ccccc} 2 & \leq & x_1 & \leq & 5\ , \\ 4 & \leq & x_2 & \leq & 12\ , \\ 0 & \leq & x_3 & \leq & 25\ , \\ 0 & \leq & x_4 & \leq & 18\ . \end{array}$$

Starting with initial BFS $x = (2, 10, 0, 6)^T$ and index partition

$$J_B = \{2, 4\},\ \ J_N^l = \{1, 3\},\ \ J_N^u = \{\}$$

we obtain the initial simplex tableau shown in Table 4.20.

				x	2	10	0	6	
Iteration 1				l	2	4	0	0	
				u	5	12	25	18	
				P	5	1	0	0	
				D	4	2	0	0	
B	P_B	D_B		x_B	A_1	A_2	A_3	A_4	
A_2	1	2		10	5	1	1	0	$\Rightarrow$
A_4	0	0		6	4	0	-1	1	
$P(x) = 30$				Δ'	0	0	1	0	
$D(x) = 40$				Δ''	2	0	2	0	
$Q(x) = 3/4$				$\Delta(x)$	$-9/2$	0	$-1/2$	0	
							$\Uparrow$		

Table 4.20. Bounded variables example – Initial tableau.

Since the aim of this problem is maximization and the bottom row of the initial tableau contains negative $\Delta_1(x) = -6/4$ and $\Delta_3(x) = -2/4$, it means that the current BFS is not optimal. In this case, we have to choose a non-basic vector A_k with negative $\Delta_k(x)$ and enter it into the basis. Let it be vector A_3, i.e. $k = 3$. Observe that non-basic variable $x_3 = 0 = l_3$, i.e. $k \in J_N^l$. The latter means that when choosing the value of θ we have to use formula (4.53). So, we have

$$\begin{aligned} \theta &= \min\{u_3 - x_3, \frac{x_2 - l_2}{x_{13}}, \frac{x_4 - u_4}{x_{23}}\} = \\ &= \min\{25 - 0, \frac{10 - 4}{1}, \frac{6 - 18}{-1}\} = 6 . \end{aligned}$$

Hence, in accordance with (4.51) we obtain

$$\begin{aligned} x_1(\theta) &= x_1 = 2 , && (\mu \in J_N, \ \mu \neq k), \\ x_2(\theta) &= x_2 - \theta x_{13} = 10 - 6 \times 1 = 4 , && (\mu \in J_B), \\ x_3(\theta) &= x_3 + \theta = 0 + 6 = 6 , && (\mu \in J_N, \ \mu = k), \\ x_4(\theta) &= x_4 - \theta x_{23} = 6 - 6 \times (-1) = 12 , && (\mu \in J_B). \end{aligned}$$

So, the new BFS is vector $x = (2, 4, 6, 12)^T$ with index partition

$$J_B = \{3, 4\}, \ J_N^l = \{1, 2\}, \ J_N^u = \{\} .$$

Now, we have to perform a pivot transformation and re-calculate Δ'_j, Δ''_j and $\Delta_j(x)$, for all $j = 1, 2, 3, 4,$. So, we obtain the simplex tableau presented in Table 4.21.

Iteration 2			x	2	4	6	12	
			l	2	4	0	0	
			u	5	12	25	18	
			P	5	1	0	0	
			D	4	2	0	0	
B	P_B	D_B	x_B	A_1	A_2	A_3	A_4	
A_3	0	0	6	5	1	1	0	$\Rightarrow$
A_4	0	0	12	9	1	0	1	
$P(x) = 24$			Δ'	-5	-1	0	0	
$D(x) = 27$			Δ''	-4	-2	0	0	
$Q(x) = 6/7$			$\Delta(x)$	$-11/7$	$5/7$	0	0	
				$\Uparrow$				

Table 4.21. Bounded variables example – After first iteration.

In the bottom row this tableau (see Table 4.21) contains negative $\Delta_1(x) = -11/7$, which means that the current BFS is not optimal and we have to enter non-basic vector A_1 into the basis. Since non-basic variable $x_1 = 2 = l_1$, i.e. $k \in J_N^l$, this means that to choose the value of θ we have to use formula (4.53). So,

$$\begin{aligned} \theta &= \min\{u_1 - x_1, \frac{x_3 - l_3}{x_{11}}, \frac{x_4 - l_4}{x_{21}}\} = \\ &= \min\{5 - 2, \frac{6-0}{5}, \frac{12-0}{9}\} = 6/5 . \end{aligned}$$

Now, using (4.51) we can re-calculate the new BFS

$$\begin{aligned} x_1(\theta) &= x_1 + \theta = 2 + 6/5 = 16/5 , & (\mu \in J_N, \ \mu = k), \\ x_2(\theta) &= x_2 = 4 , & (\mu \in J_N, \ \mu \neq k), \\ x_3(\theta) &= x_3 - \theta x_{11} = 6 - 6/5 \times 5 = 0 , & (\mu \in J_B), \\ x_4(\theta) &= x_4 - \theta x_{21} = 12 - 6/5 \times 9 = 6/5 , & (\mu \in J_B). \end{aligned}$$

So, the new BFS is vector $x = (16/5, 4, 0, 6/5)^T$ with index partition

$$J_B = \{1,4\}, \ \ J_N^l = \{2,3\}, \ \ J_N^u = \{\} .$$

After performing pivot transformation and re-calculating Δ'_j, Δ''_j and $\Delta_j(x)$, for all $j = 1,2,3,4$, we obtain the final simplex tableau as shown in Table 4.22. Since in the bottom row of this tableau all $\Delta_j(x) \geq 0, \ j = 1,2,3,4,$ it means

Iteration 3			x	16/5	4	0	6/5
			l	2	4	0	0
			u	5	12	25	18
			P	5	1	0	0
			D	4	2	0	0
B	P_B	D_B	x_B	A_1	A_2	A_3	A_4
A_1	5	4	16/5	1	1/5	1/5	0
A_4	0	0	6/5	0	−4/5	−9/5	1
$P(x) = 30$			Δ'	0	0	1	0
$D(x) = 64/6$			Δ''	0	−6/5	4/5	0
$Q(x) = 150/164$			$\Delta(x)$	0	45/41	11/41	0

Table 4.22. Bounded variables example – Final tableau.

that the current BFS x solves the problem. So,

$$x^* = (16/5, 4, 0, 6/5)^T, \quad \text{and} \quad Q(x^*) = \frac{P(x^*)}{D(x^*)} = \frac{30}{64/6} = \frac{150}{164}.$$

Before closing this discussion of the bounded-variable simplex method, we note that this method is a full adaptation of the standard simplex algorithm, which will terminate at either an optimal or unbounded solution, increase the value of the objective function at each step (if the problem is not degenerate).

12. Discussion Questions and Exercises

4.1 Convert the given LFP problem to canonical form and then using artificial variables set up the initial simplex tableau.

$$Q(x) = \frac{5x_1 - 3x_2 + 2}{4x_1 + 1x_2 - 2} \longrightarrow \max$$

subject to

$$\begin{array}{rcrcl} x_1 & + & 2x_2 & \geq & 4\,, \\ x_1 & + & 3x_2 & \leq & 6\,, \end{array}$$

$$x_1 \geq 0, \quad x_2 \geq 0\,.$$

4.2 Perform one iteration of the simplex method to obtain the next tableau from the given tableau ($p_0 = 10, \ d_0 = 20$)

			P	5	3	2	2
			D	4	2	1	2
B	P_B	D_B	x_B	A_1	A_2	A_3	A_4
A_3			120	3	1	1	0
A_4			100	2	1	0	1
$P(x) =$			Δ'				
$D(x) =$			Δ''				
$Q(x) =$			$\Delta(x)$				

4.3 Using the Big M method solve the following LFP problem

$$Q(x) = \frac{1x_1 + 3x_2 + 6}{2x_1 + 3x_2 + 12} \longrightarrow \max$$

subject to

$$\begin{array}{rcrcl} x_1 & + & 2x_2 & \geq & 40\,, \\ 2x_1 & + & 3x_2 & \leq & 60\,, \end{array}$$

$$x_1 \geq 0, \quad x_2 \geq 0$$

4.4 Using the Two-Phase Simplex method solve the following LFP problem

$$Q(x) = \frac{4x_1 + 5x_2 + 10}{2x_1 + 3x_2 + 20} \longrightarrow \max$$

subject to

$$\begin{array}{rcrcl} 3x_1 & + & 2x_2 & \geq & 30\,, \\ 1x_1 & + & 3x_2 & \leq & 100\,, \\ 3x_1 & + & 1x_2 & \leq & 100\,, \end{array}$$

$$x_1 \geq 0, \ \ x_2 \geq 0$$

4.5 Solve the following LFP problem, noting where degeneracies occur. Sketch the set of feasible solutions, indicating the order in which the extreme points are examined by the simplex method

$$Q(x) = \frac{6x_1 + 5x_2 + 1}{2x_1 + 3x_2 + 1} \longrightarrow \max$$

subject to

$$\begin{array}{rcrcl} 2x_1 & + & 2x_2 & \leq & 10\,, \\ 3x_1 & + & 2x_2 & \leq & 15\,, \end{array}$$

$$x_1 \geq 0, \ \ x_2 \geq 0$$

4.6 Using suitable transformations and the standard simplex method solve the following LFP problem with unrestricted variables

$$Q(x) = \frac{1x_1 + 3x_2 + 6}{2x_1 + 3x_2 + 12} \longrightarrow \max$$

subject to

$$\begin{array}{rcrcl} x_1 & + & 2x_2 & \leq & 50\,, \\ x_1 & + & 2x_2 & \geq & 10\,, \\ 3x_1 & + & 1x_2 & \leq & 60\,, \end{array}$$

$$x_1 - \text{urs}, \ \ x_2 - \text{urs}.$$

4.7 Using the bounded-variable simplex method solve the following LFP problem

$$Q(x) = \frac{1x_1 + 3x_2 + 6}{2x_1 + 3x_2 + 12} \longrightarrow \max$$

subject to

$$\begin{array}{rcrcl} x_1 & + & 2x_2 & \geq & 10\,, \\ 2x_1 & + & 3x_2 & \leq & 60\,, \end{array}$$

$$5 \leq x_1 \leq 15, \ \ 4 \leq x_2 \leq 30.$$

Chapter 5

DUALITY THEORY

In accordance with the duality theory of mathematical programming every mathematical programming problem has an associated dual problem. The relationship between these two problems is very useful when investigating properties of optimal solutions of both problems. Principles of duality appear in various branches of mathematics, physics and statistics. These principles are valid in linear-fractional programming too – for any LFP problem (*primal problem*) we can formulate (construct) some other problem (*dual problem*), which is very closely connected with the original problem. These connections between primal and dual problems turn out to be of great practical use. Also, duality in LFP admits an elegant and useful economic interpretation.

1. Short overview

In this section we shortly overview several approaches to constructing the dual problem for LFP. In the 1960's and 1970's several authors proposed different types of dual problems related to the primal LFP problem consisting in maximizing or minimizing linear-fractional objective function subject to a system of linear equality and/or inequality constraints. Not all of these dual problems and associated approaches have any practical interest and may be used in practice. One of them is based on the well-known Charnes & Cooper transformation [38] (Chapter 3, Section 3) and leads to the duality theory of linear programming.

Let us consider the following LFP problem in a general form:

$$Q(x) = \frac{P(x)}{D(x)} = \frac{\sum_{j=1}^{n} p_j x_j + p_0}{\sum_{j=1}^{n} d_j x_j + d_0} \to \max, \tag{5.1}$$

subject to

$$\sum_{j=1}^{n} a_{ij} x_j \le b_i, \quad i = 1, 2, \dots, m, \tag{5.2}$$

$$x_j \ge 0, \quad j = 1, 2, \dots, n. \tag{5.3}$$

In accordance with the rules formulated in Section 3 of Chapter 3, the linear analogue of LFP problem (5.1)-(5.3) may be formulated as follows:

$$L(t) = \sum_{j=0}^{n} p_j t_j \longrightarrow \max \tag{5.4}$$

subject to

$$\sum_{j=0}^{n} d_j t_j = 1. \tag{5.5}$$

$$-b_i t_0 + \sum_{j=1}^{n} a_{ij} t_j \le 0, \quad i = 1, 2, \dots, m, \tag{5.6}$$

$$t_j \ge 0, \quad j = 0, 1, 2, \dots, n, \tag{5.7}$$

where

$$t_j = \frac{x_j}{D(x)}, \quad j = 1, 2, \dots, n, \qquad t_0 = \frac{1}{D(x)}.$$

Note that (5.4)-(5.7) is a linear programming problem. As we know from the theory of linear programming and its duality, the dual problem for general LP problem in form

$$P(x) = \sum_{j=1}^{n} p_j x_j + p_0 \longrightarrow \max \tag{5.8}$$

subject to

$$\sum_{j=1}^{n} a_{ij} x_j \le b_i, \quad i = 1, 2, \dots, m, \tag{5.9}$$

$$x_j \geq 0, \quad j = 1, 2, \ldots, n \tag{5.10}$$

may be formulated as

$$\phi(u) = \sum_{i=1}^{m} b_i u_i + p_0 \longrightarrow \min \tag{5.11}$$

subject to

$$\sum_{i=1}^{m} a_{ij} u_i \geq p_j, \quad j = 1, 2, \ldots, n, \tag{5.12}$$

$$u_i \geq 0, \quad i = 1, 2, \ldots, m. \tag{5.13}$$

From the point of view of practical usability, one of the most important results of duality theory in linear programming is the interpretation of dual variables u_i, $i = 1, 2, \ldots, m$; as *shadow prices*. Let us suppose that vector $u^* = (u_1^*, u_2^*, \ldots, u_m^*)^T$ is an optimal solution of dual LP problem (5.11)-(5.13). Optimal variable u_k^* in LP may be interpreted as fluctuation of optimal value for the objective function $P(x)$ of LP problem (5.8)-(5.10) when changing element b_k by one unit in the right-hand-side vector $b = (b_1, b_2, \ldots b_m)^T$ in constraints (5.9). This interpretation may be expressed in the form of the following formula:

$$\frac{\partial P(x^*(b))}{\partial b_k} = u_k^*, \quad k = 1, 2, \ldots, m. \tag{5.14}$$

or

$$P(x') = P(x^*) + \lambda u_k^*, \tag{5.15}$$

where λ is a small enough change of element b_k, that is $b_k' = b_k + \lambda$, and vector x' is an optimal solution of the modified LP problem (5.8)-(5.10), where the original RHS vector

$$b = (b_1, b_2, \ldots, b_k, \ldots, b_m)^T$$

is replaced with the following new one:

$$b' = (b_1, b_2, \ldots, b_k + \lambda, \ldots, b_m)^T.$$

Since linear analogue (5.4)-(5.7) is a linear programming problem, we may apply formula (5.14) (or (5.15)), but obviously, there is no reason for doing so because in the right side of conditions (5.6) there is no vector b that may be varied. It means that the shadow prices in LFP are lost and in such a way we are deprived of a very important tool for analysis in economic applications.

This is why the approach to the duality in LFP based on the use of the Charnes & Cooper transformation does not have any practical interest.

Different ideas were applied for constructing dual problems in LFP by C.R.Seshan [166] in the 1980's. In [166] the dual problem for LFP is a linear-fractional programming problem too:

$$I(u,v) = \frac{p^T u + p_0}{d^T u + d_0} \longrightarrow \min \tag{5.16}$$

subject to

$$pd^T u - dp^T u - A^T v \leq p_0 d - d_0 p, \tag{5.17}$$

$$p_0 d^T u - d_0 p^T u + b^T v \leq 0, \tag{5.18}$$

$$u \geq 0, \quad v \geq 0, \quad u \in R^n, \quad v \in R^m. \tag{5.19}$$

Seshan showed that problem (5.16)-(5.19) is a dual problem for LFP problem (5.1)-(5.3), and proved the main statements of duality theory. As we can see problem (5.16)-(5.19) contains $(m+1)$ main constraints and $(n+m)$ unknown sign-restricted variables. The practical usability of vectors $u = (u_1, u_2, \dots, u_n)^T$ and $v = (v_1, v_2, \dots, v_m)^T$ is still an open question.

Another branch of investigations is connected with C.R.Bector [23], [24], [25] who used the Charnes & Cooper transformation and standard Lagrange function

$$L(x,y) = Q(x) + \sum_{i=1}^{m} y_i \, f_i(x)$$

to construct the dual problem in (linear-) fractional programming, where objective function $Q(x)$ was considered in the form without constant terms:

$$Q(x) = \frac{P(x)}{D(x)} = \frac{\sum_{j=1}^{n} p_j x_j}{\sum_{j=1}^{n} d_j x_j}.$$

C.R.Bector presented a dual problem in three equivalent forms. One of them is nearly the same as the dual problem formulated by J.Kaška [109], the only difference is in the type of relations. Another two forms may be relatively easily converted to one another and both of them are strictly connected with the first one. A more detailed comparison of different dual models formulated by J.Kaška [109], I.C.Sharma and K.Swarup [168], and K.Swarup [175], [176], for LFP may be found in J.Abrham and S.Luthra [1].

Valuable contribution to LFP duality was made by numerous other investigators: [28], [39] [43], [161] [162], [165], [168], [175], [176].

However, one of the most fruitful directions in this area of investigations is connected with a special ratio-type Lagrange function introduced by E.G.Gol'stein [76], [77]:

$$L(x,y) \;=\; \frac{P(x) + \sum_{i=1}^{m} y_i\, f_i(x)}{D(x)},$$

where

$$f_i(x) \;=\; b_i - \sum_{j=1}^{n} a_{ij}x_j, \quad i = 1,2,\dots,m.$$

In the next sections we discuss in detail this approach based on the use of the so-called *fractional Lagrangian.*

2. Gol'stein-type Lagrangian

Let us consider the general LFP problem given in the form (5.1)-(5.3). Here and in what follows we suppose that

$$D(x) > 0, \;\; \forall x \in R^+ = \{x \in R^n : x \geq 0, \; j = 1,2,\dots,n\}. \qquad (5.20)$$

Using the methodology described in [76], [77], we construct now the dual problem for the general LFP problem (5.1)-(5.3).

We use here the following fractional Lagrangian with non-negative variables x and y

$$L(x,y) = \frac{P(x) + \sum_{i=1}^{m} y_i\,(b_i - \sum_{j=1}^{n} a_{ij}\, x_j)}{D(x)}. \qquad (5.21)$$

Let us take into consideration the following function

$$\psi(y) = \max_{x \geq 0} L(x,y) \qquad (5.22)$$

and rewrite this function $\psi(y)$ as follows:

$$\begin{aligned}\psi(y) \;&=\; \max_{x\geq 0} \frac{P(x) + \sum_{i=1}^{m} y_i\,(b_i - \sum_{j=1}^{n} a_{ij}\, x_j)}{D(x)} =\\ &=\; \max_{x\geq 0} \frac{\sum_{j=1}^{n} p_j\, x_j + p_0 + \sum_{i=1}^{m} y_i\,(b_i - \sum_{j=1}^{n} a_{ij}\, x_j)}{D(x)} =\end{aligned}$$

$$= \max_{x \geq 0} \frac{\sum_{i=1}^{m} b_i\, y_i + p_0 + \sum_{j=1}^{n} x_j\, (p_j - \sum_{i=1}^{m} a_{ij}\, y_i)}{D(x)} =$$

$$= \max_{x \geq 0} \frac{p_0(y) + \sum_{j=1}^{n} p_j(y)\, x_j}{d_0 + \sum_{j=1}^{n} d_j\, x_j}, \tag{5.23}$$

where

$$p_0(y) = \sum_{i=1}^{m} b_i y_i + p_0, \quad p_j(y) = -\sum_{i=1}^{m} a_{ij} y_i + p_j, \quad j = 1, 2, \ldots, n.$$

As we can see from formula (5.23), when fixing variable y, function $\psi(y)$ becomes a linear-fractional function that depends only on non-negative variables $x_j,\ j = 1, 2, \ldots, n$.

This is why to solve the optimization problem (5.22) we should assume that variable y is fixed and then to find the maximum value of liner-fractional function (5.23) over set

$$R^+ = \{x \in R^n : x_j \geq 0,\ j = 1, 2, \ldots, n\}.$$

Since function (5.23) is a linear-fractional function, it is a monotonic function over set R^+ (see Theorem 4.2, p.78). In accordance with our assumption (5.20) denominator $D(x) > 0,\ \ \forall x \in R^+,$ so it is obvious that fractional function $L(x, y)$ on set R^+ does not have any extreme value at interior points of the domain R^+. It means that to solve maximal problem (5.22) we have to find a maximal value of $L(x, y)$ over extreme points of domain R^+. In other words, we have to calculate value $L(x, y)$ (with fixed variables y) in the zero point 0, then we have to determine maximal values (with the same fixed values$y$) of $L(x, y)$ on each axis $0x_j,\ j = 1, 2, \ldots, n,$ and, finally, on the basis of a simple comparison we have to choose from the (n+1) values obtained the one where function $L(x, y)$ reaches its maximal value. The value obtained in such a way will be the maximal value of Lagrangian $L(x, y)$ over set R^+. So, we have

$$\begin{aligned} \psi(y) &= \max_{x \geq 0} L(x, y) = \\ &= \max_{0 \leq j \leq n} \{\frac{p_0(y)}{d_0},\ G_1^*(y),\ \ldots,\ G_n^*(y)\}, \end{aligned}$$

where

$$G_j^*(y) = \max_{x_j \geq 0} G_j(x, y), \quad j = 1, 2, \ldots, n,$$

and

$$G_j(x,y) = \frac{p_0(y) + p_j(y)\, x_j}{d_0 + d_j\, x_j}, \quad j = 1,2,\ldots,n.$$

Consider now the problem

$$G_j^*(y) \;=\; \max_{x_j \geq 0} G_j(x,y)$$

for each $j = 1,2,\ldots,n$. Note that function $G_j(x,y)$ is also a linear-fractional function, $j = 1,2,\ldots,n$, and in accordance with assumption (5.20) denominator $(d_0 + d_j x_j) > 0$ on R^+. Hence,

$$\begin{aligned} G_j^*(y) &= \max_{x_j \geq 0} \frac{p_0(y) + p_j(y)\, x_j}{d_0 + d_j\, x_j} = \\ &= \max\{\frac{p_0(y)}{d_0}, \ \max\{\frac{p_0(y)}{d_0}, \frac{p_j(y)}{d_j}\}\} = \\ &= \max\{\frac{p_0(y)}{d_0}, \frac{p_j(y)}{d_j}\}, \ j = 1,2,\ldots,n. \end{aligned} \tag{5.24}$$

Hence, taking the (5.23) and (5.24) into account, we can re-formulate maximization problem (5.22) as follows:

$$\psi(y) = \max_{x \geq 0} L(x,y) = \max_{j \in J_0}\{\frac{p_j(y)}{d_j}\}, \tag{5.25}$$

where $J_0 = \{0,1,2,\ldots,n\}$.

Let J_1 denote the following set of indices j

$$J_1 = \{j : j \in J_0, \ d_j = 0\}.$$

Generally speaking, this set J_1 is not empty. So, if there is at least one index $j \in J_1$ such that $p_j(y) > 0$, then from (5.25) it follows that

$$\psi(y) = \max_{x \geq 0} L(x,y) = \infty.$$

Since in the dual problem our aim is minimization of the objective function $\psi(y)$, this function should be considered only on such a domain of points y where the function is bounded from above. The latter means that we have to exclude from our consideration such points y where function $\psi(y)$ has no upper bound. In other words, we exclude all such points y where $p_j(y) > 0, \ j \in J_1$. It becomes obvious that the dual problem may be formulated as follows:

$$\psi(y) = \max_{j \in J_0}\{\frac{p_j(y)}{d_j}\} \longrightarrow \min \tag{5.26}$$

subject to

$$p_j(y) \leq 0, \quad j \in J_1, \quad y \geq 0.$$

Let y_0 be a new variable such that $\psi(y) = y_0$. Using this notation we obtain from (5.26) that

$$y_0 \geq \frac{p_j(y)}{d_j}, \quad j \in J_0.$$

Using the latter and taking into account (5.20) we can re-formulate the dual problem in the following form:

$$\psi(y) = y_0 \longrightarrow \min$$

subject to

$$p_j(y) \leq 0, \quad j \in J_1$$
$$d_j y_0 - p_j(y) \geq 0, \quad j \in J_0$$
$$y_i \geq 0, \quad i = 1, 2, \ldots, m.$$

Now, keeping in mind definition of set J_1 (i.e. $d_j = 0, \ \forall j \in J_1,$) we can return to the original notation and re-formulate the dual problem for the general LFP problem (5.1)-(5.3) as follows:

$$\psi(y) = y_0 \longrightarrow \min \tag{5.27}$$

subject to

$$d_0 y_0 - \sum_{i=1}^{m} b_i y_i \geq p_0, \tag{5.28}$$

$$d_j y_0 + \sum_{i=1}^{m} a_{ij} y_i \geq p_j, \quad j = 1, 2, \ldots, n, \tag{5.29}$$

$$y_i \geq 0, \quad i = 1, 2, \ldots, m. \tag{5.30}$$

Note that dual problem (5.27)-(5.30) is a linear programming problem, so its dual problem is also a linear programming problem:

$$\phi(t) = \sum_{j=0}^{n} p_j t_j \longrightarrow \max \tag{5.31}$$

subject to

$$\sum_{j=0}^{n} d_j t_j = 1. \tag{5.32}$$

$$-b_i t_0 + \sum_{j=1}^{n} a_{ij} t_j \leq 0, \quad i = 1, 2, \dots, m, \tag{5.33}$$

$$t_j \geq 0, \quad j = 0, 1, 2, \dots, n, \tag{5.34}$$

It is easy to observe that problem (5.31)-(5.34) constructed as a dual problem for dual, is nothing else than the linear analogue of an LFP problem (5.1)-(5.3) (see Chapter 3, Section 3). So we are ready to formulate very important

REMARK 5.1 *In spite of the original meaning of the term "duality" as the Latin term* duo *(i.e.* two*) in the case of linear-fractional programming we have the following three problems:*

- *the primal LFP problem (5.1)-(5.3),*
- *its dual LP problem (5.27)-(5.30) and*
- *dual for dual LP problem (at the same time a linear analogue for the primal) (5.31)-(5.34).*

In the case of a standard maximization LFP problem

$$Q(x) = \frac{P(x)}{D(x)} = \frac{\sum_{j=1}^{n} p_j x_j + p_0}{\sum_{j=1}^{n} d_j x_j + d_0} \longrightarrow \max, \tag{5.35}$$

subject to

$$\sum_{j=1}^{n} a_{ij} x_j = b_i, \quad i = 1, 2, \dots, m, \tag{5.36}$$

$$x_j \geq 0, \quad j = 1, 2, \dots, n, \tag{5.37}$$

to construct the dual problem using fractional Langrangian (5.21) we consider optimization problem

$$\psi(y) = \max_{x \geq 0} L(x, y)$$

without non-negativity assumption for variable$y$. It is clear that using the same ideas as in the previous case we will obtain the following dual problem:

$$\psi(y) = y_0 \longrightarrow \min \tag{5.38}$$

subject to

$$d_0 y_0 - \sum_{i=1}^{m} b_i y_i \geq p_0, \tag{5.39}$$

$$d_j y_0 + \sum_{i=1}^{m} a_{ij} y_i \geq p_j, \quad j = 1, 2, \ldots, n. \tag{5.40}$$

Note that this problem does not contain sign-restrictions for unknown dual variables $y_i,\ i = 1, 2, \ldots, m.$

In accordance with the duality theory of linear programming the dual problem for (5.38)-(5.40) is as follows:

$$\phi(t) = \sum_{j=0}^{n} p_j t_j \longrightarrow \max$$

subject to

$$\sum_{j=0}^{n} d_j t_j = 1.$$

$$-b_i t_0 + \sum_{j=1}^{n} a_{ij} t_j = 0, \quad i = 1, 2, \ldots, m,$$

$$t_j \geq 0, \quad j = 0, 1, 2, \ldots, n,$$

Finally, we formulate the dual problem for a common linear-fractional programming problem

$$Q(x) = \frac{P(x)}{D(x)} = \frac{\sum_{j=1}^{n} p_j x_j + p_0}{\sum_{j=1}^{n} d_j x_j + d_0} \longrightarrow \max \tag{5.41}$$

subject to

$$\left.\begin{array}{ll} \sum_{j=1}^{n} a_{ij} x_j \leq b_i, & i = 1, 2, \ldots, m_1, \\ \sum_{j=1}^{n} a_{ij} x_j = b_i, & i = m_1 + 1, m_1 + 2, \ldots, m, \end{array}\right\} \tag{5.42}$$

$$x_j \geq 0, \quad j = 1, 2, \ldots, n_1, \tag{5.43}$$

where $n_1 \leq n$ and $m_1 \leq m$.

The dual problem for a common LFP problem (5.41)-(5.43) is as follows:

$$\psi(y) = y_0 \longrightarrow \min \tag{5.44}$$

subject to

$$d_0 y_0 - \sum_{i=1}^{m} b_i y_i \geq p_0, \tag{5.45}$$

$$\left.\begin{aligned} d_j y_0 + \sum_{i=1}^{m} a_{ij} y_i \geq p_j, \quad & j = 1, 2, \ldots, n_1, \\ d_j y_0 + \sum_{i=1}^{m} a_{ij} y_i = p_j, \quad & j = n_1 + 1, n_1 + 2, \ldots, n, \end{aligned}\right\} \tag{5.46}$$

$$y_i \geq 0, \quad i = 1, 2, \ldots, m_1. \tag{5.47}$$

To illustrate how the dual problem can be formulated using the formulas described above, we consider several examples.

Example 5.1 If the primal LFP problem is

$$Q(x) = \frac{1\,x_1 + 2\,x_2 + 3}{4\,x_1 + 5\,x_2 + 6} \longrightarrow \max$$

subject to

$$\begin{array}{lrcrcl} i = 1: & 7\,x_1 & + & 8\,x_2 & \leq & 100\,, \\ i = 2: & 9\,x_1 & + & 10\,x_2 & \leq & 200\,, \\ i = 3: & 11\,x_1 & + & 12\,x_2 & \leq & 300\,, \end{array}$$

$$x_1 \geq 0, \quad x_2 \geq 0;$$

then the dual problem is

$$\psi(y) = y_0 \longrightarrow \min$$

subject to

$$\begin{array}{lrcrcrcrcl} j = 0: & 6\,y_0 & - & 100\,y_1 & - & 200\,y_2 & - & 300\,y_3 & \geq & 3\,, \\ j = 1: & 4\,y_0 & + & 7\,y_1 & + & 9\,y_2 & + & 11\,y_3 & \geq & 1\,, \\ j = 2: & 5\,y_0 & + & 8\,y_1 & + & 10\,y_2 & + & 12\,y_3 & \geq & 2\,, \end{array}$$

$$y_1 \geq 0, \quad y_2 \geq 0, \quad y_3 \geq 0.$$

Observe that in forming the dual problem, constant terms $p_0 = 3, \; d_0 = 6$ of the objective function $Q(x)$ and right-hand side entries $b_1 = 100, \; b_2 =$

200, $b_3 = 300$ became the coefficients of the dual constraint marked with index j equal to zero. Further, coefficients of the ith constraint of the primal problem became the coefficients of the variable y_i in the constraints of the dual problem. Conversely, the coefficients of primal variable x_j in constraints became the coefficients of the jth constraint in the dual problem. Moreover, coefficients $p_1 = 1$, $p_2 = 2$, and $d_1 = 4$, $d_2 = 5$, of the unknown primal variables x_1 and x_2 in the objective function $Q(x)$ became the coefficients of corresponding dual constraints in the right-hand and left-hand sides respectively. These dual constraints are marked with labels $j = 1$ and $j = 2$.

Example 5.2 If the primal LFP problem is

$$Q(x) = \frac{1\,x_1 + 2\,x_2 + 3}{4\,x_1 + 5\,x_2 + 6} \longrightarrow \min$$

subject to

$$\begin{array}{lrcrcr} i=1: & 7\,x_1 & + & 8\,x_2 & = & 100\,, \\ i=2: & 9\,x_1 & + & 10\,x_2 & \leq & 200\,, \\ i=3: & 11\,x_1 & + & 12\,x_2 & = & 300\,, \end{array}$$

$$x_1 \geq 0, \quad x_2 \geq 0;$$

then the dual problem is

$$\psi(y) = y_0 \longrightarrow \max$$

subject to

$$\begin{array}{lrcrcrcrcr} j=0: & 6\,y_0 & - & 100\,y_1 & - & 200\,y_2 & - & 300\,y_3 & \geq & 3\,, \\ j=1: & 4\,y_0 & + & 7\,y_1 & + & 9\,y_2 & + & 11\,y_3 & \geq & 1\,, \\ j=2: & 5\,y_0 & + & 8\,y_1 & + & 10\,y_2 & + & 12\,y_3 & \geq & 2\,, \end{array}$$

$$y_1 - \text{unrestricted}, \quad y_2 \geq 0, \quad y_3 - \text{unrestricted}.$$

Example 5.3 If the primal LFP problem is

$$Q(x) = \frac{1\,x_1 + 2\,x_2 + 3}{4\,x_1 + 5\,x_2 + 6} \longrightarrow \max$$

subject to

$$\begin{array}{lrcrcr} i=1: & 7\,x_1 & + & 8\,x_2 & = & 100\,, \\ i=2: & 9\,x_1 & + & 10\,x_2 & \geq & 200\,, \\ i=3: & 11\,x_1 & + & 12\,x_2 & = & 300\,, \end{array}$$

$$x_1 - \text{unrestricted}, \quad x_2 \geq 0;$$

then we have first to perform the following transformations:

- Multiply the second constraint by (-1). So, instead of the original 'greater-than' constraint

$$9\,x_1 + 10\,x_2 \geq 200$$

we obtain the following 'less-than' constraint

$$-9\,x_1 - 10\,x_2 \leq -200\,.$$

- Substitute unrestricted primal variable x_1 everywhere it appears (both in the objective function and in the constraints) with the difference $x_1 = x_1' - x_1''$, where $x_1' \geq 0,\;\; x_1'' \geq 0$.

So, instead of the original problem we obtain the following LFP problem

$$Q(x) = \frac{1\,x_1' - 1\,x_1'' + 2\,x_2 + 3}{4\,x_1' - 4\,x_1'' + 5\,x_2 + 6} \longrightarrow \max$$

subject to

$$\begin{array}{lrcrcrcr} i=1: & 7\,x_1' & - & 7\,x_1'' & + & 8\,x_2 & = & 100\,, \\ i=2: & -9\,x_1' & + & 9\,x_1'' & - & 10\,x_2 & \leq & -200\,, \\ i=3: & 11\,x_1' & - & 11\,x_1'' & + & 12\,x_2 & = & 300\,, \end{array}$$

$$x_1' \geq 0,\quad x_1'' \geq 0,\quad x_2 \geq 0\,.$$

So, we can construct the following dual problem

$$\psi(y) = y_0 \longrightarrow \min$$

subject to

$$\begin{array}{lrcrcrcrcr} j=0: & 6\,y_0 & - & 100\,y_1 & + & 200\,y_2 & - & 300\,y_3 & \geq & 3\,, \\ j=1\;(x_1'): & 4\,y_0 & + & 7\,y_1 & - & 9\,y_2 & + & 11\,y_3 & \geq & 1\,, \\ j=1\;(x_1''): & -4\,y_0 & - & 7\,y_1 & + & 9\,y_2 & - & 11\,y_3 & \geq & -1\,, \\ j=2: & 5\,y_0 & + & 8\,y_1 & - & 10\,y_2 & + & 12\,y_3 & \geq & 2\,, \end{array}$$

$$y_1 - \text{unrestricted},\quad y_2 \geq 0,\quad y_3 - \text{unrestricted.}$$

Observe that the dual constraints marked with $j = 1\;(x_1')$ and $j = 1\;(x_1'')$ may be re-written as

$$\begin{array}{lrcrcrcrcr} j=1\;(x_1'): & 4\,y_0 & + & 7\,y_1 & - & 9\,y_2 & + & 11\,y_3 & \geq & 1\,, \\ j=1\;(x_1''): & 4\,y_0 & + & 7\,y_1 & - & 9\,y_2 & + & 11\,y_3 & \leq & 1\,. \end{array}$$

The last results in

$$4\,y_0 + 7\,y_1 - 9\,y_2 + 11\,y_3 = 1\,.$$

Finally, we formulate the dual problem as follows

$$\psi(y) = y_0 \longrightarrow \min$$

subject to

$$\begin{array}{lrcrcrcrcr}
j=0: & 6\,y_0 & - & 100\,y_1 & + & 200\,y_2 & - & 300\,y_3 & \geq & 3\,, \\
j=1: & 4\,y_0 & + & 7\,y_1 & - & 9\,y_2 & + & 11\,y_3 & = & 1\,, \\
j=2: & 5\,y_0 & + & 8\,y_1 & - & 10\,y_2 & + & 12\,y_3 & \geq & 2\,,
\end{array}$$

$$y_1 - \text{unrestricted}, \quad y_2 \geq 0, \quad y_3 - \text{unrestricted}.$$

We can summarize our experience in the following way: *if some variable x_j in the primal LFP problem is unrestricted in sign, then the corresponding jth dual constraint becomes equality.*

3. Main Theorems

In this section we formulate and prove the most important statements of duality. These statements establish very close and strong inter-connections between primal and dual problems, and their optimal solutions. We will see that duality theory provides useful tools necessary for quality analysis of optimal solutions and may be helpful in a wide range of real-world application.

THEOREM 5.1 (THE WEAK DUALITY THOEREM) *If vector*

$$x = (x_1, x_2, \ldots, x_n)^T$$

is feasible solution of primal LFP problem (5.1)-(5.3) and vector

$$y = (y_0, y_1, y_2, \ldots, y_m)$$

is feasible solution of its dual problem (5.27)-(5.30), then

$$Q(x) \leq \psi(y).$$

Proof. The proof of this theorem is based on a simple chain of the following obvious equalities and inequalities:

$$\begin{aligned}
P(x) &= \sum_{j=1}^{n} p_j x_j + p_0 \leq \sum_{j=1}^{n} (d_j y_0 + \sum_{i=1}^{m} a_{ij} y_i) x_j + d_0 y_0 - \sum_{i=1}^{m} b_i y_i = \\
&= y_0 (\sum_{j=1}^{n} d_j x_j + d_0) + \sum_{i=1}^{m} y_i \sum_{j=1}^{n} a_{ij} x_j - \sum_{i=1}^{m} b_i y_i \leq \\
&\leq y_0 D(x) + \sum_{i=1}^{m} b_i y_i - \sum_{i=1}^{m} b_i y_i = \\
&= y_0 D(x).
\end{aligned}$$

Indeed, since $P(x) \leq y_0 D(x)$, and $D(x) > 0, \forall x \in S$, we have $Q(x) \leq \psi(y)$. This completes the proof of the theorem.◇

LEMMA 5.1 *If vector $x^* = (x_1^*, x_2^*, \ldots, x_n^*)^T$ is a feasible solution of primal LFP problem (5.1)-(5.3), vector $y^* = (y_0^*, y_1^*, y_2^*, \ldots, y_m^*)$ is a feasible solution of dual problem (5.27)-(5.30), and the equality*

$$Q(x^*) = \psi(y^*) \tag{5.48}$$

takes place, then vector x^* and vector y^* are optimal solutions of their problems (5.1)-(5.3) and (5.27)-(5.30), respectively.

Proof. In accordance with Weak Duality Theorem 5.1, for any feasible solution x of primal LFP problem (5.1)-(5.3) and any feasible solution y^* of dual problem (5.27)-(5.30) inequality

$$Q(x) \leq \psi(y^*)$$

takes place. By using equality (5.48), from the latest relation we obtain that

$$Q(x) \leq Q(x^*).$$

This inequality is valid for any feasible solution x of primal LFP problem (5.1)-(5.3), hence, in accordance with the definition of an optimal solution for a maximization LFP problem (see Definition 3.2, p.43) vector x^* is an optimal solution of problem (5.1)-(5.3).
Since dual problem (5.27)-(5.30) is a minimization LP problem, the optimality of vector y^* may be shown in an analogous way on the basis of a definition for the optimal solution of a minimization LP problem. ◇

The following lemma establishes a connection between the solvability of primal and dual problems.

LEMMA 5.2 *If objective function $\psi(y)$ of dual problem (5.27)-(5.30) is unbounded from below on its feasible set Y, then primal LFP problem (5.1)-(5.3) is unsolvable because its feasible set S is empty.*

Proof. Let us suppose that objective function $\psi(y)$ of dual problem (5.27)-(5.30) is unbounded from below on its feasible set Y. Then, in accordance with the duality theory of LP, problem (5.31)-(5.34) has no feasible solution, that is its feasible set T is empty. Note that LP problem (5.31)-(5.34) is a dual for problem (5.27)-(5.30).
Let us suppose that feasible set S of primal LFP problem (5.1)-(5.3) is not empty and there is at least one vector $x = (x_1, x_2, \ldots, x_n)^T$, which satisfies constraints (5.2)-(5.3). In this case we can construct vector

$$t = (\frac{1}{D(x)}, \frac{x_1}{D(x)}, \frac{x_2}{D(x)}, \ldots, \frac{x_n}{D(x)}),$$

which satisfies constraints (5.32)-(5.34) of linear analogue (5.31)-(5.34). The latter means that feasible set T of linear analogue (5.31)-(5.34) is not empty. This contradiction means that feasible set S is empty, and completes the proof of the lemma. ◇

The fact that for an LFP problem there is never a gap between the primal and dual optimal objective values is usually referred to as the *Strong Duality Theorem*:

THEOREM 5.2 (THE STRONG DUALITY THEOREM) *If primal LFP problem (5.1)-(5.3) is solvable and vector x^* is its optimal solution, then its dual problem (5.27)-(5.30) is also solvable and for any optimal solution y^* of (5.27)-(5.30) the following equality takes place*

$$Q(x^*) = \psi(y^*). \tag{5.49}$$

Conversely, if dual problem (5.27)-(5.30) is solvable and y^* is its optimal solution, then the primal problem (5.1)-(5.3) is also solvable and for any optimal solution x^* of (5.1)-(5.3) equality (5.49) takes place.

Proof. We begin with the proof of the first part of the theorem. Let us suppose that the primal problem (5.1)-(5.3) is solvable (that is its feasible set S is not empty and objective function $Q(x)$ on the set S is bounded from above) and vector x^* is its optimal solution:

$$\max_{x \in S \neq \emptyset} Q(x) = Q(x^*) = M < \infty.$$

Consider the point $c = (c_1, c_2, \ldots, c_n, c_0)$, where $c_j = p_j - Md_j$, $j = 0, 1, 2, \ldots, n$. D.B.Yudin and E.G.Gol'stein [191] (Chapter 3, Theorem 6.1) have shown that this point c belongs to such convex cone R, which

1 has a vertex in the zero point $0 = (\underbrace{0, 0, \ldots, 0}_{n+1})$ and

2 contains all vectors x that may be presented in the following form

$$x = \sum_{i=1}^{m} y_i A_i - \sum_{j=1}^{n+1} v_j e_j, \tag{5.50}$$

where

$$y_i \geq 0, \quad A_i = (a_{i1}, a_{i2}, \ldots, a_{in}, -b_i), \quad i = 1, 2, \ldots, m,$$

$$v_j \geq 0, \quad e_j = (\underbrace{0, 0, \ldots, 0, 1}_{j}, 0, 0, \ldots, 0), \quad j = 1, 2, \ldots, n, n+1.$$

In other words, point c may be presented in the form (5.50). This means that there exist such non-negative coefficients $y_i^*,\ i = 1, 2, \ldots, m$ and $v_j,\ j = 1, 2, \ldots, n, n+1,$ that the following system of equalities takes place:

$$\begin{aligned} p_j - Md_j &= \sum_{i=1}^{m} y_i^* a_{ij} - v_j, \quad j = 1, 2, \ldots, n, \\ p_0 - Md_0 &= -\sum_{i=1}^{m} y_i^* b_i - v_{n+1}. \end{aligned}$$

Let us re-write this system in the following form:

$$\begin{aligned} p_j - Md_j &\leq \sum_{i=1}^{m} y_i^* a_{ij}, \quad j = 1, 2, \ldots, n, \\ p_0 - Md_0 &\leq -\sum_{i=1}^{m} y_i^* b_i, \\ y_i^* &\geq 0, \quad i = 1, 2, \ldots, m. \end{aligned}$$

The latter means that vector $y^* = (M,\ y_1^*,\ y_2^*,\ \ldots,\ y_m^*)$ is a feasible solution of dual problem (5.27)-(5.30). Since $\psi(y^*) = M,$ we obtain from Lemma 5.1 that vector y^* is an optimal solution of dual problem (5.27)-(5.30). Thus, we have seen that if the primal LFP problem (5.1)-(5.3) is solvable, then its dual problem (5.27)-(5.30) is also solvable and their optimal objective values are the same.

To prove the second part of the theorem, we suppose that dual problem (5.27)-(5.30) is solvable (that is its feasible set Y is not empty and objective function $\psi(y)$ over this set Y is bounded from below) and vector y^* is its optimal solution:

$$\min_{y \in Y \neq \emptyset} \psi(y) = \psi(y^*) = M > -\infty.$$

Let us consider LP problem (5.31)-(5.34), which is a dual problem for the problem (5.27)-(5.30) and, at the same time, is a linear analogue of the primal problem (5.1)-(5.3).

In accordance with duality theory of LP, dual problem (5.31)-(5.34) is solvable and there is at least one such vector $t^* = (t_0^*, t_1^*, \ldots, t_n^*)$ that the following relations take place:

$$\sum_{j=0}^{n} p_j t_j^* = M, \tag{5.51}$$

$$\sum_{j=0}^{n} d_j t_j^* = 1, \tag{5.52}$$

$$-b_i t_0^* + \sum_{j=1}^{n} a_{ij} t_j^* \leq 0, \quad i = 1, 2, \ldots, m, \tag{5.53}$$

$$t_j^* \geq 0, \quad j = 0, 1, 2, \ldots, n. \tag{5.54}$$

Now, we have to show that from the solvability of problem (5.27)-(5.30) it follows that feasible set S of problem (5.1)-(5.3) is not empty and there is at least one such vector $x^* \in S$, that $Q(x^*) = M$.

To show it we have to distinguish the following two cases:

$$\textbf{A:}\ t_0^* > 0 \quad \text{and} \quad \textbf{B:}\ t_0^* = 0.$$

Case (A). It is obvious that we can construct vector

$$x^* = \left(\frac{t_1^*}{t_0^*}, \frac{t_2^*}{t_0^*}, \ldots, \frac{t_n^*}{t_0^*} \right)$$

which satisfies constraints (5.2)-(5.3) and equality $Q(x^*) = M$. In accordance with Lemma 5.1 the latter means that vectorx^* is an optimal solution of primal LFP problem (5.1)-(5.3).

Case (B). Suppose that $t_0^* = 0$. Consider the following sets of indices i and j:

$$\begin{aligned} J' &= \{j : 1 \leq j \leq n;\ t_j^* > 0\} \\ J'' &= \{j : 1 \leq j \leq n;\ t_j^* = 0\} \\ I' &= \{i : 1 \leq i \leq m;\ \sum_{j \in J'} a_{ij} t_j^* < 0\} \\ I'' &= \{i : 1 \leq i \leq m;\ \sum_{j \in J'} a_{ij} t_j^* = 0\} \end{aligned}$$

Let us choose such sequence $\{t^k\}$ of $(n+1)$ dimensional vectors t^k, that

$$\lim_{k \to \infty} t_j^k = t_j^* \quad j = 0, 1, 2, \ldots, n$$

and $t_0^k > 0$ for any finite index k. Let $\lambda_k,\ k = 1, 2, \ldots$ denote the following values:

$$\lambda_k = \frac{1}{t_0^k}, \quad k = 1, 2, \ldots$$

We will show now that if vector x^* is defined in accordance with the rule

$$x_j^* = \begin{cases} \lim_{k \to \infty} \lambda_k t_j^*, & j \in J' \\ 0, & j \in J'', \end{cases} \tag{5.55}$$

then vector x^* is a feasible solution of primal LFP problem (5.1)-(5.3). Indeed, since vector $t^* = (t_0^*, t_1^*, \dots, t_n^*)$, is a feasible solution of LP problem (5.31)-(5.34), hence, vector

$$t^*(\lambda_k) = (0,\ \lambda_k t_1^*,\ \lambda_k t_2^*,\ \dots,\ \lambda_k t_n^*)$$

satisfies constraints

$$\sum_{j\in J'} a_{ij}\lambda_k t_j^* \le 0, \quad i = 1,2,\dots,m,$$

where $\lambda_k > 0$. So for $i \in I'$ we have that:

$$\lim_{k\to\infty} \sum_{j\in J'} a_{ij}\lambda_k t_j^* = -\infty, \quad i \in I'. \tag{5.56}$$

The latter means that

$$\sum_{j=1}^{n} a_{ij}x_j^* < b_i, \quad i \in I',$$

where variables x_j^* are defined in accordance with formula (5.55), and b_i are given finite constants. In other words, it means that vector x^* satisfies those constraints of (5.2), where $i \in I'$.

Let us consider those constraints of system (5.2), whose index $i \in I''$. It is obvious that those constraints of system (5.53) whose index $i \in I''$, may be re-written in the following form:

$$\sum_{j\in J'} a_{ij}t_j^* = \lim_{k\to\infty} t_0^k\, b_i, \quad i \in I''.$$

After multiplying these equalities with λ_k we get for the case $k \to \infty$ the following:

$$\lim_{k\to\infty} \sum_{j\in J'} a_{ij}t_j^*\lambda_k = b_i, \quad i \in I'', \tag{5.57}$$

System (5.57) means that the following system of equations takes place

$$\sum_{j=1}^{n} a_{ij}x_j^* = b_i, \quad i \in I''.$$

In other words, vector x^* satisfies those constraints of system (5.2), whose index $i \in I''$.

Since elements x_j^* of vector x^* are non-negative, hence vector x^* is a feasible solution of primal LFP problem (5.1)-(5.3).

Let us show now that $Q(x^*) = M$. Indeed,

$$Q(x^*) = \frac{\sum_{j=1}^{n} p_j x_j^* + p_0}{\sum_{j=1}^{n} d_j x_j^* + d_0} = \frac{\sum_{j \in J'} p_j \lim_{k\to\infty} \lambda_k \, t_j^* + p_0}{\sum_{j \in J'} d_j \lim_{k\to\infty} \lambda_k \, t_j^* + d_0}$$

$$= \frac{\lim_{k\to\infty} \lambda_k (\sum_{j\in J'} p_j t_j^*) + p_0}{\lim_{k\to\infty} \lambda_k (\sum_{j\in J'} d_j t_j^*) + d_0}.$$

Keeping in mind the definition of set J' we get from formula (5.51) and (5.52) the following chain of equalities:

$$\frac{\lim_{k\to\infty} \lambda_k \, (\sum_{j\in J'} p_j t_j^*) + p_0}{\lim_{k\to\infty} \lambda_k \, (\sum_{j\in J'} d_j t_j^*) + d_0} = \frac{\lim_{k\to\infty} \lambda_k M + p_0}{\lim_{k\to\infty} \lambda_k + d_0} = M.$$

The latter means that $Q(x^*) = \psi(y^*)$. Hence, in accordance with Lemma 5.1, vector x^* is an optimal solution of primal LFP problem (5.1)-(5.3). This completes the proof of the theorem.◊

Let us formulate the following statements that follow from Theorem 5.2.

COROLLARY 5.1 *A necessary and sufficient condition for problems (5.1)-(5.3) and (5.27)-(5.30) to be solvable is that the both problems must have at least one feasible solution.*

COROLLARY 5.2 *In order for feasible solution x^* of problem LFP (5.1)-(5.3) and feasible solution y^* of dual problem (5.27)-(5.30) to be optimal solutions of their problems, it is necessary and sufficient that*

$$Q(x^*) = \psi(y^*).$$

COROLLARY 5.3 *If LFP problem (5.1)-(5.3) (linear analogue (5.31)-(5.34)) is solvable, then its linear analogue (5.31)-(5.34) (the original LFP problem (5.1)-(5.3)) is also solvable. Moreover, for any optimal solution x^* of LFP problem (5.1)-(5.3) and any optimal solution t^* of linear analogue (5.31)-(5.34) the following equality takes place*

$$Q(x^*) = \phi(t^*),$$

and between vectors x^* and t^* the following connection may be established

$$if \quad t_0^* > 0, \quad then \ x_j^* = \frac{t_j^*}{t_0^*}, \quad j = 1, 2, \ldots, n; \tag{5.58}$$

$$if \quad t_0^* = 0, \quad then \ x_j^* = \begin{cases} t_j^* \lim\limits_{k\to\infty} \lambda_k, & j \in J', \\ 0, & j \in J'' \end{cases} \tag{5.59}$$

These theorems and corollaries establish a connection between the solvability of three mathematical programming problems: the original LFP problem (5.1)-(5.3), the dual LP problem (5.27)-(5.30) and the linear analogue (5.31)-(5.34). Sometimes it is necessary to recover the optimal solution of a dual problem when only the optimal solution of a primal LFP problem (or the optimal solution of a linear analogue) is known. The following statements can help in this regard.

DEFINITION 5.1 *We shall call constraints (5.2) and (5.33), i.e.*

$$\sum_{j=1}^{n} a_{ij}x_j \leq b_i \quad and \quad -b_i t_0 + \sum_{j=1}^{n} a_{ij}t_j \leq 0, \quad i = 1, 2, \ldots, m,$$

for each fixed value of index i $(i = 1, 2, \ldots, m)$ and constraints (5.3), (5.34), i.e.

$$x_j \geq 0 \quad and \quad t_j \geq 0, \quad j = 1, 2, \ldots, n,$$

for each fixed value of index j $(j = 1, 2, \ldots, n)$ the ***pair of analogue constraints.***

DEFINITION 5.2 *We shall call constraints (5.2) and (5.30), i.e.*

$$\sum_{j=1}^{n} a_{ij}x_j \leq b_i \quad and \quad y_i \geq 0, \quad i = 1, 2, \ldots, m,$$

for each fixed value of index i $(i = 1, 2, \ldots, m)$ and constraints (5.3), (5.29), i.e.

$$x_j \geq 0 \quad and \quad d_j y_0 + \sum_{i=1}^{m} a_{ij}y_i \geq p_j, \quad j = 1, 2, \ldots, n,$$

for each fixed value of index j $(j = 1, 2, \ldots, n)$ the ***pair of dual constraints.***

For example, if the primal LFP problem and its dual problem are as follows:
primal LFP problem

$$Q(x) = \frac{1\,x_1 + 2\,x_2 + 3}{4\,x_1 + 5\,x_2 + 6} \longrightarrow \max$$

subject to

$$\begin{array}{lrcrcl} i=1: & 7\,x_1 & + & 8\,x_2 & \leq & 100\,, \\ i=2: & 9\,x_1 & + & 10\,x_2 & \leq & 200\,, \\ i=3: & 11\,x_1 & + & 12\,x_2 & \leq & 300\,, \end{array}$$

$$x_1 \geq 0, \quad x_2 \geq 0;$$

dual problem

$$\psi(y) = y_0 \longrightarrow \min$$

subject to

$$\begin{array}{lrcrcrcrcl} j=0: & 6\,y_0 & - & 100\,y_1 & - & 200\,y_2 & - & 300\,y_3 & \geq & 3\,, \\ j=1: & 4\,y_0 & + & 7\,y_1 & + & 9\,y_2 & + & 11\,y_3 & \geq & 1\,, \\ j=2: & 5\,y_0 & + & 8\,y_1 & + & 10\,y_2 & + & 12\,y_3 & \geq & 2\,, \end{array}$$

$$y_1 \geq 0\,, \quad y_2 \geq 0\,, \quad y_3 \geq 0\,;$$

then their pairs of dual constraints are the following:
for $i = 1, 2, 3$

$$\begin{array}{lrcrcll} i=1: & 7\,x_1 & + & 8\,x_2 & \leq & 100\,, & \Longleftrightarrow \quad y_1 \geq 0\,, \\ i=2: & 9\,x_1 & + & 10\,x_2 & \leq & 200\,, & \Longleftrightarrow \quad y_2 \geq 0\,, \\ i=3: & 11\,x_1 & + & 12\,x_2 & \leq & 300\,, & \Longleftrightarrow \quad y_3 \geq 0\,, \end{array}$$

for $j = 1, 2$

$$\begin{array}{lrcrcrcrcll} j=1: & 4\,y_0 & + & 7\,y_1 & + & 9\,y_2 & + & 11\,y_3 & \geq & 1\,, & \Longleftrightarrow \quad x_1 \geq 0\,, \\ j=2: & 5\,y_0 & + & 8\,y_1 & + & 10\,y_2 & + & 12\,y_3 & \geq & 2\,, & \Longleftrightarrow \quad x_2 \geq 0\,. \end{array}$$

DEFINITION 5.3 *We shall say that constraint (5.2) (constraint (5.3)) for a fixed value of index i (index j) is **fixed**, if for any optimal solution x^* this constraint performs as a strict equality.*

DEFINITION 5.4 *We shall say constraint (5.2) (constraint (5.3)) for a fixed value of index i (index j) is **free**, if at least for one optimal solution x^* this constraint performs as a strict inequality.*

In an analogous way, these definitions may be extended to constraints (5.28), (5.29), (5.30), i.e.

$$d_0 y_0 - \sum_{i=1}^{m} b_i y_i \geq p_0,$$

$$d_j y_0 + \sum_{i=1}^{m} a_{ij} y_i \geq p_j, \quad j = 1, 2, \ldots, n,$$

$$y_i \geq 0, \quad i = 1, 2, \ldots, m,$$

of a dual problem and constraints (5.33), (5.34), i.e.

$$-b_i t_0 + \sum_{j=1}^{n} a_{ij} t_j \leq 0, \quad i = 1, 2, \ldots, m,$$

$$t_j \geq 0, \quad j = 0, 1, 2, \ldots, n,$$

of a linear analogue.

THEOREM 5.3 (THE ANALOGUE CONSTRAINTS THEOREM) *If a primal LFP (5.1)-(5.3) problem and its linear analogue (5.31)-(5.34) are solvable, i.e. have optimal solutions, then in any pair of their analogue constraints both of them are free or fixed.*

Proof. For constraints (5.3) and (5.34) the proof of this statement follows directly from formulas (5.58) and (5.59).

Consider now constraints (5.2) and (5.33). Let vector x^* be an optimal solution of primal LFP problem (5.1)-(5.3), and vector t^* be an optimal solution of a linear analogue (5.31)-(5.34). It is obvious that if $t_0^* > 0$, then we have

$$b_i - \sum_{j=1}^{n} a_{ij} x_j^* = \frac{1}{t_0^*} (b_i t_0^* - \sum_{j=1}^{n} a_{ij} t_j^*), \quad i = 1, 2, \ldots, m. \tag{5.60}$$

If $t_0^* = 0$, then in accordance with (5.56) and (5.57) from (5.59) we obtain

$$b_i - \sum_{j=1}^{n} a_{ij} x_j^* = b_i - \sum_{j=1}^{n} a_{ij} \lim_{k \to \infty} \lambda_k t_j^* = \begin{cases} \infty, & i \in I', \\ 0, & i \in I''. \end{cases} \tag{5.61}$$

It is clear from (5.60) and (5.61) that the statement is also valid for constraints (5.2) and (5.33). Thus, the theorem is proved.◇

From Theorem 5.3 and the fact that dual problem (5.27)-(5.30) and linear analogue (5.31)-(5.34) of LFP problem (5.1)-(5.3) are both linear programming problems follow the strict connections between pairs of dual constraints.

THEOREM 5.4 (THE COMPLEMENTARY SLACKNESS THEOREM) *If a primal LFP problem and its dual problem are solvable, i.e. have optimal solutions, then in each pair of their dual constraints one of them is free and the second one is fixed.*

Note that in dual problem (5.27)-(5.30) there are $n + 1$ main constraints: namely, 1 constraint (5.28) and n constraints in the form of (5.29). Constraints

(5.29) are associated with their dual constraints (5.3), while the sign-restriction constraints (5.30) form dual pairs with constraints (5.2). The only constraint in dual problem (5.27)-(5.30) which has not any dual connection with a primal LFP problem is constraint (5.28). The following Theorem establishes a connection between a primal LFP problem and constraint (5.28) of a dual problem.

THEOREM 5.5 ([8]) *If primal LFP problem (5.1)-(5.3) and its dual problem (5.27)-(5.30) are solvable, i.e. have optimal solutions, then in order for constraint (5.28) to be fixed it is necessary and sufficient that*

$$x_j^* < \infty, \quad j = 1, 2, \dots, n, \tag{5.62}$$

at least for one optimal solution x^* of the primal LFP problem (5.1)-(5.3).

Proof. *Necessity.* We begin the proof by recalling the fact that if primal LFP problem (5.1)-(5.3) and its dual problem (5.27)-(5.30) are solvable, then linear analogue (5.31)-(5.34) is also solvable, i.e. its feasible set is not empty and objective function $\phi(t)$ is bounded from above on this set. Moreover, for any optimal solution t^* of (5.31)-(5.34) we can write that

$$t_j^* < \infty, \quad j = 0, 1, 2, \dots, n. \tag{5.63}$$

Let us suppose that constraint (5.28) is fixed, then in accordance with the duality theory of linear programming constraint $t_0 \geq 0$ of the linear analogue is free. The latter means that there exists at least one optimal solution t^* of (5.31)-(5.34) such that $t_0^* > 0$. Using this optimal solution t^* and formula (5.58) we can construct such vector x^* which is an optimal solution of the primal LFP problem (5.1)-(5.3). Since (5.63) takes place, it means that all components of vector x^* satisfy conditions (5.62).

Sufficiency. Let us suppose that vector x^* is an optimal solution of primal LFP problem (5.1)-(5.20) and for this vector x^* condition (5.62) takes place.

Let us introduce the following notation:

$$\begin{aligned} \Delta_0 &= d_0 y_0 - \sum_{i=1}^{m} b_i y_i - p_0, \\ \Delta_j &= d_j y_0 + \sum_{i=1}^{m} a_{ij} y_i - p_j, \qquad j = 1, 2, \dots, n. \end{aligned}$$

Then, objective function $Q(x)$ in this point x^* may be presented as follows

$$Q(x^*) = \frac{\sum_{j=1}^{n} p_j x_j + p_0}{\sum_{j=1}^{n} d_j x_j + d_0} =$$

$$= \frac{\sum_{j=1}^{n}(d_j y_0 + \sum_{i=1}^{m} a_{ij} y_i - \Delta_j) x_j + (d_0 y_0 - \sum_{i=1}^{m} b_i y_i - \Delta_0)}{\sum_{j=1}^{n} d_j x_j + d_0}$$

or in a shorter form

$$Q(x^*) = y_0 - \frac{\sum_{i=1}^{m} \epsilon_i y_i + \sum_{j=1}^{n} \Delta_j x_j^* + \Delta_0}{D(x)}, \quad (5.64)$$

where

$$\epsilon_i = b_i - \sum_{j=1}^{n} a_{ij} x_j^*, \quad i = 1, 2, \dots, m\ .$$

Observe that equality (5.64) is valid for any feasible solution$y$ of dual problem (5.27)-(5.30). So, without any loss of generality, we can state that equality (5.64) also holds for any optimal solution$y$ too. Furthermore, since vectorsx^* and y are feasible solutions of their problems it means that

$$\Delta_0 \geq 0, \quad \Delta_j \geq 0, \quad x_j^* \geq 0, \quad j = 1, 2, \dots, n,$$

and

$$\epsilon_i \geq 0, \quad y_i^* \geq 0, \quad i = 1, 2, \dots, m.$$

In accordance with Theorem 5.2 any optimal solution x^* and y holds the following equality

$$Q(x^*) = \psi(y_0)\ .$$

From the latter it follows that

$$\frac{\sum_{i=1}^{m} y_i\ \epsilon_i + \sum_{j=1}^{n} x_j^*\ \Delta_j + \Delta_0}{D(x^*)} = 0. \quad (5.65)$$

Since function $D(x)$ is linear and holds condition (5.62), from equality (5.65) we obtain that

$$\sum_{i=1}^{m} y_i \epsilon_i = \sum_{j=1}^{n} x_j^* \Delta_j = \Delta_0 = 0.$$

Thus, the theorem is proved.◇

To illustrate this theorem we consider the LFP problem given earlier in this section on page 149. This problem has optimal solution$x^* = (0, 0)$. So,

$$x_1^* < \infty, \quad x_2^* < \infty\ .$$

The only optimal solution of the corresponding dual problem (see page 150) is vector $y^* = (0.5, 0, 0, 0)$. Substituting these optimal values $y_0^* = 0.5$, $y_1^* = y_2^* = y_3^* = 0$ to the dual constraint marked with $j = 0$ we can easily check that the constraint is fixed, i.e. is satisfied as a strict equality

$$6y_0^* - 100y_1^* - 200y_2^* - 300y_3^* = 6 \times 0.5 - 100 \times 0 - 200 \times 0 - 300 \times 0 = 3 .$$

In this section we dealt with an LFP problem only in general form (5.1)-(5.3). It is obvious that all statements formulated and proved here are valid and may be expanded to canonical or other forms of an LFP problem, since all these forms may be easily converted to one another (see Chapter 3, Section 1.3).

4. Computational Relations Between Primal and Dual Problems

In Chapter 4 we have seen how to use the simplex method to solve a linear-fractional programming problem. In this section we show that if the LFP problem to be solved is considered as a primal problem, the simplex method allows to obtain an optimal solution for the dual problem too. This result comes from the more detailed and careful examination of the information available from the final (optimal) tableau of the simplex method.

Let us suppose that we are given a general linear-fractional programming problem, and that we have entered slack and artificial variables to convert the original problem to canonical form. So, we have our problem in the form as follows

$$Q(x) = \frac{P(x)}{D(x)} = \frac{p^T x + p_0}{d^T x + d_0} \rightarrow \max, \tag{5.66}$$

subject to

$$Ax = b \tag{5.67}$$

$$x \geq 0, \tag{5.68}$$

where $D(x) > 0, \ \forall x \in S$; $A = ||a_{ij}||_{m \times n}$ is an $m \times n$ matrix, $b = (b_1, b_2, \dots, b_m)^T$, where $b_i \geq 0, \ i = 1, 2, \dots, m$; $p = (p_1, p_2, \dots, p_n)^T$, $d = (d_1, d_2, \dots, d_n)^T$, and p_0 and d_0 are scalars.

Consider a simplex tableau constructed by the simplex method during the solution of our LFP problem (5.66)-(5.68). This tableau represents a basic feasible solution x. Let basis B associated with this feasible solution x be $B = (A_{s_1}, A_{s_2}, \dots, A_{s_m})$, where $A_j = (a_{1j}, a_{2j}, \dots, a_{mj})^T$ denotes the jth column of matrix A, $j = 1, 2, \dots, n$. Let $J_B = \{s_1, s_2, \dots, s_m\}$ denote the set of basic indices, and J_N denote the set of indices of the nonbasic variables.

Recall that all nonbasic variables x_j are set equal to zero, so we can re-write objective function $Q(x)$ as follows

$$Q(x) = Q(x_B) = \frac{p_B^T x_B + p_0}{d_B^T x_B + d_0}, \quad \text{where } x_B = (x_{s_1}, x_{s_2}, \ldots, x_{s_m})^T,$$

and $p_B = (p_{s_1}, p_{s_2}, \ldots, p_{s_m})^T, \; d_B = (d_{s_1}, d_{s_2}, \ldots, d_{s_m})^T.$

For any basic feasible solution x of LFP problem (5.66)-(5.68) with corresponding basis matrix B, we define

$$u = p_B^T B^{-1}, \quad v = d_B^T B^{-1}, \tag{5.69}$$

and

$$y_i = \begin{cases} Q(x_B), & i = 0, \\ u_i - Q(x_B)\, v_i, & i = 1, 2, \ldots, m. \end{cases} \tag{5.70}$$

The following statement establishes a computational relation between optimal solutions of primal and dual problems in linear-fractional programming.

THEOREM 5.6 *If vector x is a basic optimal solution of LFP problem (5.66)-(5.68) with basic matrix B, then vector y defined by formulas (5.70) and (5.69) is an optimal solution of a problem which is dual for (5.66)-(5.68)*

Instead of proving this theorem we illustrate how it can be used to calculate the optimal solution of the dual problem using the final simplex tableau constructed for the primal LFP problem.

Consider as our primal problem the following linear-fractional programming problem

$$Q(x) = \frac{8\,x_1 + 9\,x_2 + 4\,x_3 + 4}{2\,x_1 + 3\,x_2 + 2\,x_3 + 7} \longrightarrow \max$$

subject to

$$\begin{array}{rcrcrcr} 1\,x_1 & + & 1\,x_2 & + & 2\,x_3 & \leq & 3\,, \\ 2\,x_1 & + & 1\,x_2 & + & 4\,x_3 & \leq & 4\,, \\ 5\,x_1 & + & 3\,x_2 & + & 1\,x_3 & \leq & 15\,, \end{array}$$

$$x_j \geq 0, \quad j = 1, 2, 3.$$

Introducing the slack variables x_4, x_5 and x_6, we re-write our problem as

$$Q(x) = \frac{8\,x_1 + 9\,x_2 + 4\,x_3 + 4}{2\,x_1 + 3\,x_2 + 2\,x_3 + 7} \longrightarrow \max$$

subject to

$$\begin{array}{rcrcrcrcrcrcl} 1\,x_1 & + & 1\,x_2 & + & 2\,x_3 & + & x_4 & & & & & = & 3\,, \\ 2\,x_1 & + & 1\,x_2 & + & 4\,x_3 & & & + & x_5 & & & = & 4\,, \\ 5\,x_1 & + & 3\,x_2 & + & 1\,x_3 & & & & & + & x_6 & = & 15\,, \end{array}$$

$$x_j \geq 0, \;\; j = 1,2,3,4,5,6.$$

Solving this problem by the simplex method we obtain the sequence of simplex tableaux presented in Tables 5.1-5.3.

B	P_B	D_B	x_B	A_1	A_2	A_3	A_4	A_5	A_6	
				8	9	4	0	0	0	
				2	3	2	0	0	0	
A_4	0	0	3	1	1	2	1	0	0	
A_5	0	0	4	2	1	4	0	1	0	$\Rightarrow$
A_6	0	0	15	5	3	1	0	0	1	
$P(x) = 4$				-8	-9	-4	0	0	0	
$D(x) = 7$				-2	-3	-2	0	0	0	
$Q(x) = 4/7$				$-48/7$	$-51/7$	$-20/7$	0	0	0	
				$\Uparrow$						

Table 5.1. Primal-dual connection example – Initial tableau.

B	P_B	D_B	x_B	A_1	A_2	A_3	A_4	A_5	A_6	
				8	9	4	0	0	0	
				2	3	2	0	0	0	
A_4	0	0	1	0	$1/2$	0	1	$-1/2$	0	$\Rightarrow$
A_1	8	2	2	1	$1/2$	2	0	$1/2$	0	
A_6	0	0	5	0	$1/2$	-9	0	$-5/2$	1	
$P(x) = 20$				0	-5	12	0	4	0	
$D(x) = 11$				0	-2	2	0	1	0	
$Q(x) = 20/11$				0	$-15/11$	$92/11$	0	$24/11$	0	
					$\Uparrow$					

Table 5.2. Primal-dual connection example – After first iteration.

				8	9	4	0	0	0
				2	3	2	0	0	0
B	P_B	D_B	x_B	A_1	A_2	A_3	A_4	A_5	A_6
A_2	9	3	2	0	1	0	2	-1	0
A_1	8	2	1	1	0	2	-1	1	0
A_6	0	0	4	0	0	-9	-1	-2	1
$P(x) = 30$				0	0	12	10	-1	0
$D(x) = 15$				0	0	2	4	-1	0
$Q(x) = 2$				0	0	8	2	1	0

Table 5.3. Primal-dual connection example – Final tableau.

From the final simplex tableau (see Table 5.3) we obtain the following optimal solution:

$$B = (A_2, A_1, A_6), \quad x_B = (2,1,4)^T,$$

so

$$x^* = (1,2,0,0,0,4)^T \quad \text{and} \quad Q(x^*) = \frac{30}{15} = 2.$$

The dual problem in this example is

$$\psi(y) = y_0 \longrightarrow \min$$

subject to

$$\begin{array}{rcrcrcrcr} 7\,y_0 & - & 3\,y_1 & - & 4\,y_2 & - & 15\,y_3 & \geq & 4\,, \\ 2\,y_0 & + & 1\,y_1 & + & 2\,y_2 & + & 5\,y_3 & \geq & 8\,, \\ 3\,y_0 & + & 1\,y_1 & + & 1\,y_2 & + & 3\,y_3 & \geq & 9\,, \\ 2\,y_0 & + & 2\,y_1 & + & 4\,y_2 & + & 1\,y_3 & \geq & 4\,, \end{array}$$

$$y_1 \geq 0, \;\; y_2 \geq 0, \;\; y_3 \geq 0.$$

The optimal solution y^* for the dual problem may be found from the final simplex tableau shown in Table 5.3. In this tableau the basic variables are x_2, x_1, x_6, in that order. The associated basic vectors A_2, A_1, A_6 are

$$A_2 = \begin{pmatrix} 1 \\ 1 \\ 3 \end{pmatrix}, \quad A_1 = \begin{pmatrix} 1 \\ 2 \\ 5 \end{pmatrix}, \quad A_6 = \begin{pmatrix} 0 \\ 0 \\ 1 \end{pmatrix}.$$

So,

$$B = (A_2, A_1, A_6) = \begin{pmatrix} 1 & 1 & 0 \\ 1 & 2 & 0 \\ 3 & 5 & 1 \end{pmatrix}.$$

Since the initial basic variables are x_4, x_5 and x_6 in that order, we can find the columns of inverse matrix B^{-1} under the labels A_4, A_5 and A_6 in the final tableau (see Table 5.3). So,

$$B^{-1} = \begin{pmatrix} 2 & -1 & 0 \\ -1 & 1 & 0 \\ -1 & -2 & 1 \end{pmatrix}.$$

Then for values u_i and v_i, $i = 1, 2, 3$, from (5.69) we obtain

$$(u_1, u_2, u_3) = (9, 8, 0) \begin{pmatrix} 2 & -1 & 0 \\ -1 & 1 & 0 \\ -1 & -2 & 1 \end{pmatrix} = (10, -1, 0)\,,$$

and

$$(v_1, v_2, v_3) = (3, 2, 0) \begin{pmatrix} 2 & -1 & 0 \\ -1 & 1 & 0 \\ -1 & -2 & 1 \end{pmatrix} = (4, -1, 0)\,.$$

Hence, in accordance with (5.70) for optimal entries y_1^*, y_2^*, y_3^* of optimal solution $y^* = (y_0^*, y_1^*, y_2^*, y_3^*)^T$ of our dual problem we have

$$(y_1^*, y_2^*, y_3^*) = (10, -1, 0) - 2\,(4, -1, 0) = (2, 1, 0)\,.$$

So, $y^* = (2, 2, 1, 0)^T$.

Thus, we have shown how using optimal simplex tableaus and formulas (5.69) and (5.70) we can calculate an optimal solution for a dual problem.

5. Connection with Linear Programming

Let us consider the following primal LFP problem in general form

$$Q(x) = \frac{P(x)}{D(x)} = \frac{\sum_{j=1}^{n} p_j x_j + p_0}{\sum_{j=1}^{n} d_j x_j + d_0} \longrightarrow \max, \qquad (5.71)$$

subject to

$$\sum_{j=1}^{n} a_{ij} x_j \leq b_i, \quad i = 1, 2, \ldots, m, \qquad (5.72)$$

$$x_j \geq 0, \quad j = 1, 2, \ldots, n. \tag{5.73}$$

In previous sections we have shown that its dual problem may be formulated as follows:

$$\psi(y) = y_0 \longrightarrow \min \tag{5.74}$$

subject to

$$d_0 y_0 - \sum_{i=1}^{m} b_i y_i \geq p_0, \tag{5.75}$$

$$d_j y_0 + \sum_{i=1}^{m} a_{ij} y_i \geq p_j, \quad j = 1, 2, \ldots, n, \tag{5.76}$$

$$y_i \geq 0, \quad i = 1, 2, \ldots, m. \tag{5.77}$$

Let us suppose that in LFP problem (5.71)-(5.73)

$$d_0 = 1, \quad d_j = 0, \quad j = 1, 2, \ldots, n.$$

It means that $D(x) \equiv 1, \quad Q(x) \equiv P(x)$ for all x in feasible set S, and the LFP problem we have actually is an LP problem. In this case, its dual problem (5.74)-(5.77) may be re-written as follows

$$\psi(y) = y_0 \longrightarrow \min \tag{5.78}$$

subject to

$$y_0 - \sum_{i=1}^{m} b_i y_i \geq p_0, \tag{5.79}$$

$$\sum_{i=1}^{m} a_{ij} y_i \geq p_j, \quad j = 1, 2, \ldots, n, \tag{5.80}$$

$$y_i \geq 0, \quad i = 1, 2, \ldots, m. \tag{5.81}$$

Let problem (5.71)-(5.73) be solvable and vector x^* be its optimal solution. Since problem (5.71)-(5.73) is an LP problem for any of its optimal solution x^* we can write that

$$x_j^* < \infty, \quad j = 1, 2, \ldots, n.$$

In accordance with Theorem 5.5, the latter means that dual constraint (5.79) is fixed, i.e. for any optimal solution y^* it is realized as a strict equality

$$y_0^* - \sum_{i=1}^{m} b_i y_i^* = p_0$$

So, when considering dual problem (5.74)-(5.77) instead of inequality (5.79) we can replace it with strict equality

$$y_0 - \sum_{i=1}^{m} b_i y_i = p_0$$

From the latter we obtain that

$$y_0 = \sum_{i=1}^{m} b_i y_i + p_0.$$

Using this expression and substituting its right-hand side for y_0 in dual problem (5.78)-(5.81), we obtain the dual problem in the following form

$$\psi(y) = \sum_{i=1}^{m} b_i y_i + p_0 \longrightarrow \min.$$

subject to

$$\sum_{i=1}^{m} a_{ij} y_i \geq p_j, \quad j = 1, 2, \dots, n,$$

$$y_i \geq 0, \quad i = 1, 2, \dots, m.$$

Thus, we have shown that the duality approach in LFP based on the Gol'stein-type fractional Lagrangian (5.21) contains the LP duality as a special case.

6. Dual Variables in Stability Analysis

As we saw in previous sections (see Chapter 3, Section 5), linear-fractional programming problems may have real-world economic interpretations and may serve as a useful optimization tool in numerous areas of human activity. One of the most general examples of such applications of LFP may be the problem described in Section 5.1, Chapter 3, where one must find such a production plan of n different products that maximizes the specific profit (i.e. ratio*profit/cost*) of company, while the usage of scarce resources available to the company and used in manufacturing does not exceed the given limits.

Often in real-world applications of such types, it is very important and useful to know what happens to the optimal plan, efficiency, profit and cost values if they have been calculated once but some changes in the coefficients of the problem must be done. In this section we will deal with the situation when the right-hand side vector b (that is volumes of scarce resources in stock) must be changed.

Let us consider the following LFP problem in canonical form:

$$Q(x) = \frac{P(x)}{D(x)} = \frac{\sum_{j=1}^{n} p_j x_j + p_0}{\sum_{j=1}^{n} d_j x_j + d_0} \longrightarrow \max, \tag{5.82}$$

subject to

$$\sum_{j=1}^{n} a_{ij} x_j = b_i, \quad i = 1, 2, \dots, m, \tag{5.83}$$

$$x_j \geq 0, \quad j = 1, 2, \dots, n\,. \tag{5.84}$$

We assume that

$$D(x) > 0, \;\; \forall x \in R^+ = \{R^n | x \geq 0\}.$$

As it was shown in previous sections, objective function $Q(x)$ of LFP problem (5.82)-(5.84) may reach its maximal value over feasible set S in a finite vertex as well as in an infinite point on unbounded edge (see examples shown in Figure 3.1 and Figure 3.4, respectively). G.R.Bitran and T.L.Magnanti [28], [29] showed that in the case of the optimum on an infinite edge, some small change in RHS vector b (within the range of stability for an optimal basis) does not affect the optimal value of the objective function. So, when investigating the influence of fluctuations in RHS vector b on the optimal solution and optimal value of objective function $Q(x)$ in an LFP problem, we may restrict our consideration to dealing only with the case when the maximal value is reached in a finite vertex.

So, let vector $x^* = (x_1^*, x_2^*, \dots, x_n^*)^T$ be a non-degenerate basic optimal solution of problem (5.82)-(5.84), such that

$$x_j^* < \infty, \;\; j = 1, 2, \dots, n. \tag{5.85}$$

Without loss of generality, we may assume also that optimal basis B associated with solution x^* consists of vectors $A_1, A_2, \dots, A_m$, i.e.

$$B = (A_1, A_2, \dots, A_m),$$

and hence,

$$x^* \begin{cases} > 0, & j = 1, 2, \dots, m; \\ = 0, & j = m+1, m+2, \dots, n. \end{cases} \tag{5.86}$$

Since vectors $A_1, A_2, \dots, A_m$ are linear-independent it means that any vector $b' = (b'_1, b'_2, \dots, b'_m)^T$ may be presented as their linear combination

$$\sum_{i=1}^{m} A_i x'_i = b'. \tag{5.87}$$

Let us denote by $||e_{ij}||_{m\times m}$ the inverse of basis matrix B, i.e. $B^{-1} = ||e_{ij}||_{m\times m}$, then from (5.87) we obtain that

$$x'_i = \sum_{j=1}^{m} e_{ij} b'_j, \quad i = 1, 2, \dots, m. \tag{5.88}$$

Further, we introduce the following notations:

$$e = \max_{1\le i\le m} \sum_{j=1}^{m} |e_{ij}| \quad \text{and} \quad x_0 = \min_{1\le i\le m} x^*_i.$$

Let $LFP(b)$ denote the original LFP (5.82)-(5.84) problem with RHS vector b, and $LFP(b')$ be a new LFP problem which can be obtained from $LFP(b)$, if vector b is replaced with new vector b'.

Now, we will show that if vectors b and b' satisfy condition

$$\max_{1\le j\le m} |b'_j - b_j| \le \frac{x_0}{e}, \tag{5.89}$$

then vector $x' = (x'_1, x'_2, \dots, x'_m, 0, 0, \dots, 0)^T$ is an optimal solution of problem $LFP(b')$. Indeed, since vector x^* is a feasible solution of problem $LFP(b)$, it means that

$$\sum_{i=1}^{m} a_{ij} x^*_j = b_i, \quad i = 1, 2, \dots, m,$$

and hence, using inverse B^{-1} we obtain

$$x^*_i = \sum_{j=1}^{m} e_{ij} b_j, \quad i = 1, 2, \dots, m.$$

From the latter and formula (5.88) it follows that

$$x^*_i - x'_i = \sum_{j=1}^{m} e_{ij}(b_j - b'_{ij}), \quad i = 1, 2, \dots, m,$$

meanwhile

$$\begin{aligned}
|x_i^* - x_i'| &\leq \sum_{j=1}^{m} |e_{ij}| \cdot |b_j - b_j'| \leq \\
&\leq \max_{1\leq i\leq m} \sum_{j=1}^{m} |e_{ij}| \cdot \max_{1\leq i\leq m} |b_j - b_j'| = \\
&= e \cdot \max_{1\leq i\leq m} |b_j - b_j'| \leq \\
&\leq x_0, \quad i = 1, 2, \ldots, m.
\end{aligned}$$

So, we have

$$|x_i^* - x_i'| \leq x_0, \quad i = 1, 2, \ldots, m.$$

The latter means that

$$x_i' \geq x_i^* - x_0 \geq 0, \quad i = 1, 2, \ldots, m.$$

Thus, we have shown that vector x' satisfies the sign-restrictions (5.84) of problem $LFP(b')$.

As concerning the system of main constraints

$$\sum_{j=1}^{n} a_{ij} x_j = b_i', \quad i = 1, 2, \ldots, m, \tag{5.90}$$

which is almost the same as system (5.83) but where RHS vector$b$ is replaced with vector b', this system (5.90) is satisfied by definition of vector x' (see formula (5.87)). So, we have shown that vectorx' is a basic feasible solution of problem $LFP(b')$.

Our next aim is to show that vector x' is an optimal solution of problem $LFP(b')$. In order to show it, we formulate a dual problem for $LFP(b)$ as follows

$$\psi(y) = y_0 \longrightarrow \min \tag{5.91}$$

subject to

$$d_0 y_0 - \sum_{i=1}^{m} b_i y_i \geq p_0, \tag{5.92}$$

$$d_j y_0 + \sum_{i=1}^{m} a_{ij} y_i \geq p_j, \quad j = 1, 2, \ldots, n. \tag{5.93}$$

Since the primal LFP problem is solvable, in accordance with duality theory (see Theorem 5.2) its dual problem is also solvable. So, let us suppose that vector $y^* = (y_0^*, y_1^*, \ldots, y_m^*)^T$ is an optimal solution of dual problem (5.91)-(5.93).

In accordance with Theorem 5.4 from the structure of optimal solution x^* (see (5.86)) it follows that

$$d_j y_0^* + \sum_{i=1}^{m} a_{ij} y_i^* \begin{cases} = p_j, & j = 1, 2, \dots, m, \\ \geq p_j, & j = m+1, m+2, \dots, n, \end{cases} \tag{5.94}$$

while from Theorem 5.5 and conditions (5.85) we obtain

$$d_0 y_0 - \sum_{i=1}^{m} b_i y_i^* = p_0. \tag{5.95}$$

If we multiply jth equality of (5.94) with x_j', $j = 1, 2, \dots, m$, and then after summation all of them we add the result obtained to equality(5.95), we obtain the following expression

$$y_0^* \,(\sum_{j=1}^{m} d_j x_j' + d_0) + \sum_{i=1}^{m} y_i^* \sum_{j=1}^{m} a_{ij} x_j' - \sum_{i=1}^{m} b_i y_i^* \; = \; \sum_{j=1}^{m} p_j x_j' + p_0$$

or

$$y_0^* \; D(x') + \sum_{i=1}^{m} y_i^* (b_i' - b_i) = P(x'). \tag{5.96}$$

Now, let us formulate the dual problem for $LFP(b')$. It will be as follows

$$\psi(\tilde{y}) = \tilde{y_0} \longrightarrow \min$$

subject to

$$d_0 \tilde{y_0} - \sum_{i=1}^{m} b_i' \tilde{y_i} \geq p_0,$$

$$d_j \tilde{y_0} + \sum_{i=1}^{m} a_{ij} \tilde{y_i} \geq p_j, \quad j = 1, 2, \dots, n.$$

Concerning this dual problem in accordance with Lemma 5.1 we can state that

$$\tilde{y_0} \geq \frac{P(x')}{D(x')}.$$

Observe that for any vector $x' \in S$ we have $D(x') > 0$. So

$$\tilde{y_0} D(x') \geq P(x').$$

The latter with equality (5.96) mean that

$$\tilde{y_0} D(x') \; \geq \; y_0^* \; D(x') + \sum_{i=1}^{m} y_i^* (b_i' - b_i)$$

or

$$\tilde{y_0} \geq y_0^* + \frac{\sum_{i=1}^{m} y_i^* (b_i' - b_i)}{D(x')}. \tag{5.97}$$

It should be noted that the right-hand side expression in (5.97) gives us the exact lower bound of objective function $\psi(\tilde{y})$ over its feasible set. At the same time this expression gives also the exact upper bound for objective function of problem $LFP(b')$ over its feasible set.

Observe that from equality (5.96) it follows that

$$Q(x') = y_0^* + \frac{\sum_{i=1}^{m} y_i^* \ (b_i' - b_i)}{D(x')}. \tag{5.98}$$

The latter means that objective function $Q(x)$ reaches its maximal value over the feasible set of problem $LFP(b')$ in point x'.

Herewith, we have proved the following

THEOREM 5.7 ([9]) *If LFP problem $LFP(b)$ is solvable and has at least one non-degenerate optimal solution x^* in a finite vertex and vectors b and b' satisfy restriction (5.89), then LFP problem $LFP(b')$ is also solvable and its optimal solution x' may be obtained from formula (5.88), while the optimal value of its objective function is given by (5.98)*

Note that when proving Theorem 5.7 we did not use directly our assumption on non-degeneracy of optimal solution x^*. Nevertheless, this assumption has a very important role – if vector x^* is a degenerate optimal solution, then the corresponding dual problem has multiple optimal solutions and hence, formula (5.98) is not, strictly speaking, valid at all.

REMARK 5.2 *Condition (5.89) is only sufficient, for replacing original RHS vector b with new vector b' does not lead to any change in optimal basis B, but it is not a necessary condition for it (detailed discussion of this topic will be given in Section 6, Chapter 6).*

REMARK 5.3 *When investigating the influence of change in RHS vector b we considered LFP problems in a canonical form. It is obvious that all main results obtained (including formulas (5.88), (5.89) and (5.98)) may be applied to LFP problems formulated in general form too.*

Let us suppose that

$$b' = (b_1, b_2, \dots, b_{k-1}, b_k + 1, b_{k+1}, \dots, b_m)^T$$

and restriction (5.89) is satisfied. If we recall that $Q(x^*) = y_0^*$, then for given vector b' from formula (5.98) it follows that

$$Q(x') - Q(x^*) = \frac{y_k^*}{D(x')}.$$

This formula shows that dual variables y_i in linear-fractional programming unlike linear programming cannot be interpreted as a change of optimal value in objective function when changing RHS vector$b$. This topic well be discussed in detail in the next section.

Before closing this section we illustrate how formulas (5.88) and (5.98) may be applied. Consider the numeric example described above (see page 155)

$$Q(x) = \frac{8\,x_1 + 9\,x_2 + 4\,x_3 + 4}{2\,x_1 + 3\,x_2 + 2\,x_3 + 7} \longrightarrow \max$$

subject to

$$\begin{array}{rcrcrcr} 1\,x_1 & + & 1\,x_2 & + & 2\,x_3 & \leq & 3\,, \\ 2\,x_1 & + & 1\,x_2 & + & 4\,x_3 & \leq & 4\,, \\ 5\,x_1 & + & 3\,x_2 & + & 1\,x_3 & \leq & 15\,, \end{array}$$

$$x_j \geq 0, \quad j = 1, 2, 3.$$

After entering slack variablesx_4, x_5 and x_6, we obtain our problem in canonical form

$$Q(x) = \frac{8\,x_1 + 9\,x_2 + 4\,x_3 + 4}{2\,x_1 + 3\,x_2 + 2\,x_3 + 7} \longrightarrow \max$$

subject to

$$\begin{array}{rcrcrcrcrcrcr} 1\,x_1 & + & 1\,x_2 & + & 2\,x_3 & + & x_4 & & & & & = & 3\,, \\ 2\,x_1 & + & 1\,x_2 & + & 4\,x_3 & & & + & x_5 & & & = & 4\,, \\ 5\,x_1 & + & 3\,x_2 & + & 1\,x_3 & & & & & + & x_6 & = & 15\,, \end{array}$$

$$x_j \geq 0, \quad j = 1, 2, 3, 4, 5, 6.$$

As follows from final simplex tableau (see Table 5.3, page 157), its optimal solution is as follows:

$$B = \{A_2, A_1, A_6\}, \quad x_B = (2, 1, 4)^T,$$

so

$$x^* = (1, 2, 0, 0, 0, 4)^T \quad \text{and} \quad Q(x^*) = \frac{30}{15} = 2\,,$$

where

$$B = (A_2, A_1, A_6) = \begin{pmatrix} 1 & 1 & 0 \\ 1 & 2 & 0 \\ 3 & 5 & 1 \end{pmatrix}$$

and

$$B^{-1} = \begin{pmatrix} 2 & -1 & 0 \\ -1 & 1 & 0 \\ -1 & -2 & 1 \end{pmatrix}.$$

Further, using inverse matrix B^{-1} we calculate

$$\begin{aligned} e &= \max_{1\le i\le m} \sum_{j=1}^{m} |e_{ij}| = \\ &= \max\{|2| + |-1| + |0|,\ |-1| + |1| + |0|,\ |-1| + |-2| + |1|\} = \\ &= \max\{3, 2, 4\} = 4 \end{aligned}$$

and

$$x_0 = \min_{1\le i\le m} x_i^* = \min\{x_1^*, x_2^*, x_6^*\} = \min\{1, 2, 4\} = 1.$$

Hence, for our example condition (5.89) transforms to

$$\max_{1\le j\le m} |b_j' - b_j| \le \frac{x_0}{e} = \frac{1}{4},$$

so, for new vector b' we have the following restrictions

$$\begin{array}{rcccccccl} 2.75 &=& b_1 - 0.25 &\le& b_1' &\le& b_1 + 0.25 &=& 3.25\ , \\ 3.75 &=& b_2 - 0.25 &\le& b_2' &\le& b_2 + 0.25 &=& 4.25\ , \\ 14.75 &=& b_3 - 0.25 &\le& b_3' &\le& b_3 + 0.25 &=& 15.25\ . \end{array}$$

Let new RHS vector b' be

$$b' = (3.25,\ 3.75,\ 15.20)^T.$$

Then, using inverse matrix B^{-1} and formula (5.88) we obtain

$$\begin{array}{rcl} x_2' &=& 2 \times 3.25 + (-1) \times 3.75 + 0 \times 15.20 &=& 2.75\ , \\ x_1' &=& (-1) \times 3.25 + 1 \times 3.75 + 0 \times 15.20 &=& 0.50\ , \\ x_6' &=& (-1) \times 3.25 + (-2) \times 3.75 + 1 \times 15.20 &=& 4.45\ , \end{array}$$

hence,

$$x' = (0.5,\ 2.75,\ 0,\ 0,\ 0,\ 4.45)^T.$$

Finally, we have

$$Q(x') = \frac{P(x')}{D(x')} = \frac{8 \times 0.5 + 9 \times 2.75 + 4}{2 \times 0.5 + 3 \times 2.75 + 7} = \frac{32.75}{16.26} \approx 2.0154\ .$$

We can check this result by using formula (5.98). Indeed, the dual problem in this example is

$$\psi(y) = y_0 \longrightarrow \min \tag{5.99}$$

subject to

$$\left.\begin{array}{rcrcrcrcl} 7\,y_0 & - & 3\,y_1 & - & 4\,y_2 & - & 15\,y_3 & \geq & 4\,, \\ 2\,y_0 & + & 1\,y_1 & + & 2\,y_2 & + & 5\,y_3 & \geq & 8\,, \\ 3\,y_0 & + & 1\,y_1 & + & 1\,y_2 & + & 3\,y_3 & \geq & 9\,, \\ 2\,y_0 & + & 2\,y_1 & + & 4\,y_2 & + & 1\,y_3 & \geq & 4\,, \end{array}\right\} \tag{5.100}$$

$$y_1 \geq 0, \quad y_2 \geq 0, \quad y_3 \geq 0. \tag{5.101}$$

while its optimal solution is vector $y^* = (2,2,1,0)^T$. So, from formula (5.98) we obtain the following

$$\begin{aligned} Q(x') &= y_0^* + \frac{y_1^*(b_1' - b_1) + y_2^*(b_2' - b_2) + y_3^*(b_3' - b_3)}{D(x')} = \\ &= 2 + \frac{2\,(3.25 - 3) + 1\,(3.75 - 4) + 0\,(15.2 - 15)}{16.26} = \\ &= 2 + \frac{0.25}{16.26} \approx 2 + 0.0154 = 2.0154\,. \end{aligned}$$

Thus, we have shown that having dual variables we can predict the change in optimal value of the objective function when a sufficiently small change occurs in RHS vector b.

7. Comparative Analysis of Dual Variables in LP and LFP

As we have seen in the previous section, the dual variables of linear-fractional programming differ from those of linear programming. In this section we discuss one of the possible economic interpretations of dual variables in LFP and compare them with the dual variables of LP.

Suppose that there is a company that manufactures n different products. To produce these products the company uses m kinds of scarce resources that are available in volumes $b_1, b_2, \ldots, b_m$. Further, let p_j be the profit gained by the company from a unit of the jth product and d_j be the cost associated with the same unit, while p_0 and d_0 be some constant profit and constant cost gained, respectively, whose magnitudes are independent of the output volume. Let a_{ij} be the expenditure quota of the i-th resource for manufacturing a unit of j-th kind of the product. Denote the unknown output volume of some j-th kind of the product by x_j. Our aim is to find such a production plan $x^* = (x_1^*, x_2^*, \ldots, x_n^*)^T$ which does not exceed the given limits of resources available, and leads to maximal profit

$$P(x) = \sum_{j=1}^{n} p_j x_j + p_0$$

Using the notation introduced above we can formulate the following LP problem

$$P(x) = \sum_{j=1}^{n} p_j x_j + p_0 \longrightarrow \max, \tag{5.102}$$

subject to

$$\sum_{j=1}^{n} a_{ij} x_j \leq b_i, \quad i = 1, 2, \ldots, m, \tag{5.103}$$

$$x_j \geq 0, \quad j = 1, 2, \ldots, n, \tag{5.104}$$

Let us suppose that vector $x^* = (x_1^*, x_2^*, \ldots, x_n^*)^T$ is an optimal solution of problem (5.102)-(5.104), with basis B. Further, we replace kth entry b_k in vector of resources $b = (b_1, b_2, \ldots, b_m)^T$ with $b'_k = b_k + 1$ and assume that this replacement does not affect optimal basis B.

In accordance with the theory of linear programming, this change in resource vector b generally speaking leads to change in optimal solution of LP problem (5.102)-(5.104) and affects the optimal value of objective function $P(x)$. So, for new resource vector

$$b' = (b_1, b_2, \ldots, b_{k-1}, b_k + 1, b_{k+1}, \ldots, b_m)^T$$

we obtain a new optimal solution $x' = (x'_1, x'_2, \ldots, x'_n)^T$ and a new optimal value of objective function $P(x')$ which may be calculated using formula

$$P(x') = P(x^*) + u_k^*\,, \tag{5.105}$$

where u_k^* is the kth element of optimal solution $u^* = (u_1^*, u_2^*, \ldots, u_m^*)^T$ for dual problem

$$\phi(u) = \sum_{i=1}^{m} b_i u_i + p_0 \longrightarrow \min$$

subject to

$$\sum_{i=1}^{m} a_{ij} u_i \geq p_j, \quad j = 1, 2, \ldots, n,$$

$$u_i \geq 0, \quad i = 1, 2, \ldots, m\,. \tag{5.106}$$

Consider the economic interpretation of expression (5.105). This formula expresses the fact that if the volume of the kth resource increases by one unit then the maximal value of profit-function $P(x)$ grows by u_k^* units. In accordance with the duality theory of linear programming, if the kth condition of system

(5.103) is free, i.e. at least for one optimal solution it is satisfied as a strict inequality, then the corresponding kth dual condition (5.106) is fixed and hence, $u_k^* = 0$ and $P(x') = P(x^*) + 0$. In other words, if optimal solution x^* does not require all b_k units of the kth resource and in this way results in an overstock of resource k, then any sufficiently small change ϵ in the kth resource $b_k \rightarrow b'_k = b_k + \epsilon$ does not affect the optimal solution since the optimal value of the corresponding dual variable u_k is equal to zero and thus, the optimal value of profit $P(x)$ does not change.

Now, let us apply an LFP approach to the optimization of production activity of the same company. So, our aim now is to maximize the ratio *profit/cost* using the following LFP problem

$$Q(x) = \frac{P(x)}{D(x)} = \frac{\sum_{j=1}^{n} p_j x_j + p_0}{\sum_{j=1}^{n} d_j x_j + d_0} \longrightarrow \max, \tag{5.107}$$

subject to constraints (5.103), (5.104)

As in the previous case, we suppose that vector x^* is an optimal solution of LFP problem (5.107), (5.103), (5.104), and the kth element b_k of RHS vector b must be replaced with b'_k, where $b'_k = (b_k + 1)$. In this case, in accordance with Theorem 5.7 we have

$$Q(x') = Q(x^*) + \frac{y_k^*}{D(x')}, \tag{5.108}$$

where x' is an optimal solution of LFP problem (5.107), (5.103), (5.104) with modified RHS vector b', and y_k^* is the kth element of optimal solution for corresponding dual problem

$$\psi(y) = y_0 \longrightarrow \min \tag{5.109}$$

subject to

$$d_0 y_0 - \sum_{i=1}^{m} b_i y_i \geq p_0, \tag{5.110}$$

$$d_j y_0 + \sum_{i=1}^{m} a_{ij} y_i \geq p_j, \quad j = 1, 2, \ldots, n. \tag{5.111}$$

$$y_i \geq 0, \quad i = 1, 2, \ldots, m. \tag{5.112}$$

To clear up the possible economic interpretation of dual variable y_k^* we consider formula (5.108). Obviously, we may re-write it as follows

$$y_k^* = P(x') - Q(x^*)D(x')$$

or

$$y_k^* = D(x')(Q(x') - Q(x^*)) \tag{5.113}$$

The right-hand side of the latter allows us to formulate the economic interpretation of dual variables in LFP as follows. Using an extra unit of kth resource leads to change in profit $P(x)$, cost $D(x)$, and hence, in efficiency $Q(x)$. It is obvious that the total change of profit is

$$\begin{aligned} \Delta P &= P(x') - P(x^*) = \\ &= P(x') - Q(x^*)D(x^*) = \\ &= P(x') - Q(x^*)D(x^*) + Q(x^*)D(x') - Q(x^*)D(x') = \\ &= D(x')(Q(x') - Q(x^*)) + Q(x^*)(D(x') - D(x^*)), \end{aligned}$$

or, using (5.113)

$$\Delta P = y_k^* + Q(x^*)(D(x') - D(x^*)).$$

Suppose that $D(x^*)$ units of money have been invested into such production activity, which has efficiency $Q(x^*)$. It is clear that in this case our profit will be $Q(x^*)D(x^*)$ units. If we invest $D(x')$ units into the same activity then we obtain $Q(x^*)D(x')$ units of profit. The difference between these two values of profit is

$$Q(x^*)(D(x') - D(x^*)). \tag{5.114}$$

Assume that efficiency of production activity depends on the volume of investments, so we can affect the efficiency $Q(x)$ in such a way that when we invest $D(x^*)$ units of money the efficiency is $Q(x^*)$, and if the volume of investment is $D(x')$ then efficiency becomes $Q(x')$. It is clear that in this case we obtain

$$D(x')(Q(x') - Q(x^*)) \tag{5.115}$$

units of profit. The latter means that the total change in profit consists of two parts: one of them (namely (5.114)) occurs because of a change in the volume of investments, and the second one (namely (5.115)) arises as a result of change in efficiency. Using the terminology of economics we can call these two parts of extra profit as "*extensive*" and "*intensive*" components of extra profit ΔP.

Thus, we can formulate the main difference between the economic interpretations of dual variables in LP and LFP as follows:

if dual variable u_k^* *of linear programming determines how much profit* $P(x)$ *changes when changing kth element of RHS vector* b*, then dual variable* y_k^* *of linear-fractional programming indicates only the intensive part of change in profit* $P(x)$.

This difference between the dual variables of LP and LFP determines the usability of the dual variables of LFP in real-world applications and their importance for those optimization models where scarce resources must be used with as high efficiency as possible.

To illustrate the difference between the dual variables of LP and LFP we reconsider the numeric example from the previous section (see page 166). Let us suppose that the production activity of the company described above may be formulated as the following LP problem:

$$P(x) = 8\,x_1 + 9\,x_2 + 4\,x_3 + 4 \longrightarrow \max$$

subject to

$$\left.\begin{array}{rcrcrcr} 1\,x_1 & + & 1\,x_2 & + & 2\,x_3 & \leq & 3\,, \\ 2\,x_1 & + & 1\,x_2 & + & 4\,x_3 & \leq & 4\,, \\ 5\,x_1 & + & 3\,x_2 & + & 1\,x_3 & \leq & 15\,, \\ & & \multicolumn{5}{l}{x_j \geq 0, \;\; j = 1,2,3.} \end{array}\right\} \tag{5.116}$$

Solving this problem we obtain the following optimal solution

$$x^* = (0,3,0)^T, \quad P(x^*) = 31\,,$$

while its dual problem

$$\phi(u) = 3\,u_1 + 4\,u_2 + 15\,u_3 + 4 \longrightarrow \min$$

subject to

$$\begin{array}{rcrcrcr} 1\,u_1 & + & 2\,u_2 & + & 5\,u_3 & \geq & 8\,, \\ 1\,u_1 & + & 1\,u_2 & + & 3\,u_3 & \geq & 9\,, \\ 2\,u_1 & + & 4\,u_2 & + & 1\,u_3 & \geq & 4\,, \end{array}$$

$$u_j \geq 0, \;\; j = 1,2,3.$$

has optimal solution

$$u^* = (9,0,0)^T, \quad \phi(u^*) = 31\,.$$

Vector u^* shows that if value of $b_1 = 3$ in RHS vector $b = (3,4,15)^T$ is replaced with $b_1' = 3+1 = 4$ then the optimal value of objective function $P(x)$ for the new LP problem with modified RHS vector will be

$$P(x') = P(x^*) + u_1^* = 31 + 9 = 40\,.$$

Note that when replacing $b_1 = 3 \to b_1' = 4$ we suppose that this change is small enough and does not affect the optimal basis of the primal LP problem (we will discuss this topic in detail in Chapter 6). As regards dual variables $u_2^* = 0$ and $u_3^* = 0$, they show that some small change in values of $b_2 = 4$ and $b_3 = 15$ does not affect the optimal value of objective function $P(x)$. Indeed, in accordance with the duality theory of linear programming, if dual conditions

$$u_2 \geq 0 \quad \text{and} \quad u_3 \geq 0$$

are fixed then corresponding conditions of the primal problem

$$\begin{array}{rcrcrcl} 2\,x_1 & + & 1\,x_2 & + & 4\,x_3 & \leq & 4\,, \\ 5\,x_1 & + & 3\,x_2 & + & 1\,x_3 & \leq & 15 \end{array}$$

are free. This means that if the company operates in accordance with optimal plan $x^* = (0,3,0)^T$, it results in a surplus in the RHS vector of resources since

$$\begin{array}{rcrcrcrcl} 2\times 0 & + & 1\times 3 & + & 4\times 0 & = & 3 & < & 4\,, \\ 5\times 0 & + & 3\times 3 & + & 1\times 0 & = & 9 & < & 15\,. \end{array}$$

Suppose that function of costs for a given company may be expressed as

$$D(x) = 2\,x_1 + 3\,x_2 + 2\,x_3 + 7$$

then to optimize efficiency of production activity for this company we have to consider the following LFP problem

$$Q(x) = \frac{8\,x_1 + 9\,x_2 + 4\,x_3 + 4}{2\,x_1 + 3\,x_2 + 2\,x_3 + 7} \longrightarrow \max \tag{5.117}$$

subject to constraints (5.116).
The optimal solution of this problem is

$$x^* = (1,2,0)^T, \quad Q(x^*) = \frac{30}{15} = 2\,,$$

while its dual problem (5.99)-(5.101) has the optimal solution

$$y^* = (2,2,1,0)^T, \quad \psi(y^*) = 2\,.$$

Vector y^* indicates that if we increase the volume of the first resource from $b_1 = 3$ units to $b_1' = 4$ it will result for the company higher efficiency that in its own turn will lead to extra profit $y_1^* = 2$. Indeed, if we replace $b_1 = 3$ with $b_1' = 4$ in LFP problem (5.117), (5.116) and then solve the modified problem we obtain its optimal solution as follows

$$x' = (0,4,0)^T, \quad Q(x') = \frac{40}{19} \approx 2.1053\,.$$

The total increase of profit in this case is

$$\Delta P = P(x') - P(x^*) = 40 - 30 = 10,$$

units, where

$$Q(x^*)(D(x') - D(x^*)) = 2 \times (19 - 15) = 8$$

units of them are an extensive part of extra profit ΔP, while

$$y_k^* = D(x')(Q(x') - Q(x^*)) = 19 \times (40/19 - 30/15) = 2$$

units represent its intensive part.

It is obvious that dual variable $y_2^* = 1$ may be interpreted analogically. However, dual variable $y_3^* = 0$ indicates that some small change in the third resource $b_3 \longrightarrow b_3'$ does not affect the optimal solution and hence, does not result in any change in the optimal value of objective function $Q(x)$.

8. Discussion Questions and Exercises

5.1 For LFP problems given below formulate dual problems:

1

$$Q(x) = \frac{1\,x_1 + 2\,x_2 - 5}{3\,x_1 + 1\,x_2 + 2} \longrightarrow \max$$

subject to

$$\begin{array}{rcrcr} 1\,x_1 & + & 1\,x_2 & \leq & 3\,, \\ 2\,x_1 & + & 1\,x_2 & \leq & 4\,, \\ 5\,x_1 & + & 3\,x_2 & \leq & 15\,, \end{array}$$

$$x_1 \geq 0, \quad x_2 \geq 0.$$

2

$$Q(x) = \frac{4\,x_1 + 2\,x_2 + 2\,x_3 - 15}{6\,x_1 + 7\,x_2 + 1\,x_3 + 12} \longrightarrow \min$$

subject to

$$\begin{array}{rcrcrcr} 7\,x_1 & + & 3\,x_2 & + & 3\,x_3 & = & 10\,, \\ 9\,x_1 & + & 1\,x_2 & - & 3\,x_3 & \geq & 45\,, \\ 2\,x_1 & - & 7\,x_2 & + & 5\,x_3 & \leq & 25\,, \end{array}$$

$$x_1 \geq 0, \quad x_2 \geq 0, \quad x_3 \geq 0.$$

3

$$Q(x) = \frac{-2\,x_1 - 2\,x_3}{2\,x_2 + 4\,x_3 + 10} \longrightarrow \max$$

subject to

$$\begin{array}{rcrcrcr} 4\,x_1 & + & 8\,x_2 & + & 1\,x_3 & \geq & 1\,, \\ 2\,x_1 & & & - & 1\,x_3 & = & 15\,, \\ 1\,x_1 & + & 2\,x_2 & + & 2\,x_3 & \leq & 22\,, \end{array}$$

$$x_1 \geq 0, \quad x_2 \geq 0, \quad x_3 \geq 0.$$

5.2 For the dual problems formulated for the LFP problems given in exercise 1 construct their dual problems, i.e. linear analogues of corresponding LFP problems.

5.3 Find the dual of the following LFP problem

$$Q(x) = \frac{2\,x_1 + 3\,x_2 + 1}{1\,x_1 + 2\,x_2 + 10} \longrightarrow \max$$

subject to

$$\begin{array}{rcrcr} 3\,x_1 & + & 4\,x_2 & \leq & 36\,, \\ 4\,x_1 & + & 2\,x_2 & \leq & 20\,, \\ 1\,x_1 & + & 3\,x_2 & \leq & 30\,, \end{array}$$

$$x_1 \geq 0, \quad x_2 \geq 0.$$

Then solve both problems and for all pairs of dual constraints detect if the constraint is fixed or free.

5.4 In the LFP problem given in the previous exercise we wish to change the right-hand side vector $b = (36, 20, 30)^T$ so that

1 $b_1 = 36 \;\rightarrow\; b_1' = 40$;

2 $b_2 = 20 \;\rightarrow\; b_2' = 30$;

3 $b_3 = 30 \;\rightarrow\; b_3' = 35$;

4 $b_3 = 30 \;\rightarrow\; b_3' = 29$.

Using theorem 5.3, theorem 5.7 and formula (5.98) try to predict for each case separately if the optimal value of the objective function $Q(x)$ will change. If the change in the right-hand side vector b affects the optimal value of the objective function, then calculate this change and determine the new optimal value for the objective function.

Chapter 6

SENSITIVITY ANALYSIS

In this chapter, we discuss how changes in the coefficients (including RHS vector b and the coefficients of objective function $Q(x)$) of LFP problems affect the optimal solution. This is called *sensitivity analysis.* First, in Section 1, we illustrate the concept of sensitivity analysis through a graphical example. Then in Sections 2-6 we consider separate cases when changes occur in different parts of an LFP problem.

We deal here with the following canonical LFP problem

$$Q(x) = \frac{P(x)}{D(x)} = \frac{\sum_{j=1}^{n} p_j x_j + p_0}{\sum_{j=1}^{n} d_j x_j + d_0} \longrightarrow \max \tag{6.1}$$

subject to

$$\sum_{j=1}^{n} a_{ij} x_j = b_i, \quad i = 1, 2, \ldots, m, \tag{6.2}$$

$$x_j \geq 0, \quad j = 1, 2, \ldots, n, \tag{6.3}$$

where $D(x) > 0, \ \forall x \in S$. We assume that problem (6.1)-(6.3) is solvable and vector x^* denotes its optimal solution. Without restriction of generality we may assume also that

$$x^* = (x_1^*, x_2^*, \ldots, x_m^*, 0, 0, \ldots, 0)^T$$

and its optimal basis is $B = (A_1, A_2, \ldots, A_m)$, where $A_j = (a_{1j}, a_{2j}, \ldots, a_{mj})^T$, $j = 1, 2, \ldots, n$. In the rest of the chapter we will use the following notation: $J = \{1, 2, \ldots, n\}$, $J_B = \{1, 2, \ldots, m\}$ and $J_N = J \setminus J_B$.

1. Graphical Introduction to Sensitivity Analysis

Reconsider the numeric example of Section 2, Chapter 3, page 52

$$Q(x) = \frac{6\,x_1 + 3\,x_2 + 6}{5\,x_1 + 2\,x_2 + 5} \longrightarrow \max(\min) \tag{6.4}$$

subject to

$$\left.\begin{array}{rcrcrl} 4\,x_1 & - & 2\,x_2 & \leq & 20\,, & (i) \\ 3\,x_1 & + & 5\,x_2 & \leq & 25\,, & (ii) \\ & & \multicolumn{3}{l}{x_1 \geq 0, \;\; x_2 \geq 0.} & \end{array}\right\} \tag{6.5}$$

The optimal solution for this problem is

$$x^* = (0, 5), \;\; Q(x^*) = \frac{21}{15},$$

(point A in Figure 6.1). How would the changes in the right-hand sides or in

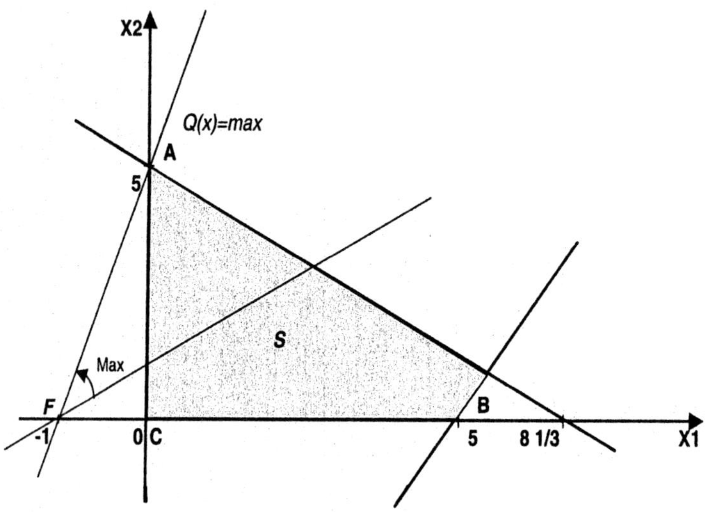

Figure 6.1. Stability – Original graphical example.

the objective function coefficients change the optimal solution of this problem?

When changing the right-hand sides of the LFP problem the analysis of the effect produced is relatively simple. Indeed, let's change the current value $b_2 = 25$ in the second constraint of (6.5) (marked by (ii)) to $b_2' = b_2 + \delta = 25+5 = 30$. It is obvious that this change does not affect focus point F but changes feasible set S as shown in Figure 6.2. From Figure 6.2, we see that in this case the optimal basis remains the same but the optimal solution of the new problem

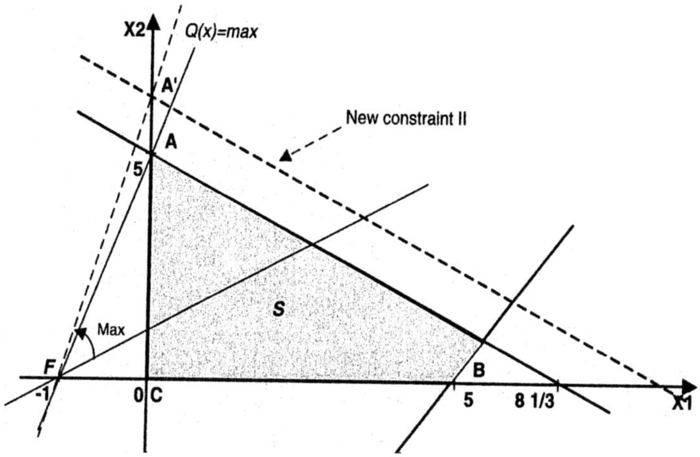

Figure 6.2. Stability – Graphical example with changed feasible set.

moves to point A' with coordinates $x' = (0,6)^T$, where $Q(x') = 24/17$. Observe that in this example we can increase the value of δ (and hence, b_2) infinitely without any change in the optimal basis. So the upper bound for δ is ∞. If we decrease b_2 the optimal basis remains stable until $b_2 + \delta \geq 0$, since for negative right-hand side b_2 the problem becomes unsolvable (infeasible). It means that the low bound of change in b_2 is $\delta \geq 0 - b_2 = -25$. Finally, we obtain the stable range for change as follows

$$-25 \leq \delta < \infty \quad \text{or} \quad 0 \leq b_2 < \infty \, .$$

Similarly, for the constraint marked by (i) we have

$$-30 \leq \delta < \infty \quad \text{and} \quad -10 \leq b_1 < \infty \, .$$

If we change the coefficients of objective function $Q(x)$ the situation becomes more complicated since in this case the change affects the focus point F and hence, may result in significant change in the properties of the problem which requires detailed investigations. Moreover, change in the coefficients of the denominator of the objective function may violate restriction $D(x) > 0 \ \forall x \in S$. This is why we defer discussion of this case to Sections 3-6 where we consider each possible change in the numerator and denominator separately. In Figure 6.3, we see how change in non-basic coefficient

$$p_1 = 6 \rightarrow p_1' = p_1 + 1 = 6 + 1 = 7$$

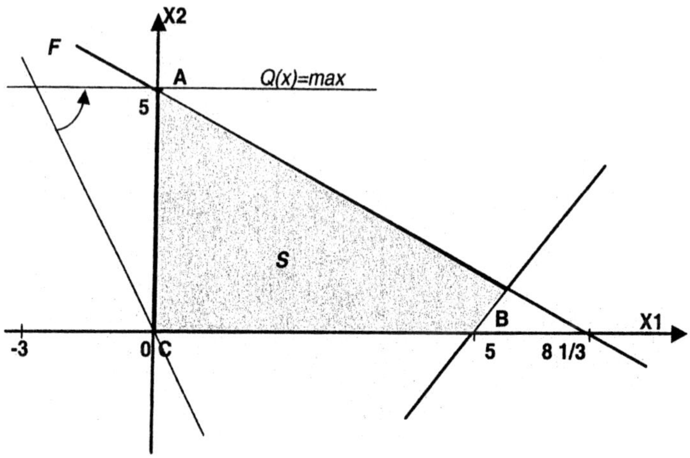

Figure 6.3. Stability – Graphical example with changed objective function.

affects the position of focus point F and the behavior of objective function $Q(x)$.

2. Change in RHS Vector b

Let us replace entry b_μ in RHS vector $b = (b_1, b_2, \ldots, b_\mu, \ldots, b_m)^T$ with $b'_\mu = b_\mu + \delta$ and investigate how this change affects optimal basis B, optimal solution x^* and the optimal value of objective function $Q(x)$.

In accordance with our assumption vector x^* is a feasible and optimal solution of canonical LFP problem (6.1)-(6.3). Since x^* is a feasible solution, it means that vector x^* satisfies conditions (6.2) and (6.3). We can re-write system (6.2) as follows

$$Bx^* = b, \tag{6.6}$$

or

$$x^* = B^{-1}b, \tag{6.7}$$

where $B^{-1} = ||e_{ij}||_{m\times m}$ denotes inverse basis matrix B (recall that matrix B consists of linear-independent column-vectors $A_j,\ j = 1, 2, \ldots, m$, so matrix B^{-1} exists).

Let us re-write equality (6.7) in the following form

$$x_i^* = \sum_{k=1}^{m} e_{ik} b_k, \quad i = 1, 2, \ldots, m. \tag{6.8}$$

Observe that when replacing $b_\mu \rightarrow b_\mu + \delta$ from formula (6.8) it follows that

$$x_i' = \sum_{k=1}^{m} e_{ik} b_k + \delta e_{i\mu} = x_i^* + \delta e_{i\mu}, \quad i = 1, 2, \ldots, m. \tag{6.9}$$

So, instead of original optimal vector x^* we obtain some other vector

$$x' = (x_1', x_2', \ldots, x_m', 0, 0, \ldots, 0).$$

Naturally, concerning this vector x' the following two questions appear:

1 Is vector x' a feasible solution of the modified problem?

2 Is vector x' an optimal solution for this problem?

Consider the first question. In accordance with the definition of feasible solution, to be a feasible solution of a modified LFP problem vector x' must satisfy conditions

$$x_i' \geq 0, \quad i = 1, 2, \ldots, m; \tag{6.10}$$

and

$$\left.\begin{array}{ll} \sum_{j=1}^{m} a_{ij} x_j' = b_i, & i = 1, 2, \ldots, m, \quad i \neq \mu; \\ \sum_{j=1}^{m} a_{ij} x_j' = b_i + \delta, & i = \mu \end{array}\right\} \tag{6.11}$$

Using (6.9) we can re-write (6.10) as follows

$$x_i' = x_i^* + \delta e_{i\mu} \geq 0, \quad i = 1, 2, \ldots, m.$$

The latter means that

$$\delta e_{i\mu} \geq -x_i^*, \quad i = 1, 2, \ldots, m,$$

or

$$\begin{array}{ll} \delta \geq -\dfrac{x_i^*}{e_{i\mu}}, & \text{for those } i \text{ that } e_{i\mu} > 0, \\ \delta \leq -\dfrac{x_i^*}{e_{i\mu}}, & \text{for those } i \text{ that } e_{i\mu} < 0. \end{array}$$

In this way, we obtain the following range:

$$\max_{\substack{1 \leq i \leq m \\ e_{i\mu} > 0}} \{-\frac{x_i^*}{e_{i\mu}}\} \leq \delta \leq \min_{\substack{1 \leq i \leq m \\ e_{i\mu} < 0}} \{-\frac{x_i^*}{e_{i\mu}}\}. \tag{6.12}$$

It is obvious that if δ is in range (6.12) then all basic entries x'_i, $i = 1, 2, \ldots, m$, are non-negative and hence, sign-restriction conditions (6.10) are satisfied.

Concerning conditions (6.11) we can assert that vector x' satisfy this system by definition, since when constructing vector x' we used basis matrix B and formula (6.6). Thus, we have shown that if δ satisfies condition (6.12) then vector x' satisfies restrictions (6.10) and (6.11), and hence, vector x' is a feasible solution of the modified LFP problem

Consider the second question: if vector x' is an optimal solution. In accordance with the theory of the simplex method and its criteria of optimality (see Chapter 4, Section 2), basic feasible vector x' is an optimal solution for the modified LFP problem if

$$\Delta_j(x') = \Delta'_j - Q(x')\Delta''_j \geq 0, \quad j = 1, 2, \ldots, n.$$

Since

$$\Delta'_j \;=\; \Delta''_j \;=\; \Delta_j(x^*) \;=\; 0, \quad \forall j \in J_B,$$

for optimality of vector x' it is enough if

$$\Delta_j(x') = \Delta'_j - Q(x')\Delta''_j \geq 0, \quad \forall j \in J_N. \tag{6.13}$$

Consider formula (6.13). Observe that reduced costs Δ'_j and Δ''_j do not depend directly on RHS vector b and vector x'. So, any change in RHS vector b may affect only the value of objective function $Q(x)$. Hence, we have

$$Q(x') \;=\; \frac{P(x^*)}{D(x^*)} = \frac{\sum\limits_{i=1}^{m} p_i x'_i + p_0}{\sum\limits_{i=1}^{m} d_i x'_i + d_0} =$$

$$\stackrel{(6.9)}{=} \frac{\sum\limits_{i=1}^{m} p_i(x^*_i + \delta e_{i\mu}) + p_0}{\sum\limits_{i=1}^{m} d_i(x^*_i + \delta e_{i\mu}) + d_0} = \frac{P(x^*) + \delta h_1}{D(x^*) + \delta h_2}, \tag{6.14}$$

where

$$h_1 = \sum_{i=1}^{m} p_i e_{i\mu}, \quad h_2 = \sum_{i=1}^{m} d_i e_{i\mu}.$$

In accordance with our assumption $D(x) > 0, \ \forall x \in S$. So, to preserve this condition we have to require that

$$D(x^*) + \delta h_2 > 0. \tag{6.15}$$

The latter gives the following restriction for δ:

$$\delta \begin{cases} > -\dfrac{D(x^*)}{h_2}, & \text{if } h_2 > 0, \\ < -\dfrac{D(x^*)}{h_2}, & \text{if } h_2 < 0. \end{cases} \tag{6.16}$$

Further, using (6.14) we can re-write (6.13) in the following form

$$\Delta'_j \geq \Delta''_j \frac{P(x^*) + \delta h_1}{D(x^*) + \delta h_2}, \quad \forall j \in J_N. \tag{6.17}$$

After transformations from (6.17) we obtain the following system

$$\delta(\Delta'_j h_2 - \Delta''_j h_1) \geq \Delta''_j P(x^*) - \Delta'_j D(x^*), \quad \forall j \in J_N,$$

which may be re-written as follows

$$\delta(\Delta'_j h_2 - \Delta''_j h_1) \geq -\Delta_j(x^*) D(x^*), \quad \forall j \in J_N.$$

The latter gives us the following restrictions

$$\max_{\substack{j \in J_N \\ g_j > 0}} \{\frac{-\Delta_j(x^*)D(x^*)}{g_j}\} \leq \delta \leq \min_{\substack{j \in J_N \\ g_j < 0}} \{\frac{-\Delta_j(x^*)D(x^*)}{g_j}\}, \tag{6.18}$$

where

$$g_j = \Delta'_j h_2 - \Delta''_j h_1, \quad j \in J_N.$$

It is obvious that if δ is in range (6.18) and satisfies restriction (6.16) then system (6.13) takes place and hence, vector x' is an optimal solution of the modified LFP problem.

Thus, we have proved the following

THEOREM 6.1 *If δ satisfies conditions (6.12), (6.16) and (6.18) then vector x' given by formula (6.9) is a feasible and optimal solution of the modified LFP problem (with replaced RHS element $b_\mu \to b'_\mu = b_\mu + \delta$).*

REMARK 6.1 *Lower and upper bounds given by expressions (6.12), (6.16) and (6.18) are a generalization of the corresponding range for the linear programming problem*

$$P(x) = \sum_{j=1}^{n} p_j x_j + p_0 \to \max$$

$$x \in S.$$

Indeed, since an LP problem is a special case of an LFP problem when $d_j = 0, j = 1, 2, \ldots, n$, *and* $d_0 = 1$, *it means that*

$$\Delta_j'' = \sum_{i=1}^{m} d_i x_{ij} - d_j = 0, \quad \textit{for all } \; j = 1, 2, \ldots, n,$$

$$h_2 = \sum_{i=1}^{m} d_i e_{i\mu} = 0, \quad D(x^*) \equiv 1,$$

$$g_j = \Delta_j' h_2 - \Delta_j'' h_1 = 0, \quad j \in J_N,$$

and

$$\Delta_j(x^*) = \Delta_j' - Q(x^*)\Delta_j'' = \Delta_j', \quad j = 1, 2, \ldots, n \, .$$

Thus, bounds (6.12) in the current form as it is are valid for an LP problem too, and guarantee non-negativity for entries x_j', $j \in J_B$, *of vector* x'. *Further, restriction (6.16) results in no bounds since* $h_2 = 0$. *Recall that originally condition (6.16) is required because of positivity for denominator* $D(x)$, *but in an LP problem* $D(x) \equiv 1 > 0$. *Analogously, expression (6.18) gives no restrictions for* δ *since all* $g_j = 0, \; j \in J_N$ *and hence, must be omitted.*

To illustrate the theoretical results described above we reconsider the numerical example from Chapter 5, page 155:

$$Q(x) = \frac{8\,x_1 + 9\,x_2 + 4\,x_3 + 4}{2\,x_1 + 3\,x_2 + 2\,x_3 + 7} \longrightarrow \max \tag{6.19}$$

subject to

$$\begin{array}{rcrcrcrcrcrcr} 1\,x_1 & + & 1\,x_2 & + & 2\,x_3 & + & x_4 & & & & & = & 3\,, \\ 2\,x_1 & + & 1\,x_2 & + & 4\,x_3 & & & + & x_5 & & & = & 4\,, \\ 5\,x_1 & + & 3\,x_2 & + & 1\,x_3 & & & & & + & x_6 & = & 15\,, \end{array} \tag{6.20}$$

$$x_j \geq 0, \quad j = 1, 2, 3, 4, 5, 6.$$

An optimal solution for this problem is given in Table 5.3 (see page 157). We have

$$B = (A_2, A_1, A_6) = \begin{pmatrix} 1 & 1 & 0 \\ 1 & 2 & 0 \\ 3 & 5 & 1 \end{pmatrix}, \quad x_B = \begin{pmatrix} x_2^* \\ x_1^* \\ x_6^* \end{pmatrix} = \begin{pmatrix} 2 \\ 1 \\ 4 \end{pmatrix},$$

so

$$x^* = (1, 2, 0, 0, 0, 4)^T \quad \text{and} \quad Q(x^*) = \frac{30}{15} = 2.$$

Using formulas (6.12), (6.16) and (6.18), and inverse matrix

$$B^{-1} = \begin{pmatrix} 2 & -1 & 0 \\ -1 & 1 & 0 \\ -1 & -2 & 1 \end{pmatrix}$$

we can determine the following ranges of stability for RHS vector$b$:

$\mu = 1$: $b_1 \to b'_1 = b_1 + \delta$. For restriction (6.12) we have

$$\max_{\substack{i\in\{2,1,6\}\\ e_{i1}>0}} \{-\frac{x_2^*}{e_{11}}\} \le \delta \le \min_{\substack{i\in\{2,1,6\}\\ e_{i1}<0}} \{-\frac{x_1^*}{e_{21}}, -\frac{x_6^*}{e_{31}}\}.$$

or

$$-1 = \max\{-\frac{2}{2}\} \le \delta \le \min\{-\frac{1}{-1}, -\frac{4}{-1}\} = 1. \qquad (6.21)$$

For restriction (6.16) we have

$$h_2 = d_2e_{11} + d_1e_{21} + d_6e_{31} = 3 \times 2 + 2 \times (-1) + 0 \times (-1) = 4,$$

hence, since $h_2 > 0$,

$$-3.75 = -\frac{15}{4} = -\frac{D(x^*)}{h_2} < \delta\,. \qquad (6.22)$$

Concerning restriction (6.18) we have

$$\max_{\substack{j\in J_N\\ g_j>0}} \frac{-\Delta_j(x^*)D(x^*)}{g_j} = \max_{j\in\{3,5\}} \frac{-\Delta_j(x^*)D(x^*)}{g_j} \le \delta\,.$$

Since

$$h_1 = p_2e_{11} + p_1e_{21} + p_6e_{31} = 9 \times 2 + 8 \times (-1) + 0 \times (-1) = 10$$

and

$$\begin{array}{lllllll} g_3 & = & \Delta'_3h_2 - \Delta''_3h_1 & = & 12 \times 4 - 2 \times 10 & = & 28\,, \\ g_4 & = & \Delta'_4h_2 - \Delta''_4h_1 & = & 10 \times 4 - 4 \times 10 & = & 0\,, \\ g_5 & = & \Delta'_5h_2 - \Delta''_5h_1 & = & (-1) \times 4 - (-1) \times 10 & = & 6\,, \end{array}$$

we have

$$\max\{\frac{-8 \times 15}{28}, \frac{-1 \times 15}{6}\} = \max\{-4\frac{2}{7}, -2\frac{1}{2}\} = -2.5 \le \delta, \qquad (6.23)$$

Finally, combining (6.21), (6.22) and (6.23) we obtain the following low and upper bounds for δ

$$-1.0 \le \delta \le 1.0$$

and for b_1

$$2.0 \le b_1 \le 4.0 \; .$$

$\mu = 2$: $b_2 \to b_2' = b_2 + \delta$. Restriction (6.12) for $\mu = 2$ gives us the following bounds

$$\max_{\substack{i \in \{2,1,6\} \\ e_{i2} > 0}} \{-\frac{x_1^*}{e_{22}}\} \le \delta \le \min_{\substack{i \in \{2,1,6\} \\ e_{i2} < 0}} \{-\frac{x_2^*}{e_{12}}, -\frac{x_6^*}{e_{32}}\}.$$

or

$$-1 = \max\{-\frac{1}{1}\} \le \delta \le \min\{-\frac{2}{-1}, -\frac{4}{-2}\} = 2. \tag{6.24}$$

From restriction (6.16) we obtain

$$h_2 = d_2 e_{12} + d_1 e_{22} + d_6 e_{32} = 3 \times (-1) + 2 \times 1 + 0 \times (-2) = -1,$$

hence, since $h_2 < 0$,

$$\delta < -\frac{D(x^*)}{h_2} = -\frac{15}{-1} = 15 \; . \tag{6.25}$$

For condition (6.18) we have

$$h_1 = p_2 e_{12} + p_1 e_{22} + p_6 e_{32} = 9 \times (-1) + 8 \times 1 + 0 \times (-2) = -1,$$

$$\begin{array}{lllllll} g_3 & = & \Delta_3' h_2 - \Delta_3'' h_1 & = & 12 \times (-1) - 2 \times (-1) & = & -10 \, , \\ g_4 & = & \Delta_4' h_2 - \Delta_4'' h_1 & = & 10 \times (-1) - 4 \times (-1) & = & -6 \, , \\ g_5 & = & \Delta_5' h_2 - \Delta_5'' h_1 & = & (-1) \times (-1) - (-1) \times (-1) & = & 0 \, . \end{array}$$

Hence,

$$\begin{aligned} \delta & \le \min_{\substack{j \in \{3,4,5\} \\ g_j < 0}} \frac{-\Delta_j(x^*) D(x^*)}{g_j} = \min_{j \in \{3,4\}} \frac{-\Delta_j(x^*) D(x^*)}{g_j} = \\ & = \min\{\frac{-8 \times 15}{-10}, \frac{-2 \times 15}{-6}\} = 5 \end{aligned} \tag{6.26}$$

Finally, combining (6.24), (6.25) and (6.26) we obtain the following low and upper bounds for δ

$$-1.0 \le \delta \le 2.0$$

and for b_2

$$3.0 \le b_2 \le 6.0 \; .$$

$\mu = 3$: $b_3 \to b_3' = b_3 + \delta$. Restriction (6.12) for $\mu = 3$ gives us the following bounds

$$\max_{\substack{i \in \{2,1,6\} \\ e_{i3} > 0}} \{-\frac{x_6^*}{e_{33}}\} \le \delta.$$

or

$$-4 = \max\{-\frac{4}{1}\} \leq \delta. \tag{6.27}$$

From restriction (6.16) we obtain

$$h_2 = d_2 e_{13} + d_1 e_{23} + d_6 e_{33} = 3 \times 0 + 2 \times 0 + 0 \times 1 = 0,$$

hence, since $h_2 = 0$,

$$-\infty \leq \delta \leq +\infty\,. \tag{6.28}$$

For condition (6.18) we have

$$h_1 = p_2 e_{13} + p_1 e_{23} + p_6 e_{33} = 9 \times 0 + 8 \times 0 + 0 \times 1 = 0,$$

$$\begin{array}{lclcl} g_3 & = & \Delta'_3 h_2 - \Delta''_3 h_1 & = & 12 \times 0 - 2 \times 0 & = & 0\,, \\ g_4 & = & \Delta'_4 h_2 - \Delta''_4 h_1 & = & 10 \times 0 - 4 \times 0 & = & 0\,, \\ g_5 & = & \Delta'_5 h_2 - \Delta''_5 h_1 & = & (-1) \times 0 - (-1) \times 0 & = & 0\,. \end{array}$$

Hence, from

$$\max_{\substack{j \in \{3,4,5\} \\ g_j > 0}} \frac{-\Delta_j(x^*) D(x^*)}{g_j} \leq \delta \leq \min_{\substack{j \in \{3,4,5\} \\ g_j < 0}} \frac{-\Delta_j(x^*) D(x^*)}{g_j}$$

we obtain

$$-\infty \leq \delta \leq +\infty\,. \tag{6.29}$$

Finally, combining (6.27), (6.28) and (6.29) we obtain the following restrictions for δ

$$-4.0 \leq \delta \leq +\infty\,.$$

The latter means that

$$11.0 \leq b_3 \leq \infty$$

Closing this section, we note that when combining restrictions (6.12) and (6.18) with condition (6.16) the latter may lead to resulting restrictions with strict inequalities '<' and '>'.

3. Change in Numerator Vector p

In this section, we consider the case when μth element p_μ of vector $p = (p_1, p_2, \dots, p_n)$ in the numerator of objective function $Q(x)$ is replaced with $p'_\mu = (p_\mu + \delta)$. Similarly to the previous section, we may assume that the optimal solution of the original LFP problem (6.1)-(6.3) is

$$x^* = (x_1^*, x_2^*, \dots, x_m^*, 0, 0, \dots, 0)^T.$$

Our goal now is to determine for δ the lower and upper bounds which guarantee that replacement $p_\mu \to p'_\mu$ does not affect the optimal basis, and the original optimal solution x^* remains feasible and optimal.

When considering this replacement $p_\mu \to p'_\mu$, we have to distinguish the following two cases:

1 $\mu \in J_N = \{m+1, m+2, \dots, n\}$, i.e. μ is a non-basic index;

2 $\mu \in J_B = \{1, 2, \dots, m\}$, i.e. μ is a basic index.

It is obvious that in both cases replacement $p_\mu \to p'_\mu$ does not affect feasibility of vector x^* since this replacement does not change feasible set S. However, it may affect the optimal value of $Q(x)$ and hence, can change the values of reduced costs $\Delta_j(x^*) = \Delta'_j - Q(x^*)\Delta''_j$. This is why when replacing $p_\mu \to p'_\mu$ the optimality of vector x^* becomes questionable. The latter means that to answer the question on the optimality of feasible vector x^* we have to investigate how replacement $p_\mu \to p'_\mu$ affects reduced costs $\Delta_j(x^*),\ j = 1, 2, \dots, n$.

Case 1 ($\mu \in J_N$)**:** First, we observe that since μ is a non-basic index it means that $x^*_\mu = 0$ and hence, the optimal value of objective function $Q(x)$ remains unchanged. Further, non-basic p_μ does not figure in the basic reduced costs $\Delta'_j,\ j = 1, 2, \dots, m$, at all, and is present in only one non-basic reduced cost Δ'_μ:

$$\Delta'_\mu = \sum_{i=1}^{m} p_i x_{i\mu} - p_\mu.$$

So, when replacing $p_\mu \to p'_\mu$ we have

$$\begin{aligned}\tilde{\Delta}'_\mu &= \sum_{i=1}^{m} p_i x_{i\mu} - p'_\mu = \sum_{i=1}^{m} p_i x_{i\mu} - (p_\mu + \delta) = \\ &= \sum_{i=1}^{m} p_i x_{i\mu} - p_\mu - \delta = \Delta'_\mu - \delta.\end{aligned}$$

Hence

$$\tilde{\Delta}_\mu(x^*) = (\Delta'_\mu - \delta) - Q(x^*)\Delta''_\mu = \Delta_\mu(x^*) - \delta.$$

The latter means that if the following condition holds,

$$\delta \le \Delta_\mu(x^*) \tag{6.30}$$

then optimal solution x^* of the original LFP problem remains optimal for the modified LFP problem (with replaced coefficient $p_\mu \to p'_\mu$) too.

Case 2 ($\mu \in J_B$)**:** Since μ is a basic index, it means that replacement $p_\mu \to p'_\mu$ affects the optimal value of $P(x)$ as well as $Q(x)$

$$\begin{aligned}\tilde{P}(x^*) &= \sum_{\substack{j \in J_B \\ j \neq \mu}} p_j x_j^* + p'_\mu x_\mu^* + p_0 = \sum_{\substack{j \in J_B \\ j \neq \mu}} p_j x_j^* + (p_\mu + \delta) x_\mu^* + p_0 = \\ &= \sum_{i=1}^{m} p_j x_j^* + p_0 + \delta x_\mu^* = P(x^*) + \delta x_\mu^*,\end{aligned}$$

and respectively,

$$\tilde{Q}(x^*) = \frac{\tilde{P}(x^*)}{D(x^*)} = \frac{P(x^*) + \delta x_\mu^*}{D(x^*)}.$$

In addition, replacement $p_\mu \to p'_\mu$ has an influence on the non-basic reduced costs $\Delta'_j,\ j \in J_N$:

$$\begin{aligned}\tilde{\Delta}'_j &= \sum_{\substack{i \in J_B \\ i \neq \mu}} p_i x_{ij} + p'_\mu x_{\mu j} - p_j = \sum_{\substack{i \in J_B \\ i \neq \mu}} p_i x_{ij} + (p_\mu + \delta) x_{\mu j} - p_j = \\ &= \sum_{i=1}^{m} p_i x_{ij} - p_j + \delta x_{\mu j} = \Delta'_j + \delta x_{\mu j},\ \ j \in J_N.\end{aligned}$$

After all these preparations, we can determine the new values of reduced costs $\Delta_j(x^*),\ j \in J_N$:

$$\begin{aligned}\tilde{\Delta}_j(x^*) &= \tilde{\Delta}'_j - \tilde{Q}(x^*)\Delta''_j = \\ &= \Delta'_j + \delta x_{\mu j} - \frac{P(x^*) + \delta x_\mu^*}{D(x^*)} \Delta''_j,\ \ j \in J_N. \qquad (6.31)\end{aligned}$$

Concerning basic indices $j \in J_B$, in this situation basic reduced costs Δ'_j do not change and they preserve their values of zero, i.e.

$$\tilde{\Delta}'_j = \Delta'_j = 0,\ \forall j \in J_B,$$

and, hence,

$$\tilde{\Delta}_j(x^*) = \Delta_j(x^*) = 0,\ \forall j \in J_B,$$

In accordance with the theory of the simplex method and its criteria of optimality, if condition

$$\tilde{\Delta}_j(x^*) \geq 0,\ \ \forall j \in J \qquad (6.32)$$

holds, then feasible vector x^* is an optimal solution. Since

$$\Delta'_j = \Delta''_j = \Delta_j(x^*) = 0,\ \ \forall j \in J_B,$$

we have to consider only those conditions of (6.32) that have non-basic indices j:

$$\tilde{\Delta}_j(x^*) \geq 0, \quad \forall j \in J_N.$$

So, from (6.31) we obtain the following system

$$\Delta'_j + \delta x_{\mu j} - \frac{P(x^*) + \delta x^*_\mu}{D(x^*)} \Delta''_j \geq 0, \quad \forall j \in J_N,$$

which gives us restrictions as follows

$$\delta(x_{\mu j} D(x^*) - \Delta''_j x^*_\mu) \geq -(D(x^*)\Delta'_j - P(x^*)\Delta''_j), \quad \forall j \in J_N,$$

or

$$\delta(x_{\mu j} D(x^*) - \Delta''_j x^*_\mu) \geq -\Delta_j(x^*) D(x^*), \quad \forall j \in J_N.$$

From the latter we obtain the following lower and upper bounds for δ:

$$\max_{\substack{j \in J_N \\ g_j > 0}} \{\frac{-\Delta_j(x^*)D(x^*)}{g_j}\} \leq \delta \leq \min_{\substack{j \in J_N \\ g_j < 0}} \{\frac{-\Delta_j(x^*)D(x^*)}{g_j}\}, \tag{6.33}$$

where

$$g_j = x_{\mu j} D(x^*) - \Delta''_j x^*_\mu, \quad \forall j \in J_N. \tag{6.34}$$

Thus, we have shown that if δ is within bounds (6.33) then optimal solution x^* of the original LFP problem also solves the modified LFP problem (with replaced coefficient $p_\mu \to p'_\mu$).

REMARK 6.2 *Expressions (6.30) and (6.33) contain the corresponding bounds for the linear programming problem*

$$P(x) = \sum_{j=1}^{n} p_j x_j + p_0 \to \max$$
$$x \in S$$

as a special case. Indeed, since an LP problem is a special case of an LFP problem when $d_j = 0, j = 1, 2, \ldots, n$, and $d_0 = 1$, it means that in an LP problem

$$\Delta''_j = 0, \quad j = 1, 2, \ldots, n, \qquad D(x^*) \equiv 1, \quad \forall x^* \in S. \tag{6.35}$$

Thus, from (6.30) for non-basic index μ we obtain

$$\delta \leq \Delta_\mu(x^*) = \Delta'_\mu - Q(x^*)\Delta''_\mu = \Delta'_\mu \, .$$

Further, if $\mu \in J_B$ *then keeping in mind (6.35) from (6.34) we obtain that*

$$g_j = x_{\mu j} D(x^*) - \Delta''_j x^*_\mu = x_{\mu j}, \quad \forall j \in J_N.$$

So, expression (6.33) gives the following range of stability for the LP problem

$$\max_{\substack{j \in J_N \\ x_{\mu j} > 0}} \{ \frac{-\Delta'_j}{x_{\mu j}} \} \le \delta \le \min_{\substack{j \in J_N \\ x_{\mu j} < 0}} \{ \frac{-\Delta'_j}{x_{\mu j}} \},$$

To illustrate how to use formulas (6.30) and (6.33), we reconsider the numeric example (6.19)-(6.20) from the previous section (see page 184). For an optimal basis $B = (A_2, A_1, A_6)$ and an optimal solution

$$x_B = \begin{pmatrix} x_2^* \\ x_1^* \\ x_6^* \end{pmatrix} = \begin{pmatrix} 2 \\ 1 \\ 4 \end{pmatrix},$$

we consider the following two cases:

Basic index $\mu \in J_B$**:** Let $\mu = 2$. First, we calculate $g_j,\ j \in J_N = \{3, 4, 5\}$:

$$\begin{array}{lllll} g_3 & = x_{23} D(x^*) - \Delta''_3 x^*_2 & = 0 \times 15 - 2 \times 2 & = & -4\,, \\ g_4 & = x_{24} D(x^*) - \Delta''_4 x^*_2 & = 2 \times 15 - 4 \times 2 & = & 22\,, \\ g_5 & = x_{25} D(x^*) - \Delta''_5 x^*_2 & = (-1) \times 15 - (-1) \times 2 & = & -13\,, \end{array}$$

and then from (6.33) we obtain

$$\max_{j \in \{4\}} \{ \frac{-\Delta_j(x^*) D(x^*)}{g_j} \} \le \delta \le \min_{j \in \{3,5\}} \{ \frac{-\Delta_j(x^*) D(x^*)}{g_j} \},$$

or

$$\max\{ \frac{-2 \times 15}{22} \} \le \delta \le \min\{ \frac{-8 \times 15}{-4}, \frac{-1 \times 15}{-13} \}.$$

Finally, we obtain the following lower and upper bounds for δ

$$-1\frac{4}{11} \le \delta \le 1\frac{2}{13}.$$

The lattet means that p_2 may vary without affecting the optimal basis within the following range of stability

$$7\frac{7}{11} \le p_2 \le 10\frac{2}{13}.$$

Non-basic index $\mu \in J_N$**:** Let $\mu = 3$. In this case, in accordance with formula (6.30) using the optimal simplex-tableau shown in Table 5.3 (see page 157) we obtain the following upper bound for δ

$$\delta \le \Delta_3(x^*) = 8,$$

which gives us the range of stability for $p_3 = 4$ as follows

$$-\infty \le p_3 \le 12.$$

4. Change in Numerator Constant p_0

In this section, we discuss how changes in coefficient p_0 of numerator $P(x)$ in objective function $Q(x)$ affect the optimal solution

$$x^* = (x_1^*, x_2^*, \dots, x_m^*, 0, 0, \dots, 0)^T$$

of the LFP problem (6.1)-(6.3).

Let us replace coefficient p_0 with $p_0' \doteq (p_0 + \delta)$. Our aim is to determine for δ the lower and upper bounds which guarantee that replacement $p_0 \to p_0'$ does not affect the optimal basis $B = (A_1, A_2, \dots, A_m)$, and the original optimal solution x^* remains feasible and optimal.

Observe that when replacing $p_0 \to p_0'$ feasible set S does not change. Hence, this replacement does not affect the feasibility of vector x^*. At the same time, since replacement $p_0 \to p_0'$ changes the value of numerator $P(x^*)$ and therefore, objective function $Q(x)$, it may result in the change of sign in reduced costs $\Delta_j(x^*)$, $j = 1, 2, \dots, n$, and in this way, violate the optimality of vector x^*.

We begin by observing that

$$\tilde{P}(x^*) = \sum_{j=1}^{n} p_j x_j^* + p_0' = \sum_{j=1}^{n} p_j x_j^* + p_0 + \delta = P(x^*) + \delta,$$

$$\tilde{Q}(x^*) = \frac{\tilde{P}(x^*)}{D(x^*)} = \frac{P(x^*) + \delta}{D(x^*)}$$

and, hence

$$\tilde{\Delta}_j(x^*) = \Delta_j' - \tilde{Q}(x^*)\,\Delta_j'' = \Delta_j' - \frac{P(x^*) + \delta}{D(x^*)}\,\Delta_j'', \quad j = 1, 2, \dots, n. \tag{6.36}$$

In accordance with the optimality criteria of the simplex method, if

$$\tilde{\Delta}_j(x^*) \geq 0, \quad \forall j \in J = \{1, 2, \dots, n\},$$

then basic feasible solution x^* is an optimal solution for the modified LFP problem (with replaced coefficient $p_0 \to p_0'$). Further, since coefficient p_0 does not figure either in the numerator's reduced costs Δ_j' or in the denominator's reduced costs Δ_j'', $j = 1, 2, \dots, n$, it means that basic reduced costs Δ_j' and Δ_j'', $j \in J_B$, preserve their zero values, i.e.

$$\Delta_j' = \Delta_j'' = 0, \quad \forall j \in J_B,$$

and, hence

$$\tilde{\Delta}_j(x^*) \;=\; \Delta'_j - \tilde{Q}(x^*)\,\Delta''_j \;=\; 0, \;\; \forall j \in J_B.$$

So, for the optimality of vector x^* we have to require non-negativity only for non-basic reduced costs $\tilde{\Delta}_j(x^*)$, i.e.

$$\tilde{\Delta}_j(x^*) \geq 0, \;\; \forall j \in J_N.$$

Hence, using (6.36) we have the following condition

$$\Delta'_j - \frac{P(x^*)+\delta}{D(x^*)}\,\Delta''_j \geq 0, \;\; \forall j \in J_N.$$

The latter after necessary transformations gives the following system

$$\delta\,\Delta''_j \leq \Delta'_j\,D(x^*) - P(x^*)\Delta'_j, \;\; \forall j \in J_N$$

Finally, from the latter we obtain the following bounds forδ

$$\max_{\substack{j\in J_N\\ \Delta''_j<0}}\{\frac{\Delta_j(x^*)D(x^*)}{\Delta''_j}\} \leq \delta \leq \min_{\substack{j\in J_N\\ \Delta''_j>0}}\{\frac{\Delta_j(x^*)D(x^*)}{\Delta''_j}\}, \qquad (6.37)$$

Thus, if δ is within bounds (6.37), basic optimal solution x^* of the original LFP problem (6.1)-(6.3) solves the modified LFP problem (with a replaced coefficient $p_0 \to p'_0$) too.

REMARK 6.3 *Observe that range (6.37) may be considered as a generalization of the corresponding range for the linear programming problem*

$$P(x) = \sum_{j=1}^{n} p_j x_j + p_0 \to \max$$
$$x \in S.$$

Indeed, in an LP problem we have $d_j = 0, j = 1, 2, \ldots, n,$ *and* $d_0 = 1$. *So, keeping in mind that* $\Delta''_j = 0$ *for all* $j = 1, 2, \ldots, n,$ *and* $D(x^*) \equiv 1$ *for any* $x^* \in S$, *from (6.37) we obtain*

$$-\infty \leq \delta \leq +\infty\,.$$

The latter means that replacement $p_0 \to p'_0$ *does not affect the optimal solution of LP problem at all.*

To illustrate the usage of expression (6.37), we reconsider numeric example (6.19)-(6.20) from the previous section (see page 184). Having optimal basis $B = (A_2, A_1, A_6)$ and optimal solution

$$x_B = \begin{pmatrix} x_2^* \\ x_1^* \\ x_6^* \end{pmatrix} = \begin{pmatrix} 2 \\ 1 \\ 4 \end{pmatrix},$$

we replace coefficient $p_0 = 4$ in the numerator of objective function $Q(x)$ with $p'_0 = (p_0 + \delta)$ and determine the lower and upper bounds (6.37). Keeping in mind that $J_N = \{3, 4, 5\}$ and

$$\begin{array}{lll} \Delta_3(x^*) = 8, & \Delta_4(x^*) = 2, & \Delta_5(x^*) = 1, \\ \Delta''_3 = 2, & \Delta''_4 = 4, & \Delta''_5 = -1, \\ & D(x^*) = 15, & \end{array}$$

we obtain the following range for δ

$$\max_{j\in\{5\}}\{\frac{\Delta_5(x^*)D(x^*)}{\Delta''_5}\} \leq \delta \leq \min_{j\in\{3,4\}}\{\frac{\Delta_3(x^*)D(x^*)}{\Delta''_3}, \frac{\Delta_4(x^*)D(x^*)}{\Delta''_4}\}$$

or

$$-15 = \max\{\frac{1 \times 15}{-1}\} \leq \delta \leq \min\{\frac{8 \times 15}{2}, \frac{2 \times 15}{4}\} = \min\{60, 7.5\} = 7.5\ .$$

Finally, for p_0 we have the following range of stability

$$-11 = 4 - 15 \leq p_0 \leq 4 + 7.5 = 11.5\ .$$

5. Change in Denominator Vector d

In this section, we consider the case when μth element d_μ of vector $d = (d_1, d_2, \ldots, d_n)$ in the denominator of objective function $Q(x)$ is replaced with $d'_\mu = (d_\mu + \delta)$. Similar to previous sections, we assume that vector

$$x^* = (x_1^*, x_2^*, \ldots, x_m^*, 0, 0, \ldots, 0)^T$$

is a basic optimal solution of the original LFP problem (6.1)-(6.3) with basis $B = (A_1, A_2, \ldots, A_m)$. Our goal now is to determine for δ the lower and upper bounds which guarantee that replacement $d_\mu \rightarrow d'_\mu$ does not affect the optimal basis, and the original optimal solution x^* remains feasible and optimal.

As in Section 3, when considering replacement $d_\mu \rightarrow d'_\mu$, we have to distinguish the following two cases:

- $\mu \in J_N = \{m + 1, m + 2, \ldots, n\}$, i.e. μ is a non-basic index;
- $\mu \in J_B = \{1, 2, \ldots, m\}$, i.e. μ is a basic index.

First of all, we observe that in both cases replacement $d_\mu \rightarrow d'_\mu$ does not affect feasibility of vector x^* since this replacement does not change feasible set S of original LFP problem (6.1)-(6.3). Thus, vector x^* is a basic feasible solution for the modified LFP problem (with replaced entry $d_\mu \rightarrow d'_\mu$). At the same time, replacement $d_\mu \to d'_\mu$ may affect the optimal values of functions $D(x)$ and $Q(x) = P(x)/D(x)$ (depending on index μ) and as a result, can change the values of reduced costs $\Delta_j(x^*) = \Delta'_j - Q(x^*)\Delta''_j$. This is why when replacing $d_\mu \to d'_\mu$ the optimality of vector x^* can be violated. The latter means that to answer the question on optimality of feasible vector x^* for the new LFP problem (with replaced entry $d_\mu \to d'_\mu$), we have to investigate how this change in denominator $D(x)$ affects reduced costs $\Delta_j(x^*),\ j = 1, 2, \dots, n.$

Case 1 ($\mu \in J_N$): Since μ is a non-basic index it means that $x^*_\mu = 0$ and hence, the optimal value of objective function $Q(x)$ does not change. Observe that in this case denominator $D(x)$ also preserves its positive optimal value $D(x^*)$.

Further, since non-basic d_μ does not figure in the basic reduced costs $\Delta''_j,\ j = 1, 2, \dots, m,$ at all, it means that all $\Delta''_j,\ j \in J_B,$ remain unchanged. However, coefficient d_μ figures in non-basic reduced cost

$$\Delta''_\mu = \sum_{i=1}^{m} d_i x_{i\mu} - d_\mu$$

and affects its value, and hence, affects the value of reduced cost $\Delta_\mu(x^*)$. Thus, when replacing $d_\mu \to d'_\mu$ we have

$$\begin{aligned}\tilde{\Delta}''_\mu &= \sum_{i=1}^{m} d_i x_{i\mu} - d'_\mu = \sum_{i=1}^{m} d_i x_{i\mu} - (d_\mu + \delta) = \\ &= \sum_{i=1}^{m} d_i x_{i\mu} - d_\mu - \delta = \Delta''_\mu - \delta\end{aligned}$$

and

$$\begin{aligned}\tilde{\Delta}_\mu(x^*) &= \Delta'_\mu - Q(x^*)\tilde{\Delta}''_\mu = \\ &= \Delta'_\mu - Q(x^*)(\Delta''_\mu - \delta) = \Delta_\mu(x^*) + Q(x^*)\,\delta.\end{aligned}$$

The latter means that if δ satisfies condition

$$\Delta_\mu(x^*) + Q(x^*)\,\delta \geq 0 \tag{6.38}$$

then optimal solution x^* of the original LFP problem also solves the modified LFP problem (with replaced coefficient $d_\mu \to d'_\mu,\ \mu \in J_N$). Finally, from (6.38) we formulate the following restrictions for δ

$$\delta \begin{cases} \geq \dfrac{-\Delta_\mu(x^*)}{Q(x^*)}, & \text{if } Q(x^*) > 0, \\ \leq \dfrac{-\Delta_\mu(x^*)}{Q(x^*)}, & \text{if } Q(x^*) < 0, \\ \text{unlimited}, & \text{if } Q(x^*) = 0. \end{cases} \tag{6.39}$$

Case 2 ($\mu \in J_B$): In this case, since μ is a basic index it means that replacement $d_\mu \to d'_\mu$ changes the value of $D(x^*)$:

$$\begin{aligned}\tilde{D}(x^*) &= \sum_{\substack{j \in J_B \\ j \neq \mu}} d_j x_j^* + d'_\mu x_\mu^* + d_0 = \sum_{\substack{j \in J_B \\ j \neq \mu}} d_j x_j^* + (d_\mu + \delta) x_\mu^* + d_0 = \\ &= D(x^*) + \delta x_\mu^*,\end{aligned}$$

and in this way, affects the value of $Q(x^*)$:

$$\tilde{Q}(x^*) = \frac{P(x^*)}{\tilde{D}(x^*)} = \frac{P(x^*)}{D(x^*) + \delta x_\mu^*}.$$

To preserve the strict positivity of denominator $D(x)$, we have to require that

$$\tilde{D}(x^*) = D(x^*) + \delta x_\mu^* > 0.$$

The latter gives the following restriction for δ

$$\delta > \frac{-D(x^*)}{x_\mu^*} \tag{6.40}$$

for the case if vector x^* is non-degenerate and hence, $x_\mu^* > 0$. Obviously, if vector x^* is a degenerate one and $x_\mu^* = 0$ then replacement $d_\mu \to d'_\mu$, $\mu \in J_B$, cannot affect the positivity of denominator $D(x)$. Thus, in this case, δ in unlimited, since for any δ denominator $D(x)$ preserves its strictly positive value.

Further, replacement $d_\mu \to d'_\mu$, $\mu \in J_B$, affects the non-basic reduced costs Δ''_j, $j \in J_N$, as follows

$$\begin{aligned}\tilde{\Delta}''_j &= \sum_{\substack{i \in J_B \\ i \neq \mu}} d_i x_{ij} + d'_\mu x_{\mu j} - d_j = \sum_{\substack{i \in J_B \\ i \neq \mu}} d_i x_{ij} + (d_\mu + \delta) x_{\mu j} - d_j = \\ &= \sum_{i=1}^{m} d_i x_{ij} - d_j + \delta x_{\mu j} = \Delta''_j + \delta x_{\mu j}, \quad j \in J_N.\end{aligned}$$

Observe that this replacement does not change either basic reduced costs Δ''_j or $\Delta_j(x^*)$, $\forall j \in J_B$, so they preserve their zero values, i.e.

$$\tilde{\Delta}''_j = \Delta''_j = 0 \text{ and } \tilde{\Delta}_j(x^*) = \Delta_j(x^*) = 0, \quad \forall j \in J_B. \tag{6.41}$$

After all these preliminaries, we can determine the new values of reduced costs $\Delta_j(x^*)$, $j \in J_N$:

$$\begin{aligned}\tilde{\Delta}_j(x^*) &= \Delta'_j - \tilde{Q}(x^*) \tilde{\Delta}''_j = \\ &= \Delta'_j - \frac{P(x^*)}{D(x^*) + \delta x_\mu^*} (\Delta''_j + \delta x_{\mu j}), \quad j \in J_N.\end{aligned} \tag{6.42}$$

In accordance with the theory of the simplex method and its criteria of optimality, basic feasible solution x^* of the modified LFP problem is optimal if $\tilde{\Delta}_j(x^*) \geq 0, \;\; \forall j \in J = \{1, 2, \dots, n\}$. Keeping in mind (6.41) we require the non-negativity only for non-basic reduced costs $\tilde{\Delta}_j(x^*)$. So, using (6.42) we obtain the following restriction

$$\Delta'_j - \frac{P(x^*)}{D(x^*) + \delta x^*_\mu} (\Delta''_j + \delta x_{\mu j}) \geq 0, \;\; \forall j \in J_N,$$

which gives us condition

$$\Delta'_j \geq \frac{P(x^*)}{D(x^*) + \delta x^*_\mu} (\Delta''_j + \delta x_{\mu j}), \;\; \forall j \in J_N. \tag{6.43}$$

Further, taking into account restriction (6.40) we can re-write condition (6.43) in the form as follows

$$\delta\, (\Delta'_j x^*_\mu - P(x^*)\, x_{\mu j}) \geq -(D(x^*)\Delta'_j - P(x^*)\Delta''_j), \quad \forall j \in J_N,$$

or

$$\delta\, (\Delta'_j x^*_\mu - P(x^*)\, x_{\mu j}) \geq -\Delta_j(x^*) D(x^*), \quad \forall j \in J_N.$$

The latter gives the following lower and upper bounds for δ

$$\max_{\substack{j \in J_N \\ g_j > 0}} \{\frac{-\Delta_j(x^*)D(x^*)}{g_j}\} \leq \delta \leq \min_{\substack{j \in J_N \\ g_j < 0}} \{\frac{-\Delta_j(x^*)D(x^*)}{g_j}\}, \tag{6.44}$$

where

$$g_j = \Delta'_j x^*_\mu - P(x^*)\, x_{\mu j}, \quad \forall j \in J_N.$$

Summarizing, we give the lower and upper bounds for δ in the non-degenerate case as follows: if δ satisfies restrictions (6.40) and (6.44) then basic optimal solution x^* of the original LFP problem also solves the modified LFP problem (with replaced entry $d_\mu \rightarrow d'_\mu, \; \mu \in J_B$). If vector x^* is a degenerate one and $x^*_\mu = 0$ then restriction (6.40) must be omitted.

REMARK 6.4 *In the case of an LP problem, expressions (6.40) and (6.44) have not any sense and result in no restrictions for δ since in an LP problem all coefficients d_j in denominator $D(x)$ are strictly fixed as follows*

$$d_j = 0, \;\; j = 1, 2, \dots, n, \quad \textit{and} \quad d_0 = 1$$

and cannot be changed.

As in the previous section, to illustrate the usage of expressions (6.40) and (6.44) we reconsider numeric example (6.19)-(6.20) (see page 184). Having optimal basis $B = (A_2, A_1, A_6)$ and optimal solution

$$x_B = \begin{pmatrix} x_2^* \\ x_1^* \\ x_6^* \end{pmatrix} = \begin{pmatrix} 2 \\ 1 \\ 4 \end{pmatrix},$$

we consider the following two cases:

Basic index $\mu \in J_B$**:** Let $\mu = 2$. First, since vector x^* is non-degenerate, from (6.40) we obtain

$$\delta > -\frac{D(x^*)}{x_2^*} = -\frac{15}{2} = -7.5\ . \qquad (6.45)$$

Further, we calculate $g_j,\ j \in J_N = \{3, 4, 5\}$:

$$\begin{array}{lclcl} g_3 & = & \Delta'_3 x_2^* - P(x^*)x_{23} & = & 12 \times 2 - 30 \times 0 & = & 24\ , \\ g_4 & = & \Delta'_4 x_2^* - P(x^*)x_{24} & = & 10 \times 2 - 30 \times 2 & = & -40\ , \\ g_5 & = & \Delta'_5 x_2^* - P(x^*)x_{25} & = & (-1) \times 2 - 30 \times (-1) & = & 28\ , \end{array}$$

and then from (6.44) we obtain

$$\max_{j \in \{3,5\}} \{\frac{-\Delta_j(x^*)D(x^*)}{g_j}\} \le \delta \le \min_{j \in \{4\}} \{\frac{-\Delta_j(x^*)D(x^*)}{g_j}\},$$

or

$$\max\{\frac{-8 \times 15}{24}, \frac{-1 \times 15}{28}\} \le \delta \le \min\{\frac{-2 \times 15}{-40}\}\ .$$

Finally, combining (6.45) with the latter we obtain the following lower and upper bounds for δ

$$-\frac{15}{28} \le \delta \le \frac{3}{4}\ .$$

The latter means that d_2 may vary without affecting the optimal basis within the following range of stability

$$2\frac{13}{28} \le d_2 \le 3\frac{3}{4}\ .$$

Non-basic index $\mu \in J_N$**:** Let $\mu = 3$. In this case, in accordance with expression (6.39) we obtain the following lower bound for δ

$$\delta \ge -\frac{\Delta_3(x^*)}{Q(x^*)} = -\frac{8}{2} = -4,$$

which gives us the range of stability for $d_3 = 2$ as follows

$$-2 \le d_3 \le \infty\ .$$

6. Change in Denominator Constant d_0

In this section, our aim is to investigate the effect of changing coefficient d_0 in denominator $D(x)$ of objective function $Q(x)$ to the optimal basis of LFP problem (6.1)-(6.3) and its optimal solution. Similar to the previous sections, we assume that vector

$$x^* = (x_1^*, x_2^*, \dots, x_m^*, 0, 0, \dots, 0)^T$$

with basis $B = (A_1, A_2, \dots, A_m)$ is a basic optimal solution of the original LFP problem.

First, we observe that replacement $d_0 \rightarrow d_0' = (d_0 + \delta)$ does not affect feasible set S of LFP problem (6.1)-(6.3), so basic optimal solution x^* of the original LFP problem is a basic feasible solution for the modified LFP problem (with replaced coefficient $d_0 \rightarrow d_0'$). However, replacement $d_0 \rightarrow d_0'$ changes the optimal value of function $D(x)$ and hence, the optimal value of objective function $Q(x) = P(x)/D(x)$. Thus, any change in coefficient d_0 may result in change in the values of reduced costs $\Delta_j(x^*) = \Delta_j' - Q(x^*)\Delta_j''$ and in this way can affect the optimality of vector x^*.

So we have

$$\tilde{D}(x^*) = \sum_{j=1}^{n} d_j x_j^* + d_0' = \sum_{j=1}^{n} d_j x_j^* + d_0 + \delta = D(x^*) + \delta,$$

$$\tilde{Q}(x^*) = \frac{P(x^*)}{\tilde{D}(x^*)} = \frac{P(x^*)}{D(x^*) + \delta}$$

and, hence

$$\tilde{\Delta}_j(x^*) = \Delta_j' - \tilde{Q}(x^*)\,\Delta_j'' = \Delta_j' - \frac{P(x^*)}{D(x^*) + \delta}\,\Delta_j'', \quad j = 1, 2, \dots, n. \qquad (6.46)$$

To preserve the strict positivity of denominator $D(x)$, we have to require that $D(x^*) + \delta > 0$. The latter gives the following restriction for δ

$$\delta > -D(x^*). \qquad (6.47)$$

Further, in accordance with the theory of the simplex method and its criteria of optimality, if

$$\tilde{\Delta}_j(x^*) \geq 0, \quad \forall j \in J = \{1, 2, \dots, n\},$$

then basic feasible solution x^* of the modified LFP problem (with replaced coefficient $d_0 \rightarrow d_0'$) is also its optimal solution.

Further, since coefficient d_0 does not figure either in the numerator's reduced costs Δ'_j or in the denominator's reduced costs Δ''_j, $j = 1, 2, \ldots, n$, it means that basic reduced costs Δ'_j and Δ''_j, $j \in J_B$, preserve their zero values, i.e.

$$\Delta'_j = \Delta''_j = 0, \quad \forall j \in J_B,$$

and, hence

$$\tilde{\Delta}_j(x^*) = \Delta'_j - \tilde{Q}(x^*)\,\Delta''_j = 0, \ \forall j \in J_B.$$

So, for the optimality of vector x^* we have to require non-negativity only for non-basic reduced costs $\tilde{\Delta}_j(x^*)$, i.e.

$$\tilde{\Delta}_j(x^*) \geq 0, \ \forall j \in J_N.$$

Thus, using (6.46) we obtain the following condition

$$\Delta'_j - \frac{P(x^*)}{D(x^*) + \delta}\,\Delta''_j \geq 0, \ \forall j \in J_N. \tag{6.48}$$

Taking into account restriction (6.47) we can re-write (6.48) in the following form

$$\delta \Delta'_j \geq -D(x^*)(\Delta'_j - Q(x^*)\Delta''_j), \ \forall j \in J_N,$$

or

$$\delta \Delta'_j \geq -D(x^*)\Delta_j(x^*), \quad \forall j \in J_N.$$

Finally, we obtain the following upper and lower bounds for

$$\max_{\substack{j \in J_N \\ \Delta'_j > 0}} \{ \frac{-\Delta_j(x^*)D(x^*)}{\Delta'_j} \} \leq \delta \leq \min_{\substack{j \in J_N \\ \Delta'_j < 0}} \{ \frac{-\Delta_j(x^*)D(x^*)}{\Delta'_j} \}, \tag{6.49}$$

Summarizing the results obtained, we can formulate restrictions for δ as follows: if δ is within bounds given by restrictions (6.47) and (6.49) the basic optimal solution x^* of the original LFP problem also solves the modified LFP problem (with replaced coefficient $d_0 \to d'_0$)

REMARK 6.5 *In the case of an LP problem, restrictions (6.47) and (6.49) cannot be applied since in an LP problem all coefficients d_j in denominator $D(x)$ are strictly fixed as follows*

$$d_j = 0, \quad j = 1, 2, \ldots, n, \quad \textit{and} \quad d_0 = 1$$

and cannot be changed.

As in the previous section, to illustrate the usage of restrictions (6.47) and (6.49) we reconsider numeric example (6.19)-(6.20) (see page 184). Having optimal basis $B = (A_2, A_1, A_6)$ and optimal solution

$$x_B = \begin{pmatrix} x_2^* \\ x_1^* \\ x_6^* \end{pmatrix} = \begin{pmatrix} 2 \\ 1 \\ 4 \end{pmatrix},$$

we replace coefficient d_0 with $d_0' = d_0 + \delta$ and determine bounds (6.47) and (6.49) for δ. First, using the optimal simplex-tableau shown in Table 5.3 (see page 157) we obtain that

$$D(x^*) = 15, \quad J_B = \{1, 2, 6\}, \quad J_N = \{3, 4, 5\}$$

and

$$\Delta_3' = 12, \ \ \Delta_4' = 10, \ \ \Delta_5' = -1,$$
$$\Delta_3(x^*) = 8, \ \ \Delta_4(x^*) = 2, \ \ \Delta_5(x^*) = 1.$$

Thus, restriction (6.47) gives

$$\delta > -15, \tag{6.50}$$

while from (6.49) we have the following lower and upper bounds

$$\max_{j \in \{3,4\}} \{\frac{-\Delta_j(x^*)D(x^*)}{\Delta_j'}\} \leq \delta \leq \min_{j \in \{5\}} \{\frac{-\Delta_j(x^*)D(x^*)}{\Delta_j'}\}$$

or

$$\max_{j \in \{3,4\}} \{\frac{-8 \times 15}{12}, \frac{-2 \times 15}{10}\} \leq \delta \leq \min_{j \in \{5\}} \{\frac{-1 \times 15}{-1}\}.$$

Finally, combining the latter with (6.50) we obtain the following range of stability for δ

$$-3 \leq \delta \leq 15$$

and for $d_0 = 7$

$$4 \leq d_0 \leq 22 \ .$$

7. Discussion Questions and Exercises

6.1 For numerical example (6.19)-(6.20) from Section 3, page 184, determine the ranges of stability for

- basic coefficients $p_1, \ p_6$ in numerator $P(x)$;
- non-basic coefficients $p_4, \ p_5$ in numerator $P(x)$;
- basic coefficients $d_1, \ d_6$ in denominator $D(x)$;
- non-basic coefficients $d_4, \ d_5$ in denominator $D(x)$.

6.2 Solve the following LFP problem with 3 variables and 2 main constraints

$$Q(x) = \frac{2.5\,x_1 + 1\,x_2 + 2\,x_3 + 10}{5\,x_1 + 2\,x_2 + 4\,x_3 + 50} \longrightarrow \max$$

subject to

$$\begin{array}{rcrcrcl} 1\,x_1 & + & 2\,x_2 & + & 2\,x_3 & \leq & 20\,, \\ 3\,x_1 & + & 3\,x_2 & + & 2\,x_3 & \leq & 30\,, \end{array}$$

$$x_j \geq 0, \;\; j = 1, 2, 3.$$

and then using the optimal simplex-tableau obtained determine the ranges of stability for

- basic and non-basic coefficients p_j in numerator $P(x)$;
- coefficient p_0 in numerator $P(x)$;
- basic and non-basic coefficients d_j in denominator $D(x)$;
- coefficient d_0 in denominator $D(x)$;
- right-hand side vector $b = (20, 30)^T$.

6.3 In the LFP problem given in the previous exercise we wish to change the right-hand side vector $b = (20, 30)^T$ so that

- $b_1 = 20 \;\; \rightarrow \;\; b_1' = 40$;
- $b_1 = 20 \;\; \rightarrow \;\; b_2' = 30$;
- $b_2 = 30 \;\; \rightarrow \;\; b_3' = 35$;
- $b_2 = 30 \;\; \rightarrow \;\; b_3' = 19$.

Using theorem 5.3, theorem 5.7 and formulas (5.98), (6.12), (6.16) and (6.18) try to predict separately for each case if the optimal value of objective function $Q(x)$ will change. If the change in the right-hand side vector b affects the optimal value of the objective function, then calculate this change and determine the new optimal value for the objective function.

6.1 Check if the optimal solution to LFP problem (6.19)-(6.20) from Section 3, page 184 changes, if

- basic coefficient $p_1 = 8$ in numerator $P(x)$ is changed to 7;
- basic coefficient $p_1 = 8$ in numerator $P(x)$ is changed to 9.5;
- basic coefficient $d_1 = 2$ in denominator $D(x)$ is changed to 1.5;
- basic coefficient $d_1 = 2$ in denominator $D(x)$ is changed to 2.5;
- non-basic coefficient $p_3 = 4$ in numerator $P(x)$ is changed to 3;
- non-basic coefficient $p_3 = 4$ in numerator $P(x)$ is changed to 15;

- non-basic coefficient $d_3 = 2$ in denominator $D(x)$ is changed to 1;
- non-basic coefficient $d_3 = 2$ in denominator $D(x)$ is changed to 5.

6.2 What will be the optimal solution to the problem in numerical example (6.19)-(6.20) from Section 3, page 184, if

- right-hand side entry $b_1 = 3$ is changed to 1.5 ?
- right-hand side entry $b_2 = 4$ is changed to 5.5 ?
- right-hand side entry $b_3 = 15$ is changed to 25 ?

6.3 Suppose that in LFP problem (6.1)-(6.3)

$$p_j = 0, \quad j = 1, 2, \dots, n, \quad \text{and} \quad p_0 = 1 \,.$$

Re-formulate restrictions (6.33) and (6.30) adapting them to this special case.

6.4 Re-formulate restrictions (6.33) and (6.30) adapting them to the following special case

$$p_j = 0, \quad j = 1, 2, \dots, n, \quad \text{and} \quad p_0 = \lambda \,.$$

6.5 Suppose that in LFP problem (6.1)-(6.3)

$$d_j = 0, \quad j = 1, 2, \dots, n, \quad \text{and} \quad d_0 = 1 \,.$$

Re-formulate restrictions (6.39), (6.40) and (6.44) adapting them to this special case.

6.6 Re-formulate restrictions (6.39), (6.40) and (6.44) adapting them to the following special case

$$d_j = 0, \quad j = 1, 2, \dots, n, \quad \text{and} \quad d_0 = \lambda \,.$$

Chapter 7

INTERCONNECTION BETWEEN LFP AND LP

As we have seen in Chapter 5, dual variables of LFP indicate if a small change in RHS vector b alters the optimal value of numerator $P(x)$ of objective function $Q(x)$ and hence, affects the optimal value of $Q(x)$. It means that the economic interpretation of dual variables in LFP and in LP may be very much alike. However, as is shown by formulas (5.98) and (5.113), dual variables of LFP act more selectively than dual variables of LP. Moreover, if the latest may be interpreted as a total change in profit $P(x)$ when changing resource vector b, then variables $y_i,\ i = 1, 2, \dots, m$ indicate only the intensive part of a total change in profit. In this chapter we deal with the interconnection between problems of LFP and LP, and their dual variables. We will show how this close connection may be used in real-world applications.

1. Preliminaries

It is known that when considering economic interests economists usually distinguish the following three levels of economic interests:

Highest level	:	*interests of human society,*
Middle level	:	*group interests*
Lowest level	:	*economic interests of individuals.*

We will consider only the two higher levels of them: the economic interests of society and group ones. Discussion of the economic interests of the lowest level, i.e. those of individuals, is beyond the scope of this book.

Let us consider a company which manufactures some products. Suppose that the company operates in a market-oriented economy and the main aim of the company is maximization of its profit. Using such a classification we may relate the company's economic interests to the middle level.

Furthermore, let unemployment have a presence in society and the main mission of the decision maker (*DM*) will be to find a way of bringing down the level of unemployment. Here and in what follows we suppose that bringing down the level of unemployment corresponds to the economic interests of society and all other interests, i.e. those of the company must be subordinated to the interests of society.

It is well known that two or more objective functions defined on the same feasible set generally speaking lead to different optimal solutions. In the case of the company and society, this means that if some production planx' is the best from the company's point of view, i.e. this plan leads to the highest company profit, there is no guarantee that the same plan maximizes the manpower requirement of the company. Conversely, if some production planx'' maximizes the manpower requirement of the company, i.e. x'' is the best output plan from the point of view of society, there is no guarantee that the same plan maximizes the profit of the company. Hence there is no guarantee that the company will prefer output plan x'' and will organize its manufacturing in accordance with it. Moreover, in such a situation the company may prefer some other aim, for example to optimize the ratio

$$\frac{Profit}{Manpower\ \ requirement}$$

So, we have three objective functions defined on the same feasible set: *Profit*, *Manpower requirement* and the fractional function *Profit / Manpower requirement*.

Below we consider the mathematical tools which may be useful for DM when these objective functions conflict with one another.

2. Primal Problems

Consider the following linear programming and linear fractional programming problems

$$\begin{array}{c} P(x) \to \max \\ x \in S, \end{array} \tag{7.1}$$

$$\begin{array}{c} D(x) \to \max \\ x \in S, \end{array} \tag{7.2}$$

$$\begin{array}{c} Q(x) \to \max \\ x \in S, \end{array} \tag{7.3}$$

where

$$Q(x) = \frac{P(x)}{D(x)} = \frac{\sum_{j=1}^{n} p_j x_j + p_0}{\sum_{j=1}^{n} d_j x_j + d_0},$$

$D(x) > 0$ for all $x = (x_1, x_2, \ldots, x_n)^T \in S$ and feasible set S is given by system of constraints

$$\sum_{j=1}^{n} a_{ij} x_j \le b_i, \quad i = 1, 2, \ldots, m;$$

$$x_j \ge 0, \quad j = 1, 2, \ldots, n.$$

Here and in what follows we assume that all three problems are solvable.

Let vector x^* be a basic optimal solution of problem (7.2) and B be the optimal basis associated with the positive components of x^*. Without loss of generality we can assume that $x^* = (x_1^*, x_2^*, \ldots, x_m^*, 0, 0, \ldots, 0)^T$ and $B = (A_1, A_2, \ldots, A_m)$, where $A_j = (a_{1j}, a_{2j}, \ldots, a_{mj})^T$ is jth column vector of matrix $A = ||a_{ij}||_{m \times n}$.

Let us suppose that this vector x^* does not solve problem (7.1) or (7.3), or solves neither, that is objective functions $P(x)$ and/or $Q(x)$ lead to some other optimal solutions x' and x''. Our aim now is to show that for any optimal solution x^* of problem (7.2) and any vector $p = (p_0, p_1, \ldots, p_n)$ we can find such vector $t = (t_0, t_1, \ldots, t_n)$ that x^* is an optimal solution of the following problems

$$\begin{aligned} P(t, x) &\to \max \\ x &\in S, \end{aligned} \tag{7.4}$$

and

$$\begin{aligned} Q(t, x) &\to \max \\ x &\in S, \end{aligned} \tag{7.5}$$

where

$$Q(t, x) = \frac{P(t, x)}{D(x)}, \quad P(t, x) = \sum_{j=1}^{n} (p_j + t_j) x_j + (p_0 + t_0).$$

Since basic vectors A_j are linearly independent we can represent any vector A_j as their linear combination

$$A_j = \sum_{i=1}^{m} A_i x_{ij}, \quad j = 1, 2, \ldots, n,$$

and we use these coefficients x_{ij} to define the following reduced costs

$$\left.\begin{array}{rcl} \Delta'_j &=& \sum_{i=1}^{m} p_i x_{ij} - p_j \,, \\ \Delta''_j &=& \sum_{i=1}^{m} d_i x_{ij} - d_j \,, \\ \Delta_j(x^*) &=& D(x^*)\Delta'_j - P(x^*)\Delta''_j \,, \end{array}\right\} \quad j = 1, 2, \ldots, n.$$

and

$$\left.\begin{array}{rcl} \Delta'_j(t) &=& \sum_{i=1}^{m} (p_i + t_i) x_{ij} - (p_j + t_j) \,, \\ \Delta_j(t, x^*) &=& D(x^*)\Delta'_j(t) - P(t, x^*)\Delta''_j \,, \end{array}\right\} \quad j = 1, 2, \ldots, n. \tag{7.6}$$

Further, the values $\Delta'_j(t)$ and $\Delta_j(t, x^*)$ can also be put in the form

$$\Delta'_j(t) = \sum_{i=1}^{m} t_i x_{ij} - t_j + \Delta'_j, \quad j = 1, 2, \ldots, n,$$

$$\Delta_j(t, x^*) = \sum_{i=1}^{m} t_i R_{ij} - t_j D(x^*) - t_0 \Delta''_j + \Delta_j(x^*), \quad j = 1, 2, \ldots, n,$$

where $R_{ij} = D(x^*) x_{ij} - \Delta''_j x^*_i, \;\; i = 1, 2, \cdots, m, \;\; j = 1, 2, \ldots, n.$

Since vector x^* is an optimal solution of problem (7.2), in accordance with the theory of linear programming we have, [69],

$$\Delta''_j \begin{cases} = 0, & j = 1, 2, \ldots, m \,, \\ \geq 0, & j = m+1, m+2, \ldots, n \,. \end{cases} \tag{7.7}$$

As in [69], [131] and [132] the basis of LP problem (7.4) and LFP problem (7.5) with fixed vector t is optimal in original form if $\Delta'_j(t) \geq 0$ for all indices j and $\Delta_j(t, x^*) \geq 0$ for all j respectively but we require only to consider $j = m+1, m+2, \ldots, n$ because

$$\Delta'_j(t) = \Delta_j(t, x^*) = 0 \,, \;\; j = 1, 2, \ldots, m.$$

The correctness of the following assertion is obvious.

THEOREM 7.1 ([13]) *If vector* $t = (t_0, t_1, \cdots, t_n)$ *satisfies conditions*

$$\left.\begin{array}{l} \sum_{i=1}^{m} t_i x_{ij} - t_j \geq -\Delta'_j \,, \\ \sum_{i=1}^{m} t_i R_{ij} - t_j D(x^*) - t_0 \Delta''_j \geq -\Delta_j(x^*) \,, \end{array}\right\} \quad j = m+1, m+2 \ldots, n, \tag{7.8}$$

then x^* is an optimal solution of LP problem (7.4) and LFP problem (7.5).

We denote the set of vectors t which satisfy the inequalities (7.8) by H. It is obvious that set $H \neq \emptyset$. Indeed, if $t = \lambda d - p$ where $\lambda \geq 0$ and $d = (d_0, d_1, \cdots, d_n)$, then using (7.6) and (7.7) we get

$$\left.\begin{array}{l} \Delta'_j(t) = \lambda\Delta''_j \geq 0\,, \\ \Delta_j(t, x^*) = D(x^*)(\lambda\Delta''_j) - (\lambda D(x^*))\Delta''_j = 0\,, \end{array}\right\} \quad j = m{+}1, m{+}2\ldots, n.$$

It means that set H contains at least vectors $\lambda d - p$, where $\lambda \geq 0$.

Thus, we have shown that for any solvable problems (7.1), (7.2) and (7.3) we can find such a vector $t = (t_0, t_1, \ldots, t_n)$ that vector x^* maximizes objective functions $P(t, x)$ and $Q(t, x)$ over feasible set S. This means that the set of problems (7.1), (7.2) and (7.3) which have at least one common optimal solution is non-empty.

Consider now function

$$T(x) \;=\; \sum_{j=1}^{n} t_j x_j + t_0\,.$$

Observe that if $t = \lambda d - p$ then for vector x^* we have

$$\begin{aligned} T(x^*) &= \sum_{i=1}^{m}(\lambda d_i - p_i)x_i^* + (\lambda d_i - p_0) = \\ &= \lambda(\sum_{i=1}^{m} d_i x_i^* + d_0) - (\sum_{i=1}^{m} p_i x_i^* + p_0) = \\ &= \lambda D(x^*) - P(x^*). \end{aligned} \tag{7.9}$$

If $Q(x^*) \geq 0$ then we may choose $\lambda = Q(x^*)$. So in this case from (7.9) we have

$$T(x^*) = Q(x^*)D(x^*) - P(x^*) = 0. \tag{7.10}$$

In other words, if $Q(x^*) \geq 0$ we can find such a vector t that $T(x^*) = 0$.

3. Stability

Let us suppose that vector $x^* = (x_1^*, x_2^*, \cdots, x_m^*, 0, 0, \cdots, 0)^T$ is a common basic optimal solution of problems (7.1), (7.2) and (7.3). We now proceed to consider the effect on the solution if any coefficient $p_j,\ j = 1, 2, \dots, n,$ is changed.

Consider the new objective functions

$$\bar{P}(x) = \sum_{\substack{j=1 \\ j \neq k}}^{n} p_j x_j + p_0 + \bar{p}_k x_k, \qquad \text{and} \quad \bar{Q}(x) = \frac{\bar{P}(x)}{D(x)},$$

where $\bar{p}_k = p_k + \varepsilon$.

Note that in this case our original common optimal solution x^* remains unaffected for problem (7.2), i.e. it remains feasible, basic and optimal. But for problems (7.1) and (7.3) vector x^* remains just a basic feasible solution and it is not necessary that it remains an optimal one for the new problems.

In accordance with the theory of linear programming, [51], [52], [69], for the new problem

$$\bar{P}(x) \rightarrow \max$$
$$x \in S$$

we have:

Case 1. if $1 \leq k \leq m$, i.e. index *k* is basic, then optimality of the solution x^* is unaffected within the following limits:

$$\max_{\substack{x_{kj}>0 \\ m+1 \leq j \leq n}} \frac{-\Delta'_j}{x_{kj}} \leq \varepsilon \leq \min_{\substack{x_{kj}<0 \\ m+1 \leq j \leq n}} \frac{-\Delta'_j}{x_{kj}};$$

Case 2. if $m+1 \leq k \leq n$, i.e. index *k* is non-basic, then x^* remains optimal if

$$\varepsilon \leq \Delta'_k.$$

Analogously, for the new problem

$$\bar{Q}(x) \rightarrow \max$$
$$x \in S$$

in accordance with [2], [28], [29] we have:

Case 1. if $1 \leq k \leq m$, i.e. index *k* is basic, then optimality of the solution x^* is unaffected if

$$\max_{\substack{g_{kj}>0 \\ m+1 \leq j \leq n}} \frac{-\Delta_j(x^*)}{g_{kj}} \leq \varepsilon \leq \min_{\substack{g_{kj}<0 \\ m+1 \leq j \leq n}} \frac{-\Delta_j(x^*)}{g_{kj}},$$

where $g_{kj} = D(x^*)x_{kj} - \Delta'_j x^*_k, \;\; j = 1, 2, \dots, n;$

Case 2. if $m+1 \leq k \leq n$, i.e. index k is non-basic, then x^* remains optimal if

$$\varepsilon \;\leq\; \frac{\Delta_k(x^*)}{D(x^*)}.$$

Thus, it is obvious that in the case of a change in basic p_k if value of ε is within the limits

$$\max\{\underline{f_1},\;\; \underline{f_2}\} \leq \;\varepsilon\; \leq\; \min\{\bar{f_1},\;\; \bar{f_2}\}, \tag{7.11}$$

then x^* remains an optimal solution for all three problems, where

$$\underline{f_1} = \max_{\substack{x_{kj}>0 \\ m+1\leq j\leq n}} \frac{-\Delta'_j}{x_{kj}}, \qquad \underline{f_2} = \max_{\substack{g_{kj}>0 \\ m+1\leq j\leq n}} \frac{-\Delta_j x^*}{g_{kj}},$$

$$\bar{f_1} = \min_{\substack{x_{kj}<0 \\ m+1\leq j\leq n}} \frac{-\Delta'_j}{x_{kj}}, \qquad \bar{f_2} = \min_{\substack{g_{kj}<0 \\ m+1\leq j\leq n}} \frac{-\Delta_j(x^*)}{g_{kj}}.$$

In the case of non-basic index k the limits are the following:

$$-\infty \leq\; \varepsilon\; \leq \min\{\Delta'_k,\;\; \frac{\Delta_k(x^*)}{D(x^*)}\}. \tag{7.12}$$

4. Dual Problems

Let us now consider the dual problems corresponding to primal problems (7.1), (7.2) and (7.3) respectively, [69] and [109],

$$\left.\begin{array}{l} \varphi(u) = \sum_{i=1}^{m} b_i u_i + p_0 \longrightarrow \min \\ \text{subject to} \\ \sum_{i=1}^{m} a_{ij} u_i \geq p_j, \;\; j = 1, 2, \dots, n; \\ \quad u_i \geq 0, \;\; i = 1, 2, \dots, m; \end{array}\right\} \tag{7.13}$$

$$\left.\begin{array}{l} \phi(v) = \sum_{i=1}^{m} b_i v_i + d_0 \longrightarrow \min \\ \text{subject to} \\ \sum_{i=1}^{m} a_{ij} v_i \geq d_j, \;\; j = 1, 2, \dots, n; \\ \quad v_i \geq 0, \;\; i = 1, 2, \dots, m; \end{array}\right\} \tag{7.14}$$

$$\left.\begin{array}{c} \psi(y) = y_0 \longrightarrow \min \\ \text{subject to} \\ d_0 y_0 - \sum_{i=1}^{m} b_i y_i \geq p_0, \\ d_j y_0 + \sum_{i=1}^{m} a_{ij} y_i \geq p_j,\ j = 1, 2, \dots, n; \\ y_i \geq 0,\ i = 1, 2, \dots, m. \end{array}\right\} \tag{7.15}$$

The next theorem indicates an important relationship between the optimal solutions of these dual problems.

THEOREM 7.2 ([13]) *If LP problems (7.1), (7.2) and LFP problem (7.3) have at least one common non-degenerate optimal solution x^*, then the following relation takes place*

$$u_i^* = y_i^* + Q(x^*)v_i^*, \quad i = 1, 2, \dots, m, \tag{7.16}$$

where vectors

$$u^* = (u_1^*, u_2^*, \dots, u_m^*)^T, \quad v^* = (v_1^*, v_2^*, \dots, v_m^*)^T$$

and

$$y^* = (y_0^*, y_1^*, y_2^*, \dots, y_m^*)^T$$

are optimal solutions of dual problems (7.13), (7.14) and (7.15) respectively.

Proof. Suppose that vector x^* is a common non-degenerate optimal solution of (7.1), (7.2) and (7.3). Let us replace the k-th element b_k of vector b by $b_k + \varepsilon$. Here and in what follows this replacement is claimed to effect no change in the basis of the optimal solution. In accordance with LP theory [69], for the new optimal solution $x' = (x_1', x_2', \dots, x_m', 0, 0, \dots, 0)^T$ we have

$$P(x') = P(x^*) + \varepsilon u_k^*, \tag{7.17}$$

$$D(x') = D(x^*) + \varepsilon v_k^*. \tag{7.18}$$

Analogously, in accordance with Theorem 5.7 and formula (5.98) for LFP problem (7.3) we get

$$Q(x') = Q(x^*) + \frac{\varepsilon y_k^*}{D(x')}.$$

Let us rewrite this equation in the following form

$$P(x') = Q(x^*)D(x') + \varepsilon y_k^*.$$

A comparison of the latter with (7.17) makes us infer that

$$P(x^*) + \varepsilon u_k^* = Q(x^*)D(x') + \varepsilon y_k^*.$$

Making use of equation (7.18) in the latter we find that

$$\varepsilon u_k^* = \varepsilon y_k^* + Q(x^*)\varepsilon v_k^*.$$

It means that formula (7.16) is valid, which was to be proved. ◇

Note that when proving this theorem we did not exploit directly the fact that vector x^* is non-degenerate but non-degeneracy of optimal vector x^* is a very important condition: degeneracy of optimal solution x^* usually leads to multiple solutions for dual problems (7.13), (7.14) and (7.15), so in the case of degenerate vector x^* formula (7.16) generally speaking is not valid.

5. Economic Interpretation

Let us now focus on the economic interpretation of the results described above. Let a certain company manufacture n different kinds of a certain product. Further, let p_j be the profit gained by the company from a unit of the j-th kind of the product, p_0 be some constant profit gained whose magnitude is independent of the output volume, b_i be the volume of some resource i available to the company and a_{ij} be the expenditure quota of the i-th resource for manufacturing a unit of j-th kind of the product. Denote the unknown output volume of some jth kind of the product by x_j. If $D(x)$ is a manpower requirement of the company, then under certain assumptions we can say that problem (7.2) corresponds to the economic interests of society. If the company's aim is maximization of its profit $P(x)$ and/or production efficiency $Q(x)$ calculated as a profit per unit of used manpower, then problems (7.1) and (7.3) correspond to the company's economic interests. Suppose that vector x^* maximizes manpower requirement function $D(x)$ on the feasible set S, i.e. x^* is the best output plan from society's point of view.

If profit vector p satisfies conditions

$$\left.\begin{array}{rcl} \Delta_j' & = & \sum\limits_{i=1}^{m} p_i x_{ij} - p_j \geq 0 \,, \\ \Delta_j(x^*) & = & \sum\limits_{i=1}^{m} p_i R_{ij} - p_j D(x^*) - p_0 \Delta_j'' \geq 0 \,, \end{array}\right\} \quad j = m+1, m+2, \ldots, n, \tag{7.19}$$

then vector x^* maximizes the company's profit $P(x)$ as well as production efficiency $Q(x)$. It means that to maximize its profit and/or production efficiency the company ought to organize its manufacturing in accordance with an output plan x^* which conforms to the economic interests of society in the best way. In this case we will say that the economic interests of the company conform to the economic interests of society.

Let us suppose now that profit vector p does not satisfy the conditions (7.19). It means that production plan x^* is not the best from the point of view of the company because x^* does not maximize profit $P(x)$ of the company or its production efficiency $Q(x)$ or does not maximize either. In this case the company would prefer some other production plan x', which corresponds to its interests in the best way. In accordance with Theorem 7.1 in this situation if the DM has a right to introduce into practice subsidies and taxes then using system (7.8) and vector $t = (t_0, t_1, \dots, t_n)$, it is possible to replace the profit vector $p = (p_0, p_1, \cdots, p_n)$ by the new vector

$$p' = p + t = (p_0 + t_0, p_1 + t_1, \dots, p_n + t_n),$$

and, in this a way, to re-orientate the objective functions of the company (problems (7.4) and (7.5)). The new objective functions

$$P'(x) = \sum_{j=1}^{n} p'_j x_j + p'_0 \quad \text{and} \quad Q'(x) = \frac{P'(x)}{D(x)}$$

will lead the company to optimal solution x^*, which in this situation is the best from all points of view:

for society this output plan x^* provides the highest manpower requirement of the company, so it may have a positive effect from the point of view of unemployment in the region;

for the company this plan x^* provides the highest profit and production efficiency calculated as a profit per unit of used manpower. Hence, it conforms to the economic interests of the profit-oriented company.

We should just remark that t_j may be interpreted as the tax if $t_j < 0$ or as the subsidy if $t_j > 0$. Moreover, as follows from (7.10) for a profitable company, i.e. in the case of $Q(x^* \geq 0$ we can find such a vector t of taxes and subsidies that the total sum of these is equal to zero, i.e. $T(x^*) = 0$.

Further, let the economic interests of the company conform to those of society and x^* be a common optimal solution of problems (7.1), (7.2) and (7.3). In this case, some little change ε in the profit p_k gained by the company does not affect the optimality of production plan x^* if this change is within the limits given by

(7.11) or (7.12). In accordance with Theorem 7.2, in this case relation (7.16) takes place. It is obvious that (7.16) may be interpreted in the following way: if the volume of resource i increases by one unit, the profit of the company rises by u_i^* units. Furthermore, y_i^* units of them are created by more intensive production, whereas $Q(x^*)v_i^*$ units by more extensive production, where v_i^* is the increase of manpower requirement.

This formula may prove to be useful if scarce resources are distributed among producers in a centralized way. Indeed, let us suppose that the company has made a request to be allocated certain extra units of the i-th resource. From the point of view of society it would be reasonable to satisfy the request if and only if $v_i^* > 0$ because it is the very case when the use of an additional volume of the i-th resource brings about an extra manpower requirement for the company.

Another way of using (7.16) is to use $Q(x^*)v_i^*$ as extra charge for an extra unit of the i-th resource. Indeed, in this case if the use of an extra unit of the i-th resource does not lead to an increase in efficiency and $y_i^* = 0$ then the extra profit of the company is equal to zero, too. It means that these extra charges will create an interest in increasing the use primarily of a resource whose index i_0 is defined from the equation

$$i_0 = ind \max_{1 \le i \le m} y_i^*$$

since in this case the extra profit is the largest. So if these extra charges have been introduced into practice they will be favorable for the intensification of production and for a more efficient use of manpower.

6. Numeric Example

Let the following feasible set S be

$$\begin{array}{rcrcrcl} 1x_1 & + & 2x_2 & + & 4x_3 & \le & 24\,, \\ 4x_1 & + & 2x_2 & + & 1x_3 & \le & 12\,, \end{array}$$

$$x_1 \ge 0, \quad x_2 \ge 0, \quad x_3 \ge 0\,.$$

Consider the following problems :

$$P(x) = 2.5x_1 + 1x_2 + 0.5x_3 + 4 \longrightarrow \max \\ x \in S, \tag{7.20}$$

$$D(x) = 1x_1 + 1.5x_2 + 2x_3 + 6 \longrightarrow \max \\ x \in S, \tag{7.21}$$

$$Q(x) = \frac{P(x)}{D(x)} \longrightarrow \max \\ x \in S. \tag{7.22}$$

By using WinGULF [14] (see Chapter 13), a program package for solving linear and linear-fractional programming problems, it is easy to show that vector $x^* = (0,4,4)^T$ solves problem (7.21) and vector $x' = (3,0,0)^T$ solves problems (7.20) and (7.22). That is these problems have no common optimal solution. Let us consider vector

$$t = (t_0, t_1, t_2, t_3),$$

where

$$t_j = \lambda d_j - p_j, \quad j = 0,1,2,3,$$

and

$$\lambda = Q(x^*) = \frac{P(x^*)}{D(x^*)} = \frac{10}{20} = 0.5 \ ,$$

that is vector

$$t = (-1, -2, -0.25, 0.5) \ ,$$

and replace vector $p = (4, 2.5, 1, 0.5)$ with vector

$$\begin{aligned} p + t &= (4 + (-1), 2.5 + (-2), 1 + (-0.25), 0.5 + 0.5) = \\ &= (3, 0.5, 0.75, 1). \end{aligned}$$

Thus, we have the following two new problems:

$$\begin{array}{c} P'(x) = 0.5x_1 + 0.75x_2 + 1x_3 + 3 \longrightarrow \max \\ x \in S, \end{array} \tag{7.23}$$

$$\begin{array}{c} Q'(x) = \dfrac{P'(x)}{D(x)} \longrightarrow \max \\ x \in S \ . \end{array} \tag{7.24}$$

These problems (7.23) and (7.24) have the same optimal solution $x^* = (0,\ 4,\ 4)^T$ as problem (7.21) does. Note that

$$P'(x^*) = 10.00, \ \ D(x^*) = 20.00, \ \ Q'(x^*) = \frac{10}{20} = 0.5$$

and

$$T(x^*) = 0 \times (-2) + 4 \times (-0.25) + 4 \times 0.5 - 1 = 0.$$

Solving the dual problems

$$\varphi(u) = 24u_1 + 12u_2 + 3 \rightarrow \min$$

subject to

$$\begin{aligned} 1u_1 + 4u_2 &\geq 0.5 \\ 2u_1 + 2u_2 &\geq 0.75 \\ 4u_1 + 1u_2 &\geq 1 \\ u_1 \geq 0,\ u_2 &\geq 0, \end{aligned}$$

$$\phi(v) = 24v_1 + 12v_2 + 6 \rightarrow \min$$

subject to

$$\begin{aligned} 1v_1 + 4v_2 &\geq 1 \\ 2v_1 + 2v_2 &\geq 1.5 \\ 4v_1 + 1v_2 &\geq 2 \\ v_1 \geq 0,\ v_2 &\geq 0 \end{aligned}$$

and

$$\psi(y) = y_0 \rightarrow \min$$

subject to

$$\begin{array}{rcrcrcr} 6y_0 & - & 24y_1 & - & 12y_2 & \geq & 3 \\ 1y_0 & + & 1y_1 & + & 4y_2 & \geq & 0.5 \\ 1.5y_0 & + & 2y_1 & + & 2y_2 & \geq & 0.75 \\ 2y_0 & + & 4y_1 & + & 1y_2 & \geq & 1 \\ & & y_1 \geq 0, & & y_2 \geq 0 & & \end{array}$$

we obtain the following relation for the dual variables

i	u_i^*	$=$	y_i^*	$+$	$Q(x^*)$	$\times$	v_i^*
$i = 1$	0.2083	$=$	0.0	$+$	0.5	$\times$	0.4167
$i = 2$	0.1667	$=$	0.0	$+$	0.5	$\times$	0.3333

We can summarize our results as follows: using Theorem 7.1 we can find such a vector $t = (t_0, t_1, \cdots, t_n)$ of taxes and subsidies that the economic interests of the company will conform to the economic interests of society in the best way. This means that trying to maximize its profit and/or production efficiency calculated as profit per unit of manpower requirement the company will automatically maximize its manpower requirement and, in this way, will be favorable for bringing down the level of unemployment in society. Theorem 7.2 allows us to indicate such resources for which the use of an additional volume of these brings about an extra manpower requirement for the company.

7. Discussion Questions and Exercises

7.1 Consider the following three problems

$$P(x) = 1x_1 + 2x_2 + 3.5x_3 + 1x_4 + 1 \rightarrow \max$$
$$x \in S,$$

$$D(x) = 2x_1 + 2x_2 + 3.5x_3 + 3x_4 + 4 \rightarrow \max$$
$$x \in S,$$

$$Q(x) = \frac{P(x)}{D(x)} \rightarrow \max$$
$$x \in S,$$

where feasible set S is given by the following constraints

$$\begin{array}{ccccccccc} 2x_1 & + & 1x_2 & + & 3x_3 & + & 3x_4 & \leq & 10\,, \\ 1x_1 & + & 2x_2 & + & 1x_3 & + & 1x_4 & \leq & 14\,, \end{array}$$

$$x_1 \geq 0, \;\; x_2 \geq 0, \;\; x_3 \geq 0, \;\; x_4 \geq 0,$$

compare their optimal solutions and then check for these problems if the interconnection (7.16) is valid.

7.2 For the numerical example given in previous exercise determine stability ranges (7.11) for basic coefficients p_j and stability ranges (7.12) for non-basic coefficients p_j.

7.3 In the numerical example given in exercise 7.1 replace function $P(x)$ with

$$P(x) = 3x_1 + 1x_2 + 2.5x_3 + 2x_4 + 5\,,$$

then using Theorem 7.1 construct restrictions (7.8) and determine such vector $t = (t_0, t_1, t_2, t_3, t_4)$ of corrections for coefficients

$$p_0 = 5, \;\; p_1 = 3, \;\; p_2 = 1, \;\; p_3 = 2.5, \;\; p_4 = 2,$$

which re-directs all three objective functions to the common optimal solution.

Chapter 8

INTEGER LFP

In some practical applications the unknown variables of a linear-fractional problem are constrained to a discrete set. Such a discrete set may consist of enumerated arbitrary (integer or real) values. If this set of feasible values for unknown variables consists of integer values, such a class of LFP problems is usually called **integer linear-fractional programming** or **ILFP**. Examples of applications of integer LFP are mostly in the field of economics and engineering, where it is very important to find such solution that provides the biggest value of a ratio expressed as an objective function - it may be a ratio of revenues and allocations subject to restriction on the availability of the goods involved in the location problem, or a maximal density of integrated elements of an electronic chip designed, etc. If the set of enumerated feasible values for unknown variables contains not only integers but real values too, we usually refer to such of LFP problems as **discrete LFP**. For example, the following LFP problem

$$Q(x) = \frac{P(x)}{D(x)} = \frac{2x_1 + 3x_2 + 10}{3x_1 - x_2 + 1} \longrightarrow \max$$

subject to

$$x_1 + x_2 \leq 35,$$

$$x_1 \in \{1.03,\ 22.5,\ 30.75,\ 50.5\}, \quad x_2 \in \{1.5,\ 2.5,\ 3.5,\ 10.25\}$$

is a discrete LFP problem.

So discrete LFP may be considered as a generalization of ILFP. Another special class of discrete LFP is such a subclass of ILFP problems where all unknown variables are restricted to have a value of 0 or 1. In such cases it is conventional to call these problems **0/1 linear-fractional programming** problems.

The versatility of the integer optimization model stems from the fact that in many practical problems, activities and resources, such as machines, airplanes, people, are indivisible. Also, many problems have only a finite number of alternative choices and consequently can appropriately be formulated as an optimization problem with integer unknown variables – the word*integer* referring to the fact that only integer values of variables are acceptable as feasible and optimal solution of the problem. Integer programming models are often referred to as *combinatorial optimization* or *combinatorial programming* models where programming refers to "planning" so that these are models used in planning where some or all of the decisions can take only a finite number of alternative possibilities.

An integer programming problem in which all variables are required to be integers is called **pure integer programming problem**. For example,

$$Q(x) = \frac{P(x)}{D(x)} = \frac{2x_1 + 3x_2 + 10}{3x_1 - x_2 + 1} \longrightarrow \max$$

subject to

$$x_1 + x_2 \leq 35,$$

$$x_1 \geq 0, \quad x_2 \geq 0, \quad x_1, \; x_2 \text{ are integer}$$

is a pure integer LFP problem. If, on the other hand, not all variables must be integer, but only some of them, the problem is called a **mixed integer programming problem** or a **MIP** problem. For example,

$$Q(x) = \frac{P(x)}{D(x)} = \frac{2x_1 + 3x_2 + 10}{3x_1 - x_2 + 1} \longrightarrow \max$$

subject to

$$x_1 + x_2 \leq 35,$$

$$x_1 \geq 0, \quad x_2 \geq 0, \quad x_1 \text{ is integer}$$

is a mixed integer LFP problem.

The general form of integer LFP problems is as follows

$$Q(x) = \frac{P(x)}{D(x)} = \frac{\sum\limits_{j \in J_1} p_j x_j + \sum\limits_{j \in J_2} p_j x_j + \sum\limits_{j \in J_3} p_j x_j + p_0}{\sum\limits_{j \in J_1} d_j x_j + \sum\limits_{j \in J_2} d_j x_j + \sum\limits_{j \in J_3} d_j x_j + d_0} \longrightarrow \max$$

subject to

$$\sum_{j \in J_1} a_{ij} x_j + \sum_{j \in J_2} a_{ij} x_j + \sum_{j \in J_3} a_{ij} x_j \; (=, \leq, \geq) \; b_i, \quad i = 1, 2, \dots, m;$$

$$x_j \geq 0, \ j \in J_1 \cup J_2 \cup J_3$$

$$x_j = 0/1, \ \forall j \in J_1; \quad x_j - \text{integer}, \ \forall j \in J_2; \quad x_j - \text{real}, \ \forall j \in J_3$$

$$D(x) > 0, \ \text{for all feasible } x.$$

Solving integer programming problems, i.e. finding an optimal solution (or optimal solutions) to such problems, can be a difficult task. The difficulty arises from the fact that unlike "continuous" LFP, for example, whose feasible region is a convex set, in integer LFP, one must search a lattice of feasible points or, in the mixed-integer case, a set of disjoint half-lines or line segments to find an optimal solution. Thus, unlike "continuous" LFP where, due to the convexity of the problem, we can exploit the fact that any local solution is a global optimum, integer LFP problems have many local optima and finding a global optimum for the problem requires one to prove that a particular solution dominates all feasible points. This means that integer LFP problems are a lot harder to solve than problems of "continuous" LFP.

In this chapter we formulate some real-word problems of integer LFP and then consider the following two different approaches to solving integer LFP problems: (1) the technique of branch-and-bound, (2) the cutting plane algorithm. The last section deals with formulating problems of discrete LFP and explaining how the problems of different subclasses for discrete LFP may be converted to one another.

1. LFP Models with Integer Variables

Here we describe some classical integer programming models to provide both an overview of the diversity and versatility of this field and to show that the solution of large real-world instances of such problems require the solution method to exploit the specialized mathematical structure of the specific application.

1.1 The Knapsack Problem

The knapsack problem is a particularly simple integer LFP problem: it has only one constraint. Moreover, the coefficients of this constraint and the fractional objective function are all non-negative. In accordance with classical definitions from linear programming, we will call a **knapcsak problem** the following integer LFP problem:

$$Q(x) = \frac{P(x)}{D(x)} = \frac{\sum_{j=1}^{n} p_j x_j + p_0}{\sum_{j=11}^{n} d_j x_j + d_0} \longrightarrow \max$$

subject to

$$\sum_{j=1}^{n} a_j x_j \leq b,$$

$$x_j = 0/1, \quad j = 1, 2, \dots, n,$$

where $D(x) > 0$ for all feasible x.

The traditional story is that there is a knapsack (here of capacity b). Furthermore, there are a number of items (here there are n items), each with a weight (here of $d_j, \ j = 1, 2, \dots, n$), a size (here of $a_j, \ j = 1, 2, \dots, n$), and a value (here of $p_j, \ j = 1, 2, \dots, n$). Here, p_0 and d_0 are the value and the weight of the knapsack, respectively. The objective is to maximize the ratio *(total value)/(total weight)* for the items in the knapsack. More extended and detailed information on this problem may found in [101].

1.2 Capital Budgeting Problems

Our next example is a capital budgeting problem. Let us suppose that there is a company considering four possible investments. Investment1 will yield a net present value (NPV) of $16000; investment 2, an NPV of $22000; investment 3, an NPV of $12000; and investment 4, an NPV of $8000. Each investment requires a certain cash flow at the present time: investment1, $7000; investment 2, $5000; investment 3, $4000; and investment 4, $3000. At present, $14000 is available for investment. The company wishes to find such an investment configuration that provides the maximum efficiency for the cash invested calculated as the following ratio

$$(total\ NVP)/(total\ investment).$$

We begin by defining a variable for each possible investment opportunity. This leads us to define the following 0/1 variables:

$$x_j = \begin{cases} 1, & \text{if investment } j \text{ is made} \\ 0 & \text{otherwise} \end{cases} \qquad j = 1, 2, 3, 4.$$

The total NVP obtained by the company (in thousands of dollars) is

$$P(x) = 16x_1 + 22x_2 + 12x_3 + 8x_4.$$

The total investment may be expressed as follows

$$D(x) = 5x_1 + 7x_2 + 4x_3 + 3x_4.$$

Since at most \$14000 can be invested, unknown variables $x_1,\ x_2,\ x_3,$ and x_4 must satisfy the following condition:

$$7x_1 + 5x_2 + 4x_3 + 3x_4 \leq 14$$

This leads to the following 0/1 (or 0-1) LFP problem:

$$Q(x) = \frac{P(x)}{D(x)} = \frac{16x_1 + 22x_2 + 12x_3 + 8x_4}{5x_1 + 7x_2 + 4x_3 + 3x_4}$$

subject to

$$7x_1 + 5x_2 + 4x_3 + 3x_4 \leq 14$$

$$x_j = 0/1, \quad j = 1, 2, 3, 4.$$

The optimal solution of this problem is: $x_1 = 0,\quad x_2 = 1,\quad x_3 = 0,\quad x_4 = 0.$ In this case, the total required investment is \$5000, which leads to NVP of \$22000. The efficiency for such investment is 4.4 .

There may be a number of additional constraints the company might want to add. For example:

1 The company can invest in at most two investments.

2 If the company invests in investment 2, it must also invest in investment 1.

3 If the company invests in investment 2, they cannot invest in investment 4, etc.

1.3 Set Covering Problems

The following example is of an important class of ILFPs known as set-covering problems. There are six cities (cities 1–6) in the district. The district is reviewing the location of its fire stations. In each city the investment required to build fire stations differ from one another. The investments required in each city are shown in Table 8.1. The district must determine where to build fire stations. The fire station can be placed in any city but the district wants to build the stations such that at least one fire station be within 15 minutes (driving time) of each city and the total cost invested per number of stations built be minimal. The times (in minutes) required to drive between the cities in the district are shown in Table 8.2. For each city we must define one variable $x_j = 0/1,\ j = 1, 2, \ldots, 6$. This variable x_j will be 1 if we place a station in

	$City\ 1$	$City\ 2$	$City\ 3$	$City\ 4$	$City\ 5$	$City\ 6$
$Million\$$	12	14	9	16	5	8

Table 8.1. Set covering problem – Investments.

	$City\ 1$	$City\ 2$	$City\ 3$	$City\ 4$	$City\ 5$	$City\ 6$
$City\ 1$	0	10	20	30	30	20
$City\ 2$	10	0	25	35	20	10
$City\ 3$	20	25	0	15	30	20
$City\ 4$	30	35	15	0	15	25
$City\ 5$	30	20	30	15	0	14
$City\ 6$	20	10	20	25	14	0

Table 8.2. Set covering problem – Driving time in minutes.

the City j, and will be 0 otherwise. Then the total number of fire stations that must be built is given by

$$D(x) = x_1 + x_2 + x_3 + x_4 + x_5 + x_6$$

and the total investment required will be

$$P(x) = 12x_1 + 14x_2 + 9x_3 + 16x_4 + 5x_5 + 8x_6.$$

This leads to the following formulation of problem

$$Q(x) = \frac{P(x)}{D(x)} = \frac{12x_1 + 14x_2 + 9x_3 + 16x_4 + 5x_5 + 8x_6}{x_1 + x_2 + x_3 + x_4 + x_5 + x_6} \longrightarrow \min$$

subject to

$$\begin{array}{cccccccc}
x_1 & +x_2 & & & & & \geq & 1 \\
x_1 & +x_2 & & & & +x_6 & \geq & 1 \\
 & & x_3 & +x_4 & & & \geq & 1 \\
 & & x_3 & +x_4 & +x_5 & & \geq & 1 \\
 & & & +x_4 & +x_5 & +x_6 & \geq & 1 \\
 & +x_2 & & & +x_5 & +x_6 & \geq & 1
\end{array}$$

$$x_j = 0/1, \quad j = 1, 2, 3, 4, 5, 6.$$

The first constraint states that there must be a fire station either in City 1 or in City 2. The next constraint is for City 2 and so on. Notice that the constraints coefficient a_{ij} is 1 if City i is adjacent to City j or if $i = j$ and 0 otherwise. The jth column of the constraint matrix represents the set of cities that can be served by a fire station in City j. We asked to find a set of such subsets j that covers the set of all cities in the sense that every city appears in the service subset associated with at least on fire station. One optimal solution of this problem is:

$$x_1 = 1, \quad x_2 = 0, \quad x_3 = 1, \quad x_4 = 0, \quad x_5 = 1, \quad x_6 = 1$$

This is an example of the set covering problem. The set covering problem is characterized by having $0/1$ (binary) variables, "great or equal" constraints each with the right-hand side of 1, and having simply sums of variables as the left-hand side of the constraints. Set covering problems have many applications in areas such as airline crew scheduling, political districting, airline scheduling, truck routing, etc.

The set covering problem of linear-fractional programming in a more general form

$$Q(x) = \frac{P(x)}{D(x)} = \frac{\sum_{j=1}^{n} p_j x_j}{\sum_{j=11}^{n} d_j x_j + d_0} \longrightarrow \min$$

subject to

$$\sum_{j=1}^{n} a_{ij} x_j \geq 1, \quad i = 1, 2, \dots, m,$$

$$x_j = 0/1, \quad j = 1, 2, \dots, n,$$

was considered in [4] and [5] where the authors presented a special technique for solving such types of ILPF. For further information on this topic see also [102].

1.4 The Traveling Salesperson Problem

One of the most famous and intensively investigated integer programming problems is the traveling salesman problem (TSP). Let us consider this problem in the form of a linear-fractional programming problem. Suppose that there is a traveling salesperson who must visit each of n cities before returning home. There is a known matrix $||d_{ij}||_{n \times n}$ of distances between each of the cities. For each pair of cities (i, j) let p_{ij} be the costs of going from city i to cite j, or from city j to cite i. Define variables $x_{ij}, i = 1,2,\ldots,n;\ j = 1,2,\ldots,n,$ where x_{ij} be 1 if the salesperson travels between cities i and j, and 0 otherwise. Here, we suppose that $P_{ij} = p_{ji},\ d_{ij} = d_{ji},$ and $x_{ij} = x_{ji},$ for all indexes i and j. This problem is known as the **symmetric TSP**. The objective is to minimize the travel cost per one kilometer:

$$Q(x) = \frac{P(x)}{D(x)} = \frac{\sum_{i=1}^{n}\sum_{j=1}^{n} p_{ij}x_{ij} + p_0}{\sum_{i=1}^{n}\sum_{j=1}^{n} d_{ij}x_{ij} + d_0} \longrightarrow \min$$

where p_0 and d_0 are some fixed cost and fixed distance, respectively, which do not depend on the path selected. The constraints for this problem are as follows:

$$\sum_{\substack{j=1\\ j\neq i}}^{n} x_{ij} = 2, \quad i = 1,2,\ldots,n,$$

These constraints say that every city must be visited. However, these constraints are not enough, since it is possible to have multiple cycles (*subtours*), rather than one big cycle (*tour*) through all the cities. To handle this condition, we have to use the following set of *subtour elimination constraints*:

$$\sum_{i\in I}\sum_{j\notin I} x_{ij} \geq 2, \quad \text{for all } I \subset N = \{1,2,\ldots,n\}$$

This set of constraints states that for any subset I of cities, the tour must enter and exit that set. These, together with

$$x_{ij} = 0/1, \quad i = 1,2,\ldots,n, \ j = 1,2,\ldots,n,$$

are sufficient to formulate the traveling salesperson problem as an integer linear-fractional problem. For detailed information on the TSP problem in conventional form of linear programming see e.g. [126].

2. The Branch-and-Bound Method

In practice, most integer programming problems are usually solved by using a special enumerative technique known as *branch-and-bound*. Methods based on this technique mainly divide the problem to be solved into smaller (sub)problems ('branching') and try to solve these smaller (sub)problems 'keeping in mind' the best objective value obtained earlier ('bound'). The method was first put forward in the early 1960's by A.H.Land and A.G.Doig in [123], where it was applied to problems of integer linear programming. Later in 1965 it was generalized in [50] for non-fractional programming problems. Pure integer and mixed LFP problems was studied by S.Chandra and M.Chandramohan in [35] and [36], D.Granot and F.Granot in [84], O.M.Saad in [156], M.Sniedovich in [172], V.Verma, H.C.Bakshi, and M.C.Puri in [185], etc.

Consider the following pure integer LFP problem

$$Q(x) = \frac{P(x)}{D(x)} = \frac{\sum_{j=1}^{n} p_j x_j + p_0}{\sum_{j=1}^{n} d_j x_j + d_0} \longrightarrow \max \tag{8.1}$$

subject to

$$\sum_{j=1}^{n} a_{ij} x_j \le b_i, \quad i = 1, 2, \dots, m, \tag{8.2}$$

$$x_j \ge 0, \quad j = 1, 2, \dots, n, \tag{8.3}$$

$$x_j - \text{ integer}, \quad j = 1, 2, \dots, n, \tag{8.4}$$

where $D(x) > 0, \ \forall x \in S_0$ and S_0 denotes a continuous feasible set determined by constraints (8.2)-(8.3).

DEFINITION 8.1 *The continuous LFP problem obtained by omitting all integer constraints (8.4) is called the* ***LFPT relaxation*** *of the ILFP (8.1)-(8.4).*

We assume that relaxation problem (8.1)-(8.3) is solvable and vector$x^{(0)}$ denotes its optimal solution.

Before explaining the principles of the branch-and-bound method, we need to make the following elementary but very important observations:

- If we solve the LFP relaxation problem (8.1)-(8.3) and obtain a solution in which all variables $x_j^{(0)}$ are integers, $j = 1, 2, \dots, n,$ then the optimal solution obtained is also the optimal solution for the pure integer LFP problem (8.1)-(8.4).
- If we add a new constraint of type

  $$x_j \le (\ge) K, \quad j \in J = \{1, 2, \dots, n\}, \tag{8.5}$$

  where K is an arbitrary constant, to the system of constraints (8.2)-(8.3) and denote it with S_1, then $S_0 \supseteq S_1$. Moreover,

  $$\max_{x \in S_0} Q(x) \ge \max_{x \in S_1} Q(x).$$

Continuing the process of adding new constraints, we can introduce set S_2 obtained by adding a new constraint of type (8.5) to set S_1, then using set S_2 we can construct set S_3 which differs from set S_2 with an extra constraint of type (8.5), etc. This generalization results in

$$S_0 \supseteq S_1 \supseteq S_2 \supseteq S_3 \supseteq \dots ,$$

and

$$\max_{x \in S_0} Q(x) \geq \max_{x \in S_1} Q(x) \geq \max_{x \in S_2} Q(x) \geq \max_{x \in S_3} Q(x) \geq \dots . \tag{8.6}$$

Now, the branch-and-bound method may be described as follows.

Step 1 (Relaxation). First, we solve the relaxation LFP problem (8.1)-(8.3). If in its optimal solution $x^{(0)}$ all variables $x_j^{(0)}$ have an integer value, then the process must be terminated, since vector $x^{(0)}$ is an optimal solution for ILFP problem (8.1)-(8.4). If not, let $Bound := -\infty$, then go to Step 2.

Step 2 (Branching). If among values $x_j^{(0)}$ there is at least one non-integer value, we choose one of them, say $x_{j_1}^{(0)}$, and partition feasible set S_0 onto two parts (S_1 and S_2) by adding to S_0 new constraints

$$x_{j_1} \leq [x_{j_1}^{(0)}], \quad \text{and} \quad x_{j_1} \leq [x_{j_1}^{(0)}] + 1, \tag{8.7}$$

respectively, where $[x_{j_1}^{(0)}]$ denotes an integer part of value $x_{j_1}^{(0)}$. It is conventional to refer to constraints (8.7) as **branching constraints**. Obviously, the variable selected must be a basic variable; otherwise its value would be equal to zero, i.e. an integer value.

Step 3 (Formation of new subproblems, or nodes). We construct the following two new mixed integer problems

$$\text{Subproblem 1}: \quad \max_{x \in S_1} Q(x) \tag{8.8}$$

and

$$\text{Subproblem 2}: \quad \max_{x \in S_2} Q(x), \tag{8.9}$$

and then solve each of these subproblems as a continuous LFP problem.

Step 4 (Termination test). Each of the nodes (subproblems) constructed in Step 3 should be marked as a terminal node (subproblem) if one of the following conditions holds.

1 The subproblem has no feasible solutions. In this case we label this node as terminal and say that the problem (or the corresponding node) is **fathomed**.

2 The subproblem has an integer optimal solution with objective value Q^*. In this case, we label this node as terminal and check the following condition: if $Q^* > Bound$ then set $Bound := Q^*$ and mark this node as a **candidate**; otherwise $Bound$ remains unchanged.

All non-terminal subproblems have to be added to the list of dangling subproblems. After performing this test go to Step 5.

Step 5 (Node selection).

1 Check if the list of dangling subproblems contains such problems that have a non-integer optimal solution with objective value$Q^* \leq Bound.$ In accordance with (8.6) we cannot expect from such a problem that any further maintenance (partitioning and branching) of corresponding node would give us an integer solution with an objective value better than $Bound.$ Since these subproblems cannot yield an integer solution better that $Bound$, the further branching on these nodes will yield no useful information about the optimal solution of the original ILFP problem. Thus, we truncate these nodes and remove the corresponding subproblems from the list of dangling subproblems.

2 Check if the list of dangling subproblems is non-empty, i.e. if there are subproblems with non-integer optimal solutions and objective values better than current $Bound.$ If there is at least one such subproblem we choose one of them and apply to the selected problem the branching process described in Step 2. In other words, keeping the selected node as a current one we go to Step 2; Otherwise, i.e if the list of dangling subproblems is empty, Stop;

After termination of the method we have to check the value of $Bound$ as follows: if $Bound > -\infty$ the candidate node associated with the value of $Bound$ contains the optimal integer solution of the original ILFP problem; otherwise the original ILFP problem has no integer optimal solutions.

Note that the conception of a $Bound$ as the best objective value associated with integer optimal solutions obtained during processing the nodes allows us to avoid finding solutions for the subproblems that cannot provide an integer solution with an objective value better than the current $Bound.$ This is why from the point of view of efficiency of calculations it is very important to have a proper rule for choosing from the dangling list a subproblem to be solved next. In the Land-Doig algorithm developed for integer LP problems, the next node to be examined is always the one having the largest objective value. This rule is referred to as a **best-first search** rule. The so-called **breadth-first search** rule prescribes to solve first the problems on the given level of the tree before going deeper. While in accordance with the **depth-first search** rule we have to go deep before going wide. Other aspect that affects the efficiency of the method is choosing a non-integer variable for a branching, if more than one integer variable has a fractional value. If this is the case, the variable with the largest fractional part should be selected for branching, see e.g. [177]. Various strategies and special procedures for branching and exploring nodes in the tree of the branch-and-bound method were described in [21], [22], [74], [130] [137], [145], [159], etc. Some of these rules were adapted to ILFP problems too, see e.g. [35]. More complicated rules recommend to split the process into two phases. During the first phase we use the depth-first search rule and branch on the variable that has the smallest fractional part. The goal of this phase is to find an integer solution as soon as possible and then to use it as a bound to fathom subproblems. Once an integer solution has been found, the algorithm enters the second phase, where our goal is to prune the branch-and-bound tree as quickly as possible. This is why in this phase we (have to) use the best-first search rule and branch on the variables that have the largest fractional part.

In order to keep track of the generated branches and nodes, a tree as shown in Figure 8.1 may be used.

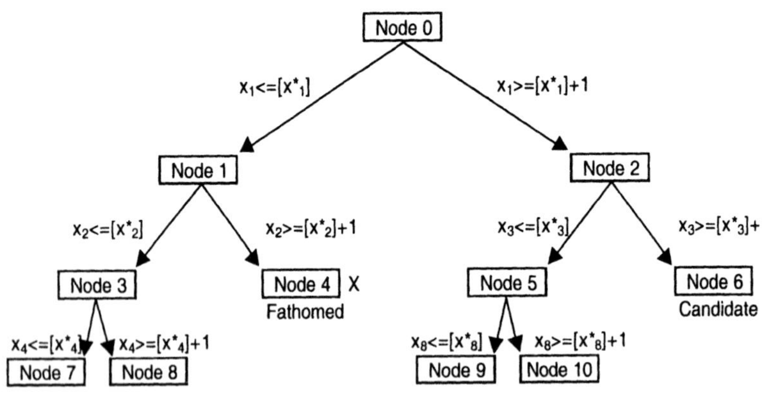

Figure 8.1. The Branch and Bound Method – A search tree.

To illustrate how the method works we consider the following pure integer ILPF numerical example:

$$Q(x) = \frac{P(x)}{D(x)} = \frac{3x_1 + 2x_2 + 2x_3 + 5}{3x_1 + 3x_2 + 2x_3 + 16} \longrightarrow \max \tag{8.10}$$

subject to

$$\left.\begin{array}{rcl} 6x_1 + 3x_2 + 5x_3 & \leq & 32, \\ 7x_1 + 4x_2 + 2x_3 & \leq & 27, \end{array}\right\} \tag{8.11}$$

$$x_j \geq 0, \quad j = 1, 2, 3, \tag{8.12}$$

$$x_j - \text{ integer}, \quad j = 1, 2, 3. \tag{8.13}$$

First, we solve the corresponding relaxation LFP problem determined as (8.10)-(8.12) and obtain the following non-integer optimal solution

$$x_1^{(0)} = 3.09, \;\; x_2^{(0)} = 0.00, \;\; x_3^{(0)} = 2.70,$$

with objective value

$$Q(x^{(0)}) = \frac{P(x^{(0)})}{D(x^{(0)})} = \frac{19.65217391}{30.65217391} \approx 0.64113475.$$

Since the optimal solution obtained is non-integer, we set $Bound := -\infty$ and choose x_3 as the branching variable, since it has the largest fractional part. The constraints to be added are

$$x_3 \leq 2, \quad \text{and} \quad x_3 \geq 3 \; .$$

To carry out Step 3 we add each of these constraints in turn to the original constraints (8.11) and in this way construct subproblems (8.8) and (8.9), respectively, with feasible sets S_1 and S_2 as follows:

$$\left.\begin{array}{rcl} 6x_1 + 3x_2 + 5x_3 & \leq & 32, \\ 7x_1 + 4x_2 + 2x_3 & \leq & 27, \\ \hline x_3 & \leq & 2, \end{array}\right\} \quad S_1$$

$$\left.\begin{array}{rcl} 6x_1 + 3x_2 + 5x_3 & \leq & 32, \\ 7x_1 + 4x_2 + 2x_3 & \leq & 27, \\ \hline x_3 & \geq & 3. \end{array}\right\} \quad S_2$$

The tree of problems which was constructed by the method for this example is given in Figure 8.2. The subproblems of type (8.8) and (8.9) we have at this point are marked in Figure 8.2 by 'Node 1' and 'Node 2', respectively. Solving these problems we obtain the following optimal solutions
for subproblem in 'Node 1':

$$x_1^{(1)} = 3.29, \;\; x_2^{(1)} = 0.00, \;\; x_3^{(1)} = 2.00, \quad Q(x^{(1)}) \approx 0.63157895,$$

for subproblem in 'Node 2':

$$x_1^{(2)} = 2.83, \;\; x_2^{(2)} = 0.00, \;\; x_3^{(2)} = 3.00, \quad Q(x^{(2)}) \approx 0.63934426 \; .$$

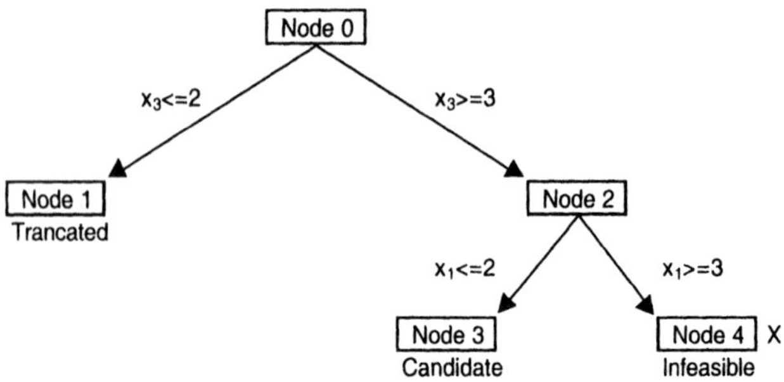

Figure 8.2. The Branch and Bound Method – Example's search tree.

Since in both nodes the optimal solutions are non-integer and the objective value in 'Node 2' is larger than in 'Node 1', we continue with subproblem in 'Node 2' recording subproblem in 'Node 1' in our list of dangling nodes. Further, in 'Node 2' we choose variable x_1 as the branching variable and construct the following constraints to be added

$$x_1 \leq 2, \quad \text{and} \quad x_1 \geq 3 .$$

The corresponding problems in 'Node 3' and 'Node 4' have feasible sets S_3 and S_4, respectively, as follows

$$\left.\begin{array}{rcl} 6x_1 + 3x_2 + 5x_3 & \leq & 32, \\ 7x_1 + 4x_2 + 2x_3 & \leq & 27, \\ \hline x_3 & \geq & 3, \\ x_1 \qquad\qquad & \leq & 2, \end{array}\right\} \quad S_3$$

$$\left.\begin{array}{rcl} 6x_1 + 3x_2 + 5x_3 & \leq & 32, \\ 7x_1 + 4x_2 + 2x_3 & \leq & 27, \\ \hline x_3 & \geq & 3, \\ x_1 \qquad\qquad & \geq & 3. \end{array}\right\} \quad S_3$$

Solving subproblems in 'Node 3' and 'Node 4' we obtain the following results. For subproblem in 'Node 3' we have

$$x_1^{(3)} = 2.00, \; x_2^{(3)} = 0.00, \; x_3^{(3)} = 4.00, \quad Q(x^{(3)}) \approx 0.63333333 .$$

Since this optimal solution is integer and $Q(x^{(3)}) > Bound$, we have to label 'Node 3' as a candidate and set $Bound := Q(x^{(3)}) = 0.63333333$. The subproblem in 'Node 4' is infeasible, i.e. it has no feasible solutions and hence, must be fathomed.

Now, performing Step 5 we have to check the list of dangling subproblems to see whether other branches of the tree must be pursued. The list contains the subproblem from 'Node 1' with objective value$Q(x^{(1)}) \approx 0.63157895$. Since

$$Q(x^{(1)}) < Bound = 0.63333333$$

branching at this dangling node will not increase the objective value. Therefore, we have to remove this subproblem from the list. Since the list is empty we terminate the branch-and-bound method. Thus, we have found an optimal solution to the original ILFP problem (8.10)-(8.13):

$$x_1^* = x_1^{(3)} = 2.00, \quad x_2^* = x_2^{(3)} = 0.00, \quad x_3^* = x_3^{(3)} = 4.00,$$
$$Q(x^*) \approx 0.63333333 .$$

Before closing this section we have to note that the branch-and-bound method may also be applied to mixed ILPF problems. Recall that in a mixed ILFP problem, some variables are required to be integers and others are allowed to be either integers or non-integers. To solve a mixed ILFP problem by the branch-and-bound method, we have to modify the method by branching only on variables that are required to be integers. Also, for a solution to a subproblem to be candidate solution, it need only assign integer values to those variables that are required to be integers.

3. The Cutting Plane Method

In the previous section we discussed how the branch-and-bound method can be used to solve ILFP problems. In this section, we discuss an alternative method, the so-called **cutting plane method**, which can be used to solve ILFP problems.

Originally, the cutting plane method was developed for integer linear programming problems by Ralph Gomory in 1958, see [80], [81], where it was shown that this process yields an integer optimal solution to the integer LP problem after a finite number of 'cuts'. Later the cutting plane method of Gomory was adapted to the class of integer LFP problems with compact feasible set (see e.g. [92], [93], [100], [156], [160]) [167].

The idea behind the algorithm is to start with a relaxation problem (i.e. ignoring the integrality constraints for variables) and solve it by the simplex method. If the resulting optimal solution is integer, the process is stopped, and we have solved the problem. Otherwise, we add to the original constraints a new one, the so-called **cutting plane**, which 'cuts off' (eliminates) a subset of feasible points, including the non-integer optimal solution just obtained. The effect of the cut can be seen in Figure 8.3. This cutting plane is constrained

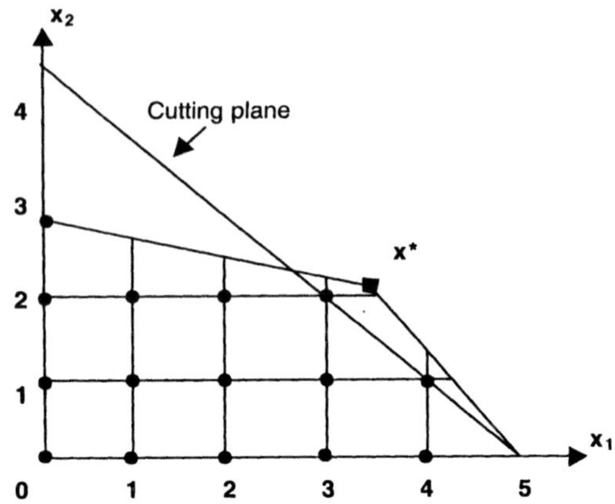

Figure 8.3. The Cutting Plane Method – Example of a cutting plane.

very carefully so it does not eliminate any integer feasible points. Then we solve the new problem and repeat this process until an integer optimal solution is obtained or the new problem is infeasible.

Consider the following pure integer LFP problem in canonical form.

$$Q(x) = \frac{P(x)}{D(x)} = \frac{\sum_{j=1}^{n} p_j x_j + p_0}{\sum_{j=1}^{n} d_j x_j + d_0} \longrightarrow \max \tag{8.14}$$

subject to

$$\sum_{j=1}^{n} a_{ij} x_j = b_i, \quad i = 1, 2, \ldots, m, \tag{8.15}$$

$$x_j \geq 0, \quad j = 1, 2, \ldots, n, \tag{8.16}$$

$$x_j - \text{ integer}, \quad j = 1, 2, \ldots, n, \tag{8.17}$$

where $D(x) > 0, \ \forall x \in S_0$ and S_0 denotes a continuous feasible set determined by constraints (8.15)-(8.16).

Here we assume that continuous feasible set S_0 of relaxation problem (8.14)-(8.16) is bounded and non-empty. Also we assume that coefficients p_j, d_j, a_{ij}, and b_i are all integers.

Let $x^{(0)}$ be an optimal non-integer solution for relaxation problem (8.14)-(8.16) and B denote the corresponding optimal basis $B = (A_{s_1}, A_{s_2}, \dots, A_{s_m})$, where $A_{s_i} = (a_{1,s_i}, a_{2,s_i}, \dots, a_{m,s_i})^T$, $i = 1, 2, \dots, m$. The ith constraint (8.15) as it appears in the final simplex tableau is

$$\sum_{j=1}^{n} x_{ij}\, x_j = x_{s_i}^{(0)}, \tag{8.18}$$

where $x_{s_i}^{(0)}$ is the optimal value of the ith basic variable and coefficients x_{ij} are defined as

$$\sum_{i=1}^{m} A_{s_i} x_{ij} = A_j, \quad j = 1, 2, \dots, n.$$

If we denote by $[x_{ij}]$ the integer part of coefficient x_{ij}, then, since $[x_{ij}] \le x_{ij}$ and $x_j \ge 0$, $j = 1, 2, \dots, n$, from (8.18) it follows that

$$\sum_{j=1}^{n} [x_{ij}]\, x_j \le x_{s_i}^{(0)}. \tag{8.19}$$

Observe that if vector $x^{(0)}$ is an integer and it satisfies condition (8.18), then it also satisfies inequality (8.19). Moreover, in this case the left-hand side of (8.19) is integer too. Thus, we can re-write (8.19) as follows

$$\sum_{j=1}^{n} [x_{ij}]\, x_j \le [x_{s_i}^{(0)}]. \tag{8.20}$$

By introducing slack variable u_i we can re-write the latter as

$$\sum_{j=1}^{n} [x_{ij}]\, x_j + u_i = [x_{s_i}^{(0)}]. \tag{8.21}$$

Further, since variable u_i expresses the difference between the integer value of the left-hand side and the integer value of the right-hand side of (8.20), it is obvious that u_i is also an integer. Thus, we have shown that any integer feasible point $x^{(0)}$ satisfies restriction (8.21) for some value of u_i.

Let us assume that $x^{(0)}$ is not an integer, and re-write (8.18) as follows

$$\sum_{j=1}^{n} ([x_{ij}] + \{x_{ij}\})\, x_j = [x_{s_i}^{(0)}] + \{x_{s_i}^{(0)}\}, \tag{8.22}$$

where $\{x_{ij}\}$ and $\{x_{s_i}^{(0)}\}$ denote the fractional part of x_{ij} and $x_{s_i}^{(0)}$, respectively. Recall that constraint (8.18) is satisfied by any optimal vector $x^{(0)}$ independently whether vector $x^{(0)}$ is an integer or is not one. At the same time, if vector $x^{(0)}$ is an integer, then it satisfies restriction (8.21) too. So, if we subtract (8.21) from (8.22), we obtain the following cutting plane constraint to be added to the constraints (8.15)

$$\sum_{j=1}^{n} \{x_{ij}\}\, x_j - u_i = \{x_{s_i}^{(0)}\}. \tag{8.23}$$

Since the coefficients x_{ij} of all basic optimal variables are equal to zero, except the ith one which is equal to 1, we can re-write (8.18) as follows

$$x_{s_i} + \sum_{j \in J_N} x_{ij}\, x_j = x_{s_i}^{(0)}, \tag{8.24}$$

where J_N denotes the set of indices of non-basic variables. Suppose that $x_{s_i}^{(0)}$ is not an integer. Constructing Gomory cutting plane (8.23) for (8.24) we obtain

$$\sum_{j \in J_N} \{x_{ij}\}\, x_j - u_i = \{x_{s_i}^{(0)}\}. \tag{8.25}$$

Taking into account that all non-basic variables of non-integer optimal solution $x^{(0)}$ are equal to zero, from (8.25) in point $x^{(0)}$ we obtain that

$$-u_i = \{x_{s_i}^{(0)}\} > 0,$$

i.e. slack variable $u_i < 0$. The latter means that point $x^{(0)}$ is no longer a feasible solution to the new LFP problem with the added constraint (8.25). So, we have 'cut off' the current optimal solution $x^{(0)}$.

So, Gomory's cutting plane method solves an integer problem by solving the corresponding relaxation problem, generating a cutting plane (if necessary) from the optimal simplex tableau, adding this additional constraint to the original system of constraints, solving the new relaxation problem, and repeating this process until an integer solution is obtained. Algorithmically, this process may be presented as follows.

Step 1 (Relaxation). First, we solve the relaxation LFP problem (8.14)-(8.16). If in its optimal solution $x^{(0)}$ all variables $x_j^{(0)}$ have an integer value, we have found an optimal solution for ILFP problem (8.14)-(8.17). Otherwise, we set $k := 0$ and proceed to Step 2.

Step 2 (Choosing fractional variable). Pick in the final simplex tableau just obtained the row whose right-hand side $x_{s_i}^{(k)}$, $i = 1, 2, \dots, m, m+1, \dots, m+k$, is fractional. The corresponding constraint should be used to generate a cutting plane constraint. Go to Step 3.

Step 3 (Cutting constraint). For the constraint identified in Step 2 write cutting plane constraint in the form of (8.23), add it to the constraints of the problem considered in Step 2, and then solve the new problem. Go to Step 4.

Step 4 (Termination test). If optimal solution $x^{(k+1)}$ obtained in Step 3 is an integer then Stop, we have found an optimal solution for original ILFP problem (8.14)-(8.17). Otherwise, set $k := k + 1$ and go to Step 2.

Several rules may be applied when we have to pick a cut plane generating row from the simplex tableau in Step 2. One of them is choosing the row whose non-integer variable $x_{s_i}^{(k)}$ has the fractional part closest to 0.5. Other rules recommend to choose the non-integer variable $x_{s_i}^{(k)}$ with the largest fractional part.

We illustrate the cutting plane algorithm by solving the following pure ILFP problem

$$Q(x) = \frac{P(x)}{D(x)} = \frac{6x_1 + 8x_2 + 3}{2x_1 + 3x_2 + 4} \longrightarrow \max \tag{8.26}$$

subject to

$$\left.\begin{array}{rcr} 1x_1 + 2x_2 & \leq & 6, \\ 4x_1 + 4x_2 & \leq & 10, \end{array}\right\} \tag{8.27}$$

$$x_j \geq 0 \text{ and integer}, \quad j = 1, 2. \tag{8.28}$$

After introducing artificial variables x_3 and x_4, and solving the canonical problem obtained, we have a final simplex tableau for the corresponding relaxation problem (8.26)-(8.28) shown in Table 8.3. Since the optimal solution

B	p_B	d_B	x_B	A_1	A_2	A_3	A_4
A_3	0	0	1	−1	0	1	−0.5
A_2	8	3	2.5	1	1	0	0.25
$P(x) = 23$				2	0	0	2
$D(x) = 11.5$				1	0	0	0.75
$Q(x) = 2$				0	0	0	0.5

Table 8.3. The Cutting Plane Method – Tableau 1.

$x^{(0)} = (0,\ 2.5)^T$ obtained is not integer, we choose in Table 8.3 row 2 associated with basic variable x_2 which has non-integer value 2.5. The corresponding constraint of type (8.18) is as follows

$$1\,x_1 + 1\,x_2 + 0\,x_3 + 0.25\,x_4 \;=\; 2.5\,. \tag{8.29}$$

Using formula (8.23) from (8.29) we obtain the following cutting constraint

$$0.25\, x_4 - u_1 \;=\; 0.5\;,$$

which may be re-written as

$$x_1 + x_2 + u_1 \;=\; 2\;, \tag{8.30}$$

since

$$4x_1 + 4x_2 + u_1 = 10.$$

Adding the cutting constraint obtained to system (8.27) and solving new LFP problem

$$Q(x) = \frac{P(x)}{D(x)} = \frac{6x_1 + 8x_2 + 3}{2x_1 + 3x_2 + 4} \longrightarrow \max$$

subject to

$$\begin{array}{rcrcrcrcrcr} 1x_1 & + & 2x_2 & + & x_3 & & & & & = & 6, \\ 4x_1 & + & 4x_2 & & & + & x_4 & & & = & 10, \\ x_1 & + & x_2 & & & & & + & u_1 & = & 2, \end{array}$$

$$x_j \geq 0 \;\; j = 1, 2, 3, 4; \quad u_1 \geq 0,$$

we have a final tableau shown in Table 8.4.

B	p_B	d_B	x_B	A_1	A_2	A_3	A_4	A_5
A_3	0	0	2	-1	0	1	0	-2
A_4	0	0	2	0	0	0	1	-4
A_2	8	3	2	1	1	0	0	1
$P(x) = 19$				2	0	0	0	8
$D(x) = 10$				1	0	0	0	3
$Q(x) = 1.9$				0.1	0	0	0	2.3

Table 8.4. The Cutting Plane Method – Tableau 2.

Table 8.4 gives an optimal solution

$$x_1^{(1)} = 0,\;\; x_2^{(1)} = 2,\;\; x_3^{(1)} = 2,\;\; x_4^{(1)} = 2,\;\; u_1 = 0.$$

Since $x_1^{(1)}$ and $x_2^{(1)}$ are integers, the optimal solution to ILFP problem (8.26)-(8.28) has been found

$$x^* = (0, 2)^T,\;\; Q(x^*) = 19/10 = 1.9\;.$$

Some minor adaptation is necessary if we apply the cutting plane method to mixed integer LFP problems. Similar to the case of pure ILFP, we start by solving a relaxation LFP problem. The optimal solution obtained will be an optimal solution of the original mixed ILPF problem if those variables which appear in the basis and are required to be integers, actually do have integer values. Otherwise we have to proceed as follows. Suppose now that basic variable x_{s_i} is required to be integer but its value $x_{s_i}^{(0)}$ is not an integer. The corresponding cutting plane constraint for x_{s_i} may be formulated in the following form

$$\sum_{j\in J_N^+} x_{ij}x_j + \frac{\{x_{s_i}^{(0)}\}}{\{x_{s_i}^{(0)}\}-1}\sum_{j\in J_N^-} x_{ij}x_j - u_i = \{x_{s_i}^{(0)}\},$$

where J_N^+ denotes the set of indices j of non-negative non-basic coefficients x_{ij}, and J_N^- is the set of indices j of negative non-basic x_{ij}, i.e.

$$J_N^+ = \{j \in J_N |\ x_{ij} \geq 0\}, \quad \text{and} \quad J_N^- = \{j \in J_N |\ x_{ij} < 0\},$$

J_N is the index set of non-basic variables.

Several remarks are in order.

In continuous LFP problems the number of positive basic optimal variables is $\leq m$ (or $= m$ in non-degenerate case). In the case of an ILFP problem this is no longer true, since the optimal solution may not occur at an extreme point of the convex feasible set S_0 of the corresponding relaxation problem.

When using the cutting plane method, in Step 3 we append a cutting plane constraint to the system of original constraints and then we have to solve the new problem obtained. Obviously, the optimal solution obtained in the previous phase in this case is no longer feasible (since it has been cut off by the cutting constraint (8.23)). Thus, to avoid solving the new problem from scratch it is highly desirable to use the so-called Dual Simplex Method in Step 3 (see Chapter 10, Section 1).

Since the cutting plane method only deals with the feasible set of a relaxation LFP problem, and the cutting plane constraint does not depend directly on the kind of objective function, it means that the method as it has just been described here may be applied to solve integer LP problems as it is, i.e. without any significant adaptations.

Unfortunately, the finite convergence of the method appears to be slow. As it was shown in [17] and [34], it is possible to improve the performance of the method if certain techniques are used, e.g. such as adding many Gomory cuts at once.

Finally, the most commonly used methods for integer programming problems, which are incorporated into professional optimization software packages, are **branch-and-cut** procedures based on the branch-and-bound method combined with the use of Gomory-type cutting planes [74].

The reader interested in different variants of the cutting plane method and their more detailed description is referred to the article [16] and book [137].

4. Formulating discrete LFP Problems

In this section, we show how discrete, integer and zero-one LFP problems may be converted to each other. Further, we show also how some practical situations may be formulated as an integer LFP problem.

4.1 Converting Problems

4.1.1 Discrete → Zero-One

Suppose we have an LFP problem where one of the variables, sayx_k has the following restriction

$$x_k \in \Lambda = \{\lambda_1, \lambda_2, \dots, \lambda_r\}, \tag{8.31}$$

i.e. is required only to take values from discrete setΛ, where $\lambda_i, i = 1, 2, \dots, r$, are arbitrary non-integer as well as integer constants. To convert this variable to zero-one form we have to substitute variablex_k with r new zero-one variables $x_k^{(i)}, i = 1, 2, \dots, r,$ as follows:

$$x_k = \sum_{i=1}^{r} x_k^{(i)} \lambda_i$$

and then append to the system of constraints of the problem the following condition

$$\sum_{i=1}^{r} x_k^{(i)} = 1.$$

Finally, instead of constraint (8.31) we obtain the following set of new constraints

$$\sum_{i=1}^{r} x_k^{(i)} = 1$$

$$x_k^{(i)} \text{ are } 0/1, \quad i = 1, 2, \dots, r.$$

4.1.2 Integer $\longrightarrow$ Zero-One

Consider the following ILPF problem

$$\max_{x \in S} Q(x) \longrightarrow \max .$$

Let us suppose that in the given ILFP problem we have to convert some integer variable, say x_k to zero-one form. We assume that variable x_k is required to be non-negative, so $x_k \geq 0$, and an integer. First, in this situation we have to find an upper bound of this variable that may involve looking for a solution for the following LP problem

$$\max_{x \in S} x_k \longrightarrow \max .$$

Let K denote the upper bound of variable x_k. Then, to convert this variable to zero-one form we have to substitute variable x_k with r new zero-one variables $x_k^{(i)}, i = 0, 1, 2, \ldots, r-1,$ as follows:

$$x_k = \sum_{i=0}^{r-1} x_k^{(i)} \lambda_i ,$$

where $\lambda_i = 2^{\,i}$ and r is determined by the following rule

- If K is equal to a power of 2, then $r = \log_2 K$;
- otherwise, r is equal to the smallest power of 2 that exceeds K, i.e. $K \in [2^{\,r-1}, 2^{\,r}]$.

For example, if $K = 9$, then $r = 4$, since $9 \in [2^{\,3}, 2^{\,4}]$. Thus, integer variable x_k has to be replaced with the following expression

$$x_k = 2^{\,0} x_k^0 + 2^{\,1} x_k^1 + 2^{\,2} x_k^2 + 2^{\,3} x_k^3 ,$$

where $x_k^i = 0/1$ for all $i = 0, 1, \ldots, r-1$.

4.2 Practical Situations

4.2.1 Either-Or Constraints

When modelling real-world problems the following situation may often occur. We have an unknown variable, say x_k which is allowed to take values only from one of the following two ranges

$$x_k \in [\lambda_1', \lambda_1''], \quad \text{or} \quad x_k \in [\lambda_2', \lambda_2''].$$

Typically, such a situation may occur if some product to be economically acceptable should be produced in an amount of at least λ_2' or should not be produced at all ($\lambda_1' = \lambda_1'' = 0$). Thus, in a more general case, we have two constraints of the form

$$f_1(x) \le 0, \quad \text{and} \quad f_2(x) \le 0, \tag{8.32}$$

and we wish to ensure that at least one of them is satisfied. Such constraints are usually referred to as **either-or constraints** [188]. In this situation we have to introduce a zero-one variable y and a large constant M such that

$$f_1(x) \le M, \quad f_2(x) \le M, \quad \forall x \in \text{ feasible set},$$

and then append to the main constraints the following two conditions

$$f_1(x) \le My, \quad f_2(x) \le M(1-y).$$

For example, consider the following situation

$$0 \le x_k \le 10, \quad \text{or} \quad 100 \le x_k \le 1000.$$

Obviously, constraints

$$x_k \ge 0, \quad \text{and} \quad x_k \le 1000,$$

are of conventional type and hence, may be maintained as they are. Further, converting remaining constraints to the form of (8.32) we obtain

$$f_1(x) = x_k - 10 \le 0, \qquad \text{for} \qquad x_k \le 10,$$
$$f_2(x) = 100 - x_k \le 0, \qquad \text{for} \qquad 100 \le x_k.$$

Thus, we have the following new constraints

$$\begin{array}{rcl} x_k - 10 & \le & My, \\ 100 - x_k & \le & M - My, \\ & y = 0/1 \,. & \end{array}$$

4.2.2 If-Then Constraints

Another type of special situation which may occur in many applications is as follows: we have an unknown variable, say x_k which may take any nonnegative value. If the value is positive, then some other variable(s) must be zero. Mathematically, restrictions of this type may be expressed as follows

$$\text{If } x_k > 0, \text{ then } x_{k+1} = 0.$$

Typically, such a situation may occur in economic problems where if product k must be produced in an amount greater than zero, then production of product $k+1$ is not allowed. Generally speaking, in such a situation we have two functions $f_1(x)$ and $f_2(x)$ which are interconnected as follows: if constraint $f_1(x) \geq 0$ is satisfied as strict inequality $f_1(x) > 0$, then the other constraint $f_2(x) \geq 0$ must be satisfied. While if $f_1(x) = 0$, then constraint $f_2(x) \geq 0$ may or may not be satisfied. Such constraints are usually referred to as**if-then constraints** [188].

Similar to the previous case, in this situation we have to introduce a zero-one variable y and a large constant M such that

$$f_1(x) \leq M, \quad -f_2(x) \leq M, \qquad \forall x \in \text{ feasible set},$$

and then append to the main constraints the following two conditions

$$f_1(x) \leq M(1-y), \quad -f_2(x) \leq My.$$

For example, consider the following situation

$$\text{If } \ x_k > 10, \ \text{ then } \ x_{k+1} \geq 50.$$

Let

$$\begin{aligned} f_1(x) &= x_k - 10 \geq 0, &\quad \text{for} \quad & x_k \geq 10, \\ f_2(x) &= x_{k+1} - 50 \geq 0, &\quad \text{for} \quad & x_k \geq 50, \end{aligned}$$

then, we have the following new constraints

$$\begin{aligned} x_k - 10 &\leq M(1-y), \\ -(x_{k+1} - 50) &\leq My, \\ y &= 0/1 \, . \end{aligned}$$

5. Discussion Questions and Exercises

8.1 Use the branch-and-bound method to solve the following pure ILFP problem

$$Q(x) = \frac{P(x)}{D(x)} = \frac{2x_1 + 4x_2 + 3x_3 + 10}{4x_1 + 5x_2 + 3x_3 + 24} \longrightarrow \max$$

subject to

$$\left. \begin{aligned} 7x_1 + 1x_2 + 5x_3 &\leq 45, \\ 5x_1 + 3x_2 + 7x_3 &\leq 38, \end{aligned} \right\}$$

$$x_j \geq 0, \text{ and integer } \ j = 1, 2, 3.$$

8.2 Examine how the flow of the branch-and-bound method changes if in the previous example we omit the integrality restriction for variablex_1.

8.3 Use the branch-and-bound method to solve the following mixed ILFP problem

$$Q(x) = \frac{P(x)}{D(x)} = \frac{3x_1 + 5x_2 + 3x_3 + 5}{5x_1 + 7x_2 + 5x_3 + 18} \longrightarrow \max$$

subject to

$$\left.\begin{array}{rcl} 7x_1 + 6x_2 + 6x_3 & \leq & 49, \\ 8x_1 + 9x_2 + 5x_3 & \leq & 47, \end{array}\right\}$$

$$x_j \geq 0, \quad j = 1, 2, 3,$$

$$x_1 \text{ - integer}, \quad x_2 \text{ - integer}.$$

8.4 Use the cutting plane method to solve the following pure integer LFP problem

$$Q(x) = \frac{P(x)}{D(x)} = \frac{6x_1 + 8x_2 + 3}{2x_1 + 3x_2 + 2} \longrightarrow \max$$

subject to

$$\begin{array}{rcl} 1x_1 + 2x_2 & \leq & 4, \\ 7x_1 + 3x_2 & \leq & 10, \end{array}$$

$$x_1 \geq 0, \quad x_2 \geq 0,$$

$$x_1 \text{ - integer}, \quad x_2 \text{ - integer}.$$

8.5 Examine how the flow of the cutting plane method changes if in the previous example variable x_1 is not required to be an integer.

Chapter 9

SPECIAL LFP PROBLEMS

In this chapter, we discuss the following special classes of linear-fractional programming problems: transportation problems, assignment problems, and transshipment problems. While each of these problems can be solved by the simplex method, there are specialized algorithms for each type of problems that are much more efficient than the simplex algorithm.

1. The Transportation Problem

In this section we deal with the so-called *transportation problem* of linear-fractional programming. First, in Section 1.1 we formulate the problem in a general form and shortly overview the main theoretical results. The main computational technique used to solve transportation LFP problems is the *Transportation Simplex Method.* It is presented in Section 1.2. Then in Section 1.3 we shortly overview the basics of the methods that allow to determine an initial basic feasible solution needed to start the transportation simplex method. Finally, in Section 1.5 we construct the dual problem of the transportation LFP problem and formulate the main statements of duality theory for this problem.

1.1 Formulation and Preliminaries

In general, an LFP transportation (LFPT) problem is specified by the following information:

1. A set of m *supply points* (or *stores*) from which a good must be shipped. Supply point i ($i = 1, 2, \dots, m$) can supply at most b_i units of the good.

2. A set of n *demand points* (or *shops*) to which the good must be shipped. Demand point j ($j = 1, 2, \ldots, n$) must obtain at least a_j units of the good.

3. Profit matrix $P = ||p_{ij}||_{m\times n}$ which determines the profit p_{ij} gained by a transportation company if a unit of the good is shipped from supply point i to demand point j.

4. Cost matrix $D = ||p_{ij}||_{m\times n}$ which determines the cost d_{ij} of shipping a unit of the good from supply point i to demand point j.

5. Constants p_0 and d_0 which determine some constant profit and cost, respectively.

If variable x_{ij} is an unknown quantity of the good shipped from supply point i to demand point j then the transportation problem of linear-fractional programming may be formulated as follows:
Given objective function

$$Q(x) = \frac{P(x)}{D(x)} = \frac{\sum_{i=1}^{m}\sum_{j=1}^{n} p_{ij}x_{ij} + p_0}{\sum_{i=1}^{m}\sum_{j=1}^{n} d_{ij}x_{ij} + d_0}, \tag{9.1}$$

which must be minimized (or maximized) subject to

$$\sum_{j=1}^{n} x_{ij} \leq b_i, \quad i = 1, 2, \ldots, m, \tag{9.2}$$

$$\sum_{i=1}^{m} x_{ij} \geq a_j, \quad j = 1, 2, \ldots, n, \tag{9.3}$$

$$x_{ij} \geq 0, \quad i = 1, 2, \ldots, m; \ j = 1, 2, \ldots, n. \tag{9.4}$$

Here and in what follows we suppose that $D(x) > 0, \ \forall x = (x_{ij}) \in S,$ where S denotes a feasible set defined by constraints (9.2)-(9.4). Further, we assume that

$$b_i > 0, \ a_j > 0, \ i = 1, 2, \ldots, m; \ j = 1, 2, \ldots, n \tag{9.5}$$

and total supply is not less than total demand, i.e.

$$\sum_{i=1}^{m} b_i \geq \sum_{j=1}^{n} a_j. \tag{9.6}$$

This form of the LFP transportation problem has the following properties:

- The problem always has a feasible solution, i.e. feasible set $S \neq \emptyset$.
- The set of feasible solutions is bounded.
- Hence, the problem is always solvable.

Indeed, let

$$x'_{ij} = \frac{b_i a_j}{K}, \quad i = 1, 2, \dots, m; \; j = 1, 2, \dots, n, \tag{9.7}$$

where

$$K = \sum_{j=1}^{n} a_j > 0.$$

Substituting x'_{ij} for constraints (9.2) and (9.3) we obtain that

$$\sum_{j=1}^{n} x'_{ij} = \sum_{j=1}^{n} \frac{b_i a_j}{K} = \frac{b_i}{K} \sum_{j=1}^{n} a_j = \frac{b_i}{K} K = b_i, \quad i = 1, 2, \dots, m,$$

and

$$\begin{aligned} \sum_{i=1}^{m} x'_{ij} &= \sum_{i=1}^{m} \frac{b_i a_j}{K} = \frac{a_j}{K} \sum_{i=1}^{m} b_i \overset{(9.6)}{\geq} \frac{a_j}{K} \sum_{j=1}^{n} a_j = \\ &= \frac{a_j}{K} K = a_j, \quad j = 1, 2, \dots, n, \end{aligned}$$

respectively. Hence, constraints (9.2) and (9.3) are satisfied by x'_{ij}. Since from (9.5) and (9.7) it follows that $x'_{ij} > 0$, $i = 1, 2, \dots, m$; $j = 1, 2, \dots, n$, it becomes obvious that $x' = (x'_{ij})$ is a feasible solution of the problem. Thus, we have shown that feasible set S is not empty.

Further, from (9.2) and (9.4) we have that

$$0 \leq x_{ij} \leq b_i, \quad i = 1, 2, \dots, m; \; j = 1, 2, \dots, n.$$

The latter means that feasible set S is bounded.

Finally, since functions $P(x)$ and $D(x)$ are linear and $D(x) > 0$ over bounded feasible set S, it means that objective function $Q(x)$ is also bounded over set S and hence, the LFP transportation problem is solvable.

DEFINITION 9.1 *If total demand equals to total supply, i.e.*

$$\sum_{i=1}^{m} b_i = \sum_{j=1}^{n} a_j, \tag{9.8}$$

then the LFPT problem is said to be a ***balanced transportation problem.***

LFPT problem (9.1)-(9.4) is also known as the *un-capacitated transportation problem* because there are no specified upper bounds on unknown variables x_{ij}, although the inequality supply constraints (9.2) and nonnegativity constraints (9.4) automatically imply an upper bound for every x_{ij}, namely

$$x_{ij} \leq \min_{1 \leq i \leq m} \{b_i\}, \; i = 1, 2, \dots, m, \; j = 1, 2, \dots, n.$$

If a transportation problem has a total supply that is strictly less than total demand, the problem has no feasible solution. In this situation it is sometimes desirable to allow the possibility of leaving some demand unmet. In such a case, we can balance a transportation problem by creating a *dummy supply point* that has a supply equal to the amount of unmet demand, and associating a penalty with it. For more information on this topic see for example [78], [188], etc.

1.2 The Transportation Simplex Method

As in the case of a general LFP problem, the solution process of an LFPT problem consists of two phases:

1 Finding an initial basic feasible solution (BFS);

2 Improving the current basic feasible solution until the optimality criterion is satisfied.

Since the process of finding initial BFS for LFPT is the same as in the LP case, we will focus mainly on the second stage.

Consider the following LFPT problem in a canonical form

$$Q(x) = \frac{P(x)}{D(x)} = \frac{\sum_{i=1}^{m}\sum_{j=1}^{n} p_{ij}x_{ij} + p_0}{\sum_{i=1}^{m}\sum_{j=1}^{n} d_{ij}x_{ij} + d_0} \longrightarrow \max \tag{9.9}$$

subject to

$$\sum_{j=1}^{n} x_{ij} = b_i, \quad i = 1, 2, \dots, m, \tag{9.10}$$

$$\sum_{i=1}^{m} x_{ij} = a_j, \quad j = 1, 2, \dots, n, \tag{9.11}$$

$$x_{ij} \geq 0, \quad i = 1, 2, \ldots, m; \; j = 1, 2, \ldots, n, \tag{9.12}$$

where $D(x) > 0, \;\; \forall x = (x_{ij}) \in S.$

This LFPT problem has the following augmented matrix of main constraints (9.10)-(9.11)

$$A\,|R = \left(\begin{array}{cccccccccccccc|c}
1 & 1 & \cdots & 1 & & & & & & & & & & & b_1 \\
 & & & & 1 & 1 & \cdots & 1 & & & & & & & b_2 \\
 & & & & & & & & \ddots & & & & & & \vdots \\
 & & & & & & & & & 1 & 1 & \cdots & 1 & & b_m \\
1 & & & & 1 & & & & & 1 & & & & & a_1 \\
 & 1 & & & & 1 & & & & & 1 & & & & a_2 \\
 & & \ddots & & & & \ddots & & \cdots & & & \ddots & & & \vdots \\
 & & & 1 & & & & 1 & & & & & 1 & & a_n
\end{array}\right),$$

where R denotes the column vector of supplies $b_i, \;\; i = 1, 2, \ldots, m,$ and demands $a_j, \;\; j = 1, 2, \ldots, n,$ i.e.

$$R = (b_1, b_2, \ldots, b_m, a_1, a_2, \ldots, a_n)^T,$$

and matrix A has $m + n$ rows and $m \times n$ columns.

Let $A_{ij}, \;\; i = 1, 2, \ldots, m, \;\; j = 1, 2, \ldots, n,$ denote column vectors of matrix A. It is obvious that vector A_{ij} contains '1's in position i and position $m + j$. All other elements of vector A_{ij} are equal to zero.

THEOREM 9.1 (REDUNDANCY) *There is exactly one redundant equality constraint in (9.10) and (9.11). When any one of the constraints in (9.10) or (9.11) is dropped, what remains is a linearly independent system of constraints. So the rank of matrix A is equal to $m + n - 1$.*

We omit the proof because it is exactly the same as in LP, see e.g., [135].

DEFINITION 9.2 *We say that system B of $m + n - 1$ vectors A_{ij} is a **basis** of LFPT problem (9.9)-(9.12), if these vectors A_{ij} are linearly independent.*

Suppose that a given system B of $m + n - 1$ vectors A_{ij} is a basis. Let J_B denote the set of the pairs of indices (ij) corresponding to basic vectors A_{ij}. If set J denotes all possible pairs (ij) of indices $i = 1, 2, \ldots, m$ and $j = 1, 2, \ldots, n,$ then set $J_N = J \setminus J_B$ denotes the indices (ij) of those vectors Aij, which are not in basis B.

DEFINITION 9.3 *We will say that* $x = (x_{ij})$ *is a* ***basic solution*** *of LFPT problem (9.9)-(9.12) if* x *satisfies system*

$$\sum_{(ij)\in J_B} A_{ij}x_{ij} = R \quad \text{and} \quad x_{ij} = 0, \ \forall (ij) \in J_N.$$

As in the case of a general LFP problem, those variables x_{ij} whose indices (ij) are in the set J_B are said to be *basic variables* or *BV's*. If variable x_{ij} is such that $(ij) \in J_N$, we say that this variable is a *nonbasic variable* or *NBV*.

The conception of a (non-)degenerate basic feasible solution in an LFPT problem has a role in the transportation simplex method as important as in the case of the common simplex method.

DEFINITION 9.4 *We will say that basic solution* x *is* ***degenerate,*** *if at least one of its basic variables is equal to zero, i.e.* $\exists (ij) : (ij) \in J_B,$ *such that* $x_{ij} = 0.$ *In the case if* $x_{ij} \neq 0, \ \forall (ij) \in J_B,$ *basic solution* x *is said to be* ***non-degenerate.***

DEFINITION 9.5 *Basic solution* $x = (x_{ij})$ *is said to be a* ***basic feasible solution (BFS)*** *of LFPT problem (9.9)-(9.12) if all elements* $x_{ij}, \ (ij) \in J_B,$ *satisfy nonnegativity constraints (9.12).*

Unlike an LFPT problem in the form (9.1)-(9.4) with assumptions (9.5) and (9.6), the canonical LFPT problem can have no solutions. The following statement establishes the necessary and sufficient condition for the canonical LFPT problem (9.9)-(9.12) to be solvable.

THEOREM 9.2 *LFPT problem (9.9)-(9.12) is solvable if and only if the following* ***balance equality*** *holds*

$$\sum_{i=1}^{m} b_i = \sum_{j=1}^{n} a_j. \tag{9.13}$$

Proof. *Necessity.* Suppose that LFPT problem (9.9)-(9.12) is solvable and x is its basic feasible solution. Adding together separately supply constraints (9.10) for all indices $i = 1, 2, \dots, m,$ and demand constraints (9.11) for all indices $j = 1, 2, \dots, n,$ we obtain

$$\sum_{i=1}^{m}\sum_{j=1}^{n} x_{ij} = \sum_{i=1}^{m} b_i, \quad \text{and} \quad \sum_{j=1}^{n}\sum_{i=1}^{m} x_{ij} = \sum_{j=1}^{n} a_j,$$

respectively. Since the left-hand sides of the equalities obtained are exactly the same, their right-hand sides are equal to one another. Hence, balance equality (9.13) holds.

Sufficiency. Suppose now that condition (9.13) holds. We have to show that in this case feasible set S of LFPT (9.9)-(9.12) problem is not empty and objective function $Q(x)$ over set S is bounded. Let

$$x'_{ij} = \frac{b_i a_j}{K}, \quad i = 1,2,\dots,m, \; j = 1,2,\dots,n,$$

where

$$K = \sum_{i=1}^{m} b_i = \sum_{j=1}^{n} a_j.$$

It is easy to show that x' is a feasible solution of (9.9)-(9.12). Indeed, we have

$$\sum_{j=1}^{n} x'_{ij} = \sum_{j=1}^{n} \frac{b_i a_j}{K} = \frac{b_i}{K} \sum_{j=1}^{n} a_j = b_i, \quad i = 1,2,\dots,m,$$

$$\sum_{i=1}^{m} x'_{ij} = \sum_{i=1}^{m} \frac{b_i a_j}{K} = \frac{a_j}{K} \sum_{i=1}^{m} b_i = a_j, \quad j = 1,2,\dots,n.$$

Further, since $a_j > 0, \; j = 1,2,\dots,n,$ and $b_i > 0, \; i = 1,2,\dots,m,$ it ensures us that $x'_{ij} \geq 0$ for all indices $i = 1,2,\dots,m,$ and $j = 1,2,\dots,n.$ Hence, we have shown that feasible set S is not empty and contains at least one feasible solution $x' = (x'_{ij})$.

From (9.10) and (9.12) we have that

$$0 \leq x_{ij} \leq b_i, \quad i = 1,2,\dots,m; \; j = 1,2,\dots,n.$$

The latter means that feasible set S is bounded.

Finally, since $P(x)$ and $D(x)$ are linear functions and $D(x) > 0$ over bounded feasible set S, it means that objective function $Q(x)$ is also bounded over set S and hence, LFPT problem (9.9)-(9.12) is solvable.◇

The following statement (formulated as a theorem with omitted proof) indicates a very useful property of an LFPT problem.

THEOREM 9.3 (INTEGER PROPERTY) *If all a_j and b_i in LFPT problem (9.9)-(9.12) are positive integers, then every basic solution of (9.9)-(9.12) is an integer vector. Hence, if all a_j and b_i in LFPT problem (9.9)-(9.12) are positive integers and balance equality (9.13) holds, then canonical balanced*

LFPT problem (9.9)-(9.12) has an optimal solution x^* *with integer elements* $x^*_{ij},\ i=1,2,\dots,m,\ j=1,2,\dots,n.$

We now show how the simplex method may be adapted to the case when an LFPT problem is to be solved. First, we have to introduce special simplex multipliers $u'_i,\ v'_j$ and $u''_i,\ v''_j$ associated with numerator $P(x)$ and denominator $D(x)$, respectively. Elements u'_i and $u''_i,\ i=1,2,\dots,m,$ correspond to m supply constraints (9.10) and elements v'_j and $v''_j,\ j=1,2,\dots,n,$ correspond to n demand constraints (9.11). We calculate these variables from the following systems of linear equations

$$u'_i + v'_j = p_{ij},\quad (ij)\in J_B\,, \tag{9.14}$$

and

$$u''_i + v''_j = d_{ij},\quad (ij)\in J_B\,. \tag{9.15}$$

Then, using these variables $u'_i,\ v'_j,\ u''_i,$ and v''_j we define the following 'reduced costs' Δ'_{ij} and Δ''_{ij}

$$\left.\begin{array}{rcl}\Delta'_{ij} &=& u'_i + v'_j - p_{ij}\\ \Delta''_{ij} &=& u''_i + v''_j - d_{ij}\end{array}\right\}\quad i=1,2,\dots,m,\ j=1,2,\dots,n. \tag{9.16}$$

Further, we define the following values

$$U_i(x) = u'_i - Q(x)\,u''_i,\quad i=1,2,\dots,m,$$

$$V_j(x) = v'_j - Q(x)\,v''_j,\quad j=1,2,\dots,n,$$

$$Z_{ij}(x) = U_i(x) + V_j(x),\quad i=1,2,\dots,m,\ j=1,2,\dots,n,$$

$$C_{ij}(x) = p_{ij} - Q(x)\,d_{ij},\quad i=1,2,\dots,m,\ j=1,2,\dots,n,$$

and, finally

$$\Delta_{ij}(x) = Z_{ij}(x) - C_{ij}(x),\quad i=1,2,\dots,m,\ j=1,2,\dots,n.$$

It is easy to show that the latter may also be expressed as follows

$$\Delta_{ij}(x) = \Delta'_{ij} - Q(x)\Delta''_{ij},\quad i=1,2,\dots,m,\ j=1,2,\dots,n. \tag{9.17}$$

Using this notation we can formulate criteria of optimality for BFS x as follows

THEOREM 9.4 (OPTIMALITY CRITERIA) *Basic feasible solution* $x=(x_{ij})$ *of LFPT problem (9.9)-(9.12) is optimal if*

$$\Delta_{ij}(x)\ge 0,\quad i=1,2,\dots,m,\ j=1,2,\dots,n. \tag{9.18}$$

Proof. Let B be a feasible basis of LFPT problem (9.9)-(9.12) and let x denote a corresponding basic feasible solution. Suppose that there is another solution x' which differs from x by only one element and may be obtained from x by entering into the basis non-basic variable x_{rk}, $(rk) \in J_N$. We have

$$P(x') = P(x) - \theta\Delta'_{rk}, \quad \text{and} \quad D(x') = D(x) - \theta\Delta''_{rk},$$

where $\theta \leq 0$ is a value associated with new basic variable x_{rk}, and reduced costs Δ'_{rk} and Δ''_{rk} are determined by (9.16).

Calculating the difference between $Q(x')$ and $Q(x)$ we obtain that

$$\begin{aligned} Q(x') - Q(x) &= \frac{P(x')}{D(x')} - \frac{P(x)}{D(x)} = \frac{P(x) - \theta\Delta'_{rk}}{D(x) - \theta\Delta''_{rk}} - \frac{P(x)}{D(x)} = \\ &= -\frac{\theta\Delta'_{rk}D(x) - \theta\Delta''_{rk}P(x)}{D(x')D(x)} = -\frac{\theta(\Delta'_{rk} - Q(x)\Delta''_{rk})}{D(x')} = \\ &= -\frac{\theta\Delta_{rk}(x)}{D(x')}, \quad \forall (rk) \in J_N. \end{aligned} \tag{9.19}$$

Since $D(x') > 0$, $\forall x' \in S$, formula (9.19) shows that difference $Q(x') - Q(x)$ can be positive in the only case if there exists such index (rk) that $\Delta_{rk}(x) < 0$. Thus, we have proven the statement.$\diamondsuit$

It is convenient to maintain the data of the problem in a transportation simplex tableau shown in Table 9.1, where T_{ij} denotes basic variable x_{ij}, if $(ij) \in J_B$, and value $\Delta_{ij}(x)$ for non-basic indices $(ij) \in J_N$. So the cells of the transportation simplex tableau are as follows

p_{ij}	
	x_{ij}
d_{ij}	

, $\forall (ij) \in J_B$,

and

p_{ij}	
	$\Delta_{ij}(x)$
d_{ij}	

or

p_{ij}		Δ'_{ij}
	$\Delta_{ij}(x)$	
d_{ij}		Δ''_{ij}

, $\forall (ij) \in J_N$.

DEFINITION 9.6 *A **circle** C (or **loop**) is an ordered subset of at least four different cells of a transportation simplex tableau if*

1 Any two consecutive cells lie in either the same row or same column,

	Shop 1	*Shop* 2	...	*Shop* n	*Supply*
Store 1	p_{11} T_{11} d_{11}	p_{12} T_{12} d_{12}	...	p_{1n} T_{1n} d_{1n}	b_1
Store 2	p_{21} T_{21} d_{21}	p_{22} T_{22} d_{22}	...	p_{2n} T_{2n} d_{2n}	b_2
⋮	⋮	⋮	⋮	⋮	⋮
Store m	p_{m1} T_{m1} d_{m1}	p_{m2} T_{m2} d_{m2}	...	p_{mn} T_{mn} d_{mn}	b_m
Demand	a_1	a_2	...	a_n	

Table 9.1. Transportation simplex tableau for an LFPT problem.

2 *No three consecutive cells lie in the same row or column,*

3 *The last cell in the sequence has a row or column in common with the first cell in the sequence.*

The circles that we are interested in are most often of a special type. One of the cells, for example (rk), is not in the current basis, i.e. $(rk) \in J_N$, while all the remaining circle cells are in the current basis. The non-basic cell(rk) is said to be the one that *forms* or *generates* the circle C. So the circles we will use in the transportation simplex method may be thought of as a closed path starting in a non-basic cell, then running through several basic cells and, finally reaching its end in the start cell.

	1	2	3	4
1	→	↓		
2		→		↓
3				
4	↑			←

	1	2	3	4	5
1			→		↓
2	↓				←
3					
4	→		↑		

Table 9.2. Transportation LFP problem – Circle examples.

Tables 9.2 and 9.3 show some examples of the preceding definition. The LFPT tableaus shown in Table 9.2 contain the following two circles:

$$(1,1) \to (1,2) \to (2,2) \to (2,4) \to (4,4) \to (4,1) \to (1,1)$$

and

$$(1,3) \to (1,5) \to (2,5) \to (2,1) \to (4,1) \to (4,3) \to (1,3).$$

Paths shown in Table 9.3 do not represent a circle. The reason is that in the left-hand side tableau the first row contains more thatn two cells, while in the right-hand side tableau the problem is that in the second and third columns we marked only one cell in each column (namely, cell $(1,2)$ in column 2 and cell $(4,3)$ in column 3).

	1	2	3	4	5
1	$\rightarrow$		$\rightarrow$		$\downarrow$
2	$\uparrow$				$\leftarrow$
3					

	1	2	3
1	$\rightarrow$	$\downarrow$	
2			
3			
4	$\uparrow$		$\leftarrow$

Table 9.3. Transportation LFP problem – Non-circle examples.

Let us suppose that B is a feasible basis and x denotes an associated BFS. To check if the current BFS x is optimal, we construct systems of equalities (9.14) and (9.15), and then solve these systems to determine variables u'_i, v'_j, u''_i, and v''_j. Note that each of systems (9.14) and (9.15) consists of $m+n-1$ equations (associated with $m+n-1$ basic variables x_{ij}, $(ij) \in J_B$) and contains $m+n$ unknown variables u'_i, v'_j and u''_i, v''_j, respectively. Since these systems have multiple solutions, we can solve them by fixing any one of the variables in both systems to an arbitrary value, e.g. zero. Having solved these systems and having known values of u'_i, v'_j, u''_i, and v''_j, we calculate reduced costs Δ'_{ij}, Δ''_{ij} and $\Delta'_{ij}(x)$ for all non-basic indices $(ij) \in J_N$. In this phase the following two situations may occur:

1 All non-basic reduced costs $\Delta_{ij}(x), \ \forall (ij) \in J_N$, are nonnegative. Since $\Delta_{ij}(x) = 0, \ \forall (ij) \in J_B$, it means that $\Delta_{ij}(x) \geq 0, \ \forall (ij) \in J$.

2 At least one of non-basic reduced costs $\Delta_{ij}(x)$ has a negative value, that is

$$J_N^- = \{(ij) | \ (ij) \in J_N, \ \Delta_{ij}(x) < 0\} \neq \emptyset.$$

In case 1 in accordance with the criteria of optimality (see Theorem 9.4) the current BFS x is an optimal solution of the LFPT problem. The process must be terminated.

In the second case we have to choose some index-pair $(rk) \in J_N^-$ and enter non-basic variable x_{rk} into the basis using the following rule:

- First, we mark cell (rk) with sign '+' and then starting from this cell (rk) we build a circle marking by turns the next cell with sign '-', then the second cell with '+' and so on. Once, we have built the circle we can determine the value of θ as

$$\theta = \min_{(ij) \in J_B^-} x_{ij} = x_{fq}, \tag{9.20}$$

  where J_B^- denotes an index set of those basic variables x_{ij} which are in the circle and are marked with sign '-'.

- Recalculate basic variables included in the circle by formula

$$x_{ij}(\theta) = \begin{cases} x_{ij} - \theta, & \text{if } (ij) \in J_B^-, \\ x_{ij} + \theta, & \text{if } (ij) \in J_B^+, \end{cases}$$

  where J_B^+ is an index set of those basic variables x_{ij} which are in the circle and are marked with sign '+'.

- All other basic variables which are not included in the circle remain unchanged.

- New basic variable $x_{rk}(\theta) = \theta$.

- Basic variable $x_{fq}(\theta) = 0$ and hence it leaves the basis.

Once we have calculated the new basic feasible solution $x(\theta)$ we have to recalculate variables $u'_i,\ v'_j,\ u''_i,\ v''_j$ and reduced costs $\Delta'_{ij},\ \Delta''_{ij}\ \ \Delta'_{ij}(x)$ in the new basis and then to check if the new BFS is optimal. Since a balanced canonical LFPT problem always has an optimal solution, the iterative process of the transportation simplex method after a finite number of iterations will terminate.

Before closing this discussion of the transportation simplex method we have to note that in a transportation LFP problem a degeneracy may occur. Suppose that when determining the value of θ we obtained in (9.20) multiple indices of minimal value, i.e.

$$\theta = \min_{(ij) \in J_B^-} x_{ij} = x_{f_1 q_1} = x_{f_2 q_2} = \ldots = x_{f_h q_h}.$$

This expression shows that after performing simplex iteration we obtain degenerate BFS $x(\theta)$ since some of its basic variables

$$x_{f_1q_1}(\theta),\ x_{f_2q_2}(\theta),\ \ldots,\ x_{f_hq_h}(\theta)$$

have a value equal to zero.

Suppose now that the current BFS x is degenerate. In this situation it may happen that we obtained such a circle that contains zero-valued basic variables in the cells marked by '-'. In other words, it means that there are such indices $(i_0 j_0) \in J_B^-$ that $x_{i_0 j_0} = 0$. Hence, in accordance with formula (9.20) we have that $\theta = 0$ and there is no change in the objective value. Obviously, in this case to avoid a possible cycling of the transportation simplex method we may apply the same special pivoting rules as we used in the simplex method for a common LFP problem (see Chapter 4, Section 9).

1.3 Determining Initial BFS

Here we briefly discuss some of the most widespread special procedures for determining an initial BFS for canonical LFPT problem (9.9)-(9.12). All of these methods originate from a transportation problem of linear programming, and may be easily adapted to the case of LFPT. Here and in what follows we assume that balance equality (9.13) holds.

1.3.1 Northwest Corner Method

This method does not use either coefficients of numerator in objective function $Q(x)$ or coefficients of the denominator. To build an initial BFS by this method, we begin in the upper left ('northwest') corner of the transportation simplex method and set variable x_{11} as large as possible, i.e. $x_{11} = \min\{b_1, a_1\}$. If $x_{11} = b_1$, we mark (cross out or put '×' against) the first row of the tableau and in this way indicate that x_{11} is an only basic variable in the first row. Also we change a_1 to $a_1 - b_1$. If $x_{11} = a_1$, we mark the first column of the tableau that will indicate that x_{11} is an only basic variable in this column. Also we have to change b_1 to $b_1 - a_1$. If $x_{11} = a_1 = b_1$, we mark either column 1 or row 1 of the tableau. If we mark column 1, we have to change b_1 to 0. If we mark row 1, we change a_1 to 0. Once we have completed the processing of x_{11}, we continue applying this procedure to the most northwest cell (it will be the next basic variable) of the tableau that does not lie in a marked row or column. We have to repeat this procedure until the last column or row is marked. An initial BFS has now been obtained.

To illustrate how the Northwest Corner Method may be used we consider the transportation tableau given in Table 9.4 (we do not indicate here the coefficients

	1	2	3	4	
1					200
2					100
3					100
4					100
	120	150	180	50	

Table 9.4. Northwest Corner Method Example – Original tableau.

of the objective function, since this method does not use these coefficients at all).

	1	2	3	4	
1	120				80
2					100
3					100
4					100
	×	150	180	50	

	1	2	3	4	
1	120	80			×
2					100
3					100
4					100
	×	70	180	50	

Table 9.5. Northwest Corner Method Example – Tableaus 1 and 2.

	1	2	3	4	
1	120	80			×
2		70			30
3					100
4					100
	×	×	180	50	

	1	2	3	4	
1	120	80			×
2		70	30		×
3					100
4					100
	×	×	150	50	

Table 9.6. Northwest Corner Method Example – Tableaus 3 and 4.

	1	2	3	4	
1	120	80			×
2		70	30		×
3			100		×
4					100
	×	×	50	50	

	1	2	3	4	
1	120	80			×
2		70	30		×
3			100		×
4			50		50
	×	×	×	50	

Table 9.7. Northwest Corner Method Example – Tableaus 5 and 6.

	1	2	3	4	
1	120	80			×
2		70	30		×
3			100		×
4			50	50	×
	×	×	×	×	

Table 9.8. Northwest Corner Method Example – Final tableau 7.

To begin, we set $x_{11} = \min\{b_1, a_1\} = \min\{200, 120\} = 120$, Table 9.5, tableau 1. Then we cross out column 1 and change $b_1 \to b_1 - a_1 = 200 - 120 = 80$. The most northwest remaining variable is x_{12}. So, we set $x_{12} = \min\{b_1, a_2\} = \min\{80, 150\} = 80$, Table 9.5, tableau 2. We continue this process until we obtain the final tableau 7, Table 9.8, which gives us the following BFS $x_{11} = 120,\ x_{12} = 80,\ x_{22} = 70,\ x_{23} = 30,\ x_{33} = 100,\ x_{34} = 50,\ x_{44} = 50$, with basic index set

$$J_B = \{(1,1), (1,2), (2,2), (2,3), (3,3), (4,3), (4,4)\},$$

Note that this BFS contains exactly $m+n-1 = 4+4-1 = 7$ basic variables.

Since this method does not use coefficients p_{ij} and d_{ij} of objective function $Q(x)$, it may result in an initial BFS that has a very high shipping cost $D(x)$ and/or very low profit $P(x)$. The following method allows us to utilize the coefficients of $Q(x)$.

1.3.2 Maximum Profit (or Minimum Cost) Method

Actually, Minimum Cost (or Maximum Profit) method used in linear-fractional programming to find an initial BFS for transportation problems descends from the well-known Minimum Cost method of linear programming, and is adapted to LFPT in two modifications: one of them to choose a basic variable utilizes like its ancestor the shipping costs d_{ij} presented in function $D(x)$ (Minimum Cost method), and the second one is based on shipping profits p_{ij} presented in function $P(x)$ (Maximum Profit method). Since these two methods differ from one another only in the rule for choosing variables, we restrict our consideration with the Maximum Profit method.

To build an initial BFS by this method we begin by finding the variable $x_{i_1 j_1}$ which corresponds to the highest profit p_{ij}. Then assign $x_{i_1 j_1}$ its largest possible value, i.e. $x_{i_1 j_1} = \min\{b_{i_1}, a_{j_1}\}$. As in the Northwest Corner method, we mark (cross out) row i_1 or column j_1 and reduce the corresponding supply or demand by the value of $x_{i_1 j_1}$. Then we repeat this procedure using only those cells that do not lie in the crossed out rows and columns. We have to continue this process until there is only one cell in the transportation tableau that can be chosen.

To illustrate this method we consider the LFPT problem given in Table 9.9. Since this method does not use coefficients d_{ij} of function $D(x)$, we give in the initial tableau only coefficients p_{ij}. The cell that contains the highest profit is $(4, 2)$, so we choose variable x_{42} and set it $x_{42} = \min\{b_4, a_2\} = \min\{100, 150\} = 100$. Then we replace $a_2 \rightarrow a_2 - b_4 = 50$ and we mark row 4 with sign '$\times$' or cross it out. It results in a tableau shown in Table 9.10. The next variable that must be chosen is x_{34}. We determine its value as $x_{34} = \min\{b_3, a_4\} = \min\{100, 50\} = 50$, then replace $b_3 \rightarrow b_3 - a_4 = 100 - 50 = 50$ and mark column 4 with '$\times$'. The result is shown in Table 9.11. We could now choose either x_{11} or x_{33}, since both of them have a profit of 8. We arbitrarily choose x_{11} and set $x_{11} = \min\{b_1, a_1\} = \min\{200, 120\} = 120$. Then we mark column 1 and change b_1 to $200 - 120 = 80$. The result is presented in Table 9.12. Table 9.13 contains results obtained after choosing variable x_{33} and setting it $x_{33} = \min\{b_3, a_3\} = \min\{50, 180\} = 50$. Repeating this process several times we obtain the final tableau shown in Table 9.14, which gives the following initial BFS $x_{11} = 120,\ x_{12} = 50,\ x_{13} = 30,\ x_{23} = 100,\ x_{33} = 50,\ x_{34} = 50,\ x_{42} = 100$ with profit value of $P(x) = 4090$. Observe that the profit value obtained is better than the one associated with the initial BFS provided by the Northwest Corner method ($P(x) = 3050$).

	1	2	3	4	
1	8	6	6	1	200
2	3	4	6	8	100
3	7	3	8	9	100
4	4	12	4	3	100
	120	150	180	50	

Table 9.9. Maximum Profit Method Example – Original tableau.

	1	2	3	4	
1	8	6	6	1	200
2	3	4	6	8	100
3	7	3	8	9	100
4	4	12 100	4	3	×
	120	50	180	50	

Table 9.10. Maximum Profit Method Example – Tableau 1.

	1	2	3	4	
1	8	6	6	1	200
2	3	4	6	8	100
3	7	3	8	9 50	50
4	4	12 100	4	3	×
	120	50	180	×	

Table 9.11. Maximum Profit Method Example – Tableau 2.

	1	2	3	4	
1	8 120	6	6	1	80
2	3	4	6	8	100
3	7	3	8	9 50	50
4	4	12 100	4	3	×
	×	50	180	×	

Table 9.12. Maximum Profit Method Example – Tableau 3.

	1	2	3	4	
1	8 120	6	6	1	80
2	3	4	6	8	100
3	7	3	8 50	9 50	×
4	4	12 100	4	3	×
	×	50	130	×	

Table 9.13. Maximum Profit Method Example – Tableau 4.

	1	2	3	4	
1	8 120	6 50	6 30	1	×
2	3	4	6 100	8	×
3	7	3	8 50	9 50	×
4	4	12 100	4	3	×
	×	×	×	×	

Table 9.14. Maximum Profit Method Example – Final tableau 5.

1.3.3 Vogel's Method

Like the method described in the previous section, Vogel's method allows for two modifications. One of them is based on computing for each row (or column) a *penalty* equal to the difference between the two smallest costs in the row (or column). The second one uses differences between the largest profits in the row (or column). Since both modifications may be applied to rows as well as to columns, we have totally four different modifications of this method. Let us restrict our consideration only with that one which utilizes shipping costs and is applied to columns.

We begin by calculating for each column a penalty equal to the difference between two smallest shipping cost in the column. Then we choose the column with the largest penalty and find in this column the variable that has the smallest shipping cost. Similar to the Northwest Corner and Maximum Profit methods, we set this variable as large as possible, change the corresponding supply or demand and mark the corresponding column or row. Now we re-calculate penalties for all columns using only cells that do not lie in the marked row or column. We have to repeat this procedure until only one cell remains in unmarked columns and rows.

We illustrate this method by finding a BFS for the LFPT problem given in Table 9.9. We suppose that values presented in this tableau are shipping costs. First, for each column we calculate penalties (see Table 9.15) and choose column 4 (or column 3) with the largest penalty of 2. Then we choose in this column variable x_{14} since it has the smallest shipping cost of 1. So we set $x_{14} = \min\{b_1, a_4\} = \min\{200, 50\} = 50$, then reduce supply $b_1 \rightarrow b_1 - a_4 = 200 - 50 = 150$, and mark column 4 with '$\times$'. After recalculating new penalties, we obtain the tableau shown in Table 9.16. The largest penalty now occurs in column 3, so we choose variable x_{43} since it is associated with the smallest shipping cost in this column, and set it $x_{43} = \min\{b_4, a_3\} = \min\{100, 180\} = 100$. Then we change $a_3 \rightarrow a_3 - b_4 = 180 - 100 = 80$ and mark row 4 with '$\times$'. The result of this procedure is shown in Table 9.17. The recalculated penalties show that we have to choose column 1 since it is associated with the largest penalty of 4. In this column we find the smallest shipping cost 3 (in row 2) and choose variable x_{21} to enter it into basis. So we set it $x_{21} = \min\{b_2, a_1\} = \min\{100, 120\} = 100$, replace $a_1 \rightarrow a_1 - b_2 = 120 - 100 = 20$ and mark row 2. After recalculating penalties we obtain the tableau presented in Table 9.18. In accordance with penalties calculated we choose on the next step column 2 since it has the largest penalty of 3, and then find the smallest shipping cost of 3 in row 3. So we set variable $x_{32} = \min\{b_3, a_2\} = \min\{100, 150\} = 100$, replace $a_2 \rightarrow a_2 - b_3 = 150 - 100 = 50$ and mark row 3. The result is given in Table 9.19. Since tableau 5 contains

	1	2	3	4	
1	8	6	6	1	200
2	3	4	6	8	100
3	7	3	8	9	100
4	4	12	4	3	100
	120	150	180	50	
	$4-3=1$	$4-3=1$	$6-4=2$	$3-1=2$	

Table 9.15. Vogel's Method Example – Tableau 1.

	1	2	3	4	
1	8	6	6	50 1	150
2	3	4	6	8	100
3	7	3	8	9	100
4	4	12	4	3	100
	120	150	180	×	
	$4-3=1$	$4-3=1$	$6-4=2$	–	

Table 9.16. Vogel's Method Example – Tableau 2.

	1	2	3	4	
1	8	6	6	50 1	150
2	3	4	6	8	100
3	7	3	8	9	100
4	4	12	100 4	3	×
	120	150	80	×	
	$7-3=4$	$4-3=1$	$8-6=2$	–	

Table 9.17. Vogel's Method Example – Tableau 3.

	1	2	3	4	
1	8	6	6	50 1	150
2	100 3	4	6	8	×
3	7	3	8	9	100
4	4	12	100 4	3	×
	20	150	80	×	
	$8-7=1$	$6-3=3$	$8-6=2$	–	

Table 9.18. Vogel's Method Example – Tableau 4.

	1	2	3	4	
1	8	6	6	50 1	150
2	100 3	4	6	8	×
3	7	100 3	8	9	×
4	4	12	100 4	3	×
	20	50	80	×	
	8	6	6	–	

Table 9.19. Vogel's Method Example – Tableau 5.

only one unmarked row, namely row 1, we use shipping costs$8,\ 6,$ and 6 of the row as penalties for columns 1, 2, and 3 respectively and set the remaining unmarked variables $x_{11} = 20,\ x_{12} = 50$, and $x_{13} = 80$. The final tableau is presented in Table 9.20. So, we have obtained the following BFS:$x_{11} = 20$, $x_{12} = 50,\ x_{13} = 80,\ x_{14} = 50,\ x_{21} = 100,\ x_{32} = 100, x_{43} = 100.$

Summarizing, we have to note that among three methods we have discussed in this section, the Northwest Corner method requires the least effort, while Maximum Profit (or Minimum Cost) method is usually more expensive, and Vogel's method usually requires the most effort. However, as shown by practice and extensive research, if Vogel's method is used to determine an initial BFS, it provides a basic feasible solution significantly more close to an optimal one. This is why the Northwest Corner and Maximum Profit (or Minimum Cost) methods are relatively rarely used to find an initial BFS for real-world large transportation problems.

1.4 Numerical Example

In this section, we illustrate how the transportation simplex method described above may be applied.

	1	2	3	4	
1	20 8	50 6	80 6	50 1	×
2	100 3	 4	 6	 8	×
3	 7	100 3	 8	 9	×
4	 4	 12	100 4	 3	×
	×	×	×	×	

Table 9.20. Vogel's Method Example – Final tableau.

Consider the following balanced LFPT problem:

$$Q(x') = \frac{P(x)}{D(x)} = \frac{\sum_{i=1}^{3}\sum_{j=1}^{4} p_{ij}x_{ij} + p_0}{\sum_{i=1}^{3}\sum_{j=1}^{4} d_{ij}x_{ij} + d_0} \longrightarrow \max \tag{9.21}$$

subject to

$$\left.\begin{aligned} x_{11} + x_{12} + x_{13} + x_{14} &\leq 150, \\ x_{21} + x_{22} + x_{23} + x_{24} &\leq 250, \\ x_{31} + x_{32} + x_{33} + x_{34} &\leq 200, \end{aligned}\right\} \tag{9.22}$$

$$\left.\begin{aligned} x_{11} + x_{21} + x_{31} &\geq 150, \\ x_{12} + x_{22} + x_{32} &\geq 250, \\ x_{13} + x_{23} + x_{33} &\geq 50, \\ x_{14} + x_{24} + x_{34} &\geq 150; \end{aligned}\right\} \tag{9.23}$$

$$x_{ij} \geq 0, \quad i = 1, 2, 3, \quad j = 1, 2, 3, 4, \tag{9.24}$$

where $p_0 = 100$, $d_0 = 120$, and coefficients p_{ij} and d_{ij} are given in the following tableaus

p_{ij}	1	2	3	4
1	10	14	8	12
2	8	12	14	8
3	9	6	15	9

d_{ij}	1	2	3	4
1	15	12	16	8
2	10	6	13	12
3	13	15	12	10

Applying the Maximum Profit Method we obtain an initial feasible solution presented in Table 9.21. When applying the Maximum Profit Method we ob-

	1	2	3	4	
1	10 0 15	14 150 12	8 16	12 8	150
2	8 10	12 100 6	14 13	8 150 12	250
3	9 150 13	6 15	15 50 12	9 10	200
	150	250	50	150	

Table 9.21. Transportation Simplex Method Example – Initial BFS.

tained only 5 cells containing a shipment, namely $x_{12} = 150$, $x_{22} = 100$, $x_{24} = 150$, $x_{31} = 150$, and $x_{33} = 50$. It means that the given feasible solution is not a BFS, since it contains not $m + n - 1 = 3 + 4 - 1 = 6$ basic variables but only 5. In this situation we enter into the basis any non-basic variable, for example $x_{11} = 0$. So the solution presented in Table 9.21 is a degenerate one with the following basic index set

$$J_B = \{(1,1),(1,2),(2,2),(2,4),(3,1),(3,3)\}.$$

For this BFS we have the following objective values

$$P(x) = 6700, \quad D(x) = 6870, \quad Q(x) = 6700/6870 \ (\approx 0.975255)\ .$$

Now using this BFS and formulas (9.14)-(9.14) we construct the following systems of linear equations

$$\left.\begin{array}{rcr} u_1' + v_1' & = & 10, \\ u_1' + v_2' & = & 14, \\ u_2' + v_2' & = & 12, \\ u_2' + v_4' & = & 8, \\ u_3' + v_1' & = & 9, \\ u_3' + v_3' & = & 15, \end{array}\right\} \qquad (9.25)$$

$$\left.\begin{array}{rcr} u_1'' + v_1'' & = & 15, \\ u_1'' + v_2'' & = & 12, \\ u_2'' + v_2'' & = & 6, \\ u_2'' + v_4'' & = & 12, \\ u_3'' + v_1'' & = & 13, \\ u_3'' + v_3'' & = & 12, \end{array}\right\} \qquad (9.26)$$

Setting $u_1' = 0$ and $u_1'' = 0$ in (9.25) and (9.26) respectively, and then solving these systems for remaining unknowns we obtain the following solutions

$$u_1' = 0,\ u_2' = -2,\ u_3' = -1, \quad v_1' = 10,\ v_2' = 14,\ v_3' = 16,\ v_4' = 10,$$

and

$$u_1'' = 0,\ u_2'' = -6,\ u_3'' = -2, \quad v_1'' = 15,\ v_2'' = 12,\ v_3'' = 14,\ v_4'' = 18.$$

We use these variables to calculate reduced costs Δ_{ij}' and Δ_{ij}'' (see formulas (9.16)) for non-basic indices

$$J_N = \{(1,3),(1,4),(2,1),(2,3),(4,2),(4,4)\}$$

as follows

$$\begin{array}{rcrcrcr} \Delta_{13}' & = & u_1' + v_3' - p_{13} & = & 0 + 16 - 8 & = & 8, \\ \Delta_{14}' & = & u_1' + v_4' - p_{14} & = & 0 + 10 - 12 & = & -2, \\ \Delta_{21}' & = & u_2' + v_1' - p_{21} & = & -2 + 10 - 8 & = & 0, \\ \Delta_{23}' & = & u_2' + v_3' - p_{23} & = & -2 + 16 - 14 & = & 0, \\ \Delta_{32}' & = & u_3' + v_2' - p_{32} & = & -1 + 14 - 6 & = & 7, \\ \Delta_{34}' & = & u_3' + v_4' - p_{34} & = & -1 + 10 - 9 & = & 0, \end{array}$$

$$\begin{array}{rcrcrcr} \Delta_{13}'' & = & u_1'' + v_3'' - d_{13} & = & 0 + 14 - 16 & = & -2, \\ \Delta_{14}'' & = & u_1'' + v_4'' - d_{14} & = & 0 + 18 - 8 & = & 10, \\ \Delta_{21}'' & = & u_2'' + v_1'' - d_{21} & = & -6 + 15 - 10 & = & -1, \\ \Delta_{23}'' & = & u_2'' + v_3'' - d_{23} & = & -6 + 14 - 13 & = & -5, \\ \Delta_{32}'' & = & u_3'' + v_2'' - d_{32} & = & -2 + 12 - 15 & = & -5, \\ \Delta_{34}'' & = & u_3'' + v_4'' - d_{34} & = & -2 + 18 - 10 & = & 6. \end{array}$$

Further, having values of non-basic Δ'_{ij} and Δ''_{ij} and using formulas (9.17) we can determine values for non-basic reduced costs $\Delta_{ij}(x)$

$$\begin{aligned}
\Delta_{13}(x) &= \Delta'_{13} - Q(x)\,\Delta''_{13} &=& \quad 9\frac{653}{687}, \\
\Delta_{14}(x) &= \Delta'_{14} - Q(x)\,\Delta''_{14} &=& \quad -11\frac{517}{687}, \\
\Delta_{21}(x) &= \Delta'_{21} - Q(x)\,\Delta''_{21} &=& \quad \frac{670}{687}, \\
\Delta_{23}(x) &= \Delta'_{23} - Q(x)\,\Delta''_{23} &=& \quad 4\frac{602}{687}, \\
\Delta_{32}(x) &= \Delta'_{32} - Q(x)\,\Delta''_{32} &=& \quad 11\frac{602}{687}, \\
\Delta_{34}(x) &= \Delta'_{34} - Q(x)\,\Delta''_{34} &=& \quad -5\frac{195}{229}.
\end{aligned}$$

Since not all non-basic reduced costs $\Delta_{ij}(x)$ are non-negative, in accordance with the criteria of optimality (see Theorem 9.4) it means that the current BFS x is not an optimal solution. So we have to choose one of non-basic variables x_{ij} associated with negative reduced cost $\Delta_{ij}(x)$ and enter this variable into the basis. Let it be variable x_{14}. Further, we enter shipment θ (in the meantime unknown) into cell $(1,4)$ and construct a circle to determine the value of shipment θ. The result of this manipulation is given in Table 9.22. Once we have

	1	2	3	4	
1	10 0 15	14 $150-\theta$ 12 ↓	8 16	12 ← θ 8	150
2	8 10	12 $100+\theta$ → 6	14 13	8 ↑ $150-\theta$ 12	250
3	9 150 13	6 15	15 50 12	9 10	200
	150	250	50	150	

Table 9.22. Transportation Simplex Method Example – Tableau 1.

constructed the circle we can determine the value ofθ as follows

$$\theta = \min\{x_{12},\ x_{24}\} = \min\{150,\ 150\} = 150.$$

Note that in the expression above we obtained multiple indices of minimal value for θ, namely $(1,2)$ and $(2,4)$. It means that one of variables $x_{12},\ x_{24}$ leaves the current basis and the second one remains in the basis with the value of zero as a degenerate variable. Let us choose variable x_{24} to leave the basis and variable x_{12} to remain in the basis, so the new basic index set J_B be as follows

$$J_B = \{(1,1),(1,2),(1,4),(2,2),(3,1),(3,3)\},$$

while

$$J_N = \{(1,3),(2,1),(2,3),(2,4),(3,2),(3,4)\}.$$

After performing the corresponding transformations we obtain the tableau given in Table 9.23. This tableau contains the new BFS

	1	2	3	4	
1	10 0 15	14 0 12	8 16	12 150 8	150
2	8 10	12 250 6	14 13	8 12	250
3	9 150 13	6 15	15 50 12	9 10	200
	150	250	50	150	

Table 9.23. Transportation Simplex Method Example – Tableau 2.

$$x_{11} = 0,\ x_{12} = 0,\ x_{14} = 150,\ x_{22} = 250,\ x_{31} = 150,\ x_{33} = 50$$

with objective values of

$$P(x) = 7000,\ \ D(x) = 5370,\ \ Q(x) = 7000/5370\ (\approx 1.303538).$$

Further, for the new basis we construct the following two systems of equations

$$\begin{array}{rcr} u_1' + v_1' & = & 10, \\ u_1' + v_2' & = & 14, \\ u_1' + v_4' & = & 12, \\ u_2' + v_2' & = & 12, \\ u_3' + v_1' & = & 9, \\ u_3' + v_3' & = & 15, \end{array}$$

and

$$\begin{array}{rcr} u_1'' + v_1'' & = & 15, \\ u_1'' + v_2'' & = & 12, \\ u_1'' + v_4'' & = & 8, \\ u_2'' + v_2'' & = & 6, \\ u_3'' + v_1'' & = & 13, \\ u_3'' + v_3'' & = & 12. \end{array}$$

Solving these systems we obtain the following values

$$u_1' = 0,\ u_2' = -2,\ u_3' = -1, \quad v_1' = 10,\ v_2' = 14,\ v_3' = 16,\ v_4' = 12,$$

and

$$u_1'' = 0,\ u_2'' = -6,\ u_3'' = -2, \quad v_1'' = 15,\ v_2'' = 12,\ v_3'' = 14,\ v_4'' = 8,$$

which allow us to re-calculate non-basic reduced costsΔ_{ij}', Δ_{ij}'', and $\Delta_{ij}(x)$, as follows

$$\begin{array}{lclclcl} \Delta_{13}' & = & u_1' + v_3' - p_{13} & = & 0 + 16 - 8 & = & 8, \\ \Delta_{21}' & = & u_2' + v_1' - p_{21} & = & -2 + 10 - 8 & = & 0, \\ \Delta_{23}' & = & u_2' + v_3' - p_{23} & = & -2 + 16 - 14 & = & 0, \\ \Delta_{24}' & = & u_2' + v_4' - p_{24} & = & -2 + 12 - 8 & = & 2, \\ \Delta_{32}' & = & u_3' + v_2' - p_{32} & = & -1 + 14 - 6 & = & 7, \\ \Delta_{34}' & = & u_3' + v_4' - p_{34} & = & -1 + 12 - 9 & = & 2, \end{array}$$

$$\begin{array}{lclclcr} \Delta_{13}'' & = & u_1'' + v_3'' - d_{13} & = & 0 + 14 - 16 & = & -2, \\ \Delta_{21}'' & = & u_2'' + v_1'' - d_{21} & = & -6 + 15 - 10 & = & -1, \\ \Delta_{23}'' & = & u_2'' + v_3'' - d_{23} & = & -6 + 14 - 13 & = & -5, \\ \Delta_{24}'' & = & u_2'' + v_4'' - d_{24} & = & -6 + 8 - 12 & = & -10, \\ \Delta_{32}'' & = & u_3'' + v_2'' - d_{32} & = & -2 + 12 - 15 & = & -5, \\ \Delta_{34}'' & = & u_3'' + v_4'' - d_{34} & = & -2 + 8 - 10 & = & -4, \end{array}$$

and, finally

$$\begin{aligned}
\Delta_{13}(x) &= \Delta'_{13} - Q(x)\,\Delta''_{13} &= 10\frac{326}{537}, \\
\Delta_{21}(x) &= \Delta'_{21} - Q(x)\,\Delta''_{21} &= 1\frac{163}{537}, \\
\Delta_{23}(x) &= \Delta'_{23} - Q(x)\,\Delta''_{23} &= 6\frac{278}{537}, \\
\Delta_{24}(x) &= \Delta'_{24} - Q(x)\,\Delta''_{24} &= 15\frac{19}{537}, \\
\Delta_{32}(x) &= \Delta'_{32} - Q(x)\,\Delta''_{32} &= 13\frac{278}{537}, \\
\Delta_{34}(x) &= \Delta'_{34} - Q(x)\,\Delta''_{34} &= 7\frac{115}{537}.
\end{aligned}$$

Since all non-basic reduced costs $\Delta_{ij}(x) \geq 0, \ (ij) \in J_N$ it means that the current BFS x is an optimal solution.

1.5 Duality Theory for the Transportation Problem

Our aim in this section is to shortly overview the main results of duality theory adapted to the LFPT problem. Duality theory was covered in Chapter 5 for the case of maximization LFP problems in a general form.

Consider the following LFPT problem

$$Q(x) = \frac{P(x)}{D(x)} = \frac{\sum_{i=1}^{m}\sum_{j=1}^{n} p_{ij}x_{ij} + p_0}{\sum_{i=1}^{m}\sum_{j=1}^{n} d_{ij}x_{ij} + d_0} \longrightarrow \max \tag{9.27}$$

subject to

$$\sum_{j=1}^{n} x_{ij} \leq b_i, \quad i = 1, 2, \ldots, m, \tag{9.28}$$

$$\sum_{i=1}^{m} x_{ij} \geq a_j, \quad j = 1, 2, \ldots, n, \tag{9.29}$$

$$x_{ij} \geq 0, \quad i = 1, 2, \ldots, m; \ j = 1, 2, \ldots, n. \tag{9.30}$$

As usual, we suppose that $D(x) > 0, \ \ \forall x = (x_{ij}) \in S,$ where S denotes a feasible set defined by constraints (9.28)-(9.30).

The dual problem for LFPT problem (9.27)-(9.30) may be formulated as follows:

$$\psi(y) = y_0 \longrightarrow \min \tag{9.31}$$

subject to

$$d_0 y_0 - \sum_{i=1}^{m} b_i u_i + \sum_{j=1}^{n} a_j v_j \geq p_0, \tag{9.32}$$

$$d_{ij} y_0 + u_i - v_j \geq p_{ij}, \quad i = 1, 2, \ldots, m, \; j = 1, 2, \ldots, n, \tag{9.33}$$

$$u_i \geq 0, \; i = 1, 2, \ldots, m, \quad v_j \geq 0, \; j = 1, 2, \ldots, n, \tag{9.34}$$

where y denotes a vector containing $n + m + 1$ components

$$y_0, \quad u_1, u_2, \ldots, u_m, \quad v_1, v_2, \ldots, v_n.$$

Since an LFPT problem is a special case of a common LFP problem, it means that all basic statements of duality theory formulated in Chapter 5 are also valid for LFPT problem (9.27)-(9.30). Here we just reformulate some of the main results described in Chapter 5.

THEOREM 9.5 (THE WEAK DUALITY THOEREM) *If x is a feasible solution of primal LFPT problem (9.27)-(9.30) and y is a feasible solution of its dual problem (9.31)-(9.34), then*

$$Q(x) \leq \psi(y).$$

LEMMA 9.1 *If x^* is a feasible solution of primal LFPT problem (9.27)-(9.30), y^* is a feasible solution of dual problem (9.31)-(9.34), and the equality*

$$Q(x^*) = \psi(y^*) \tag{9.35}$$

holds, then x^* and y^* are optimal solutions of their problems (9.27)-(9.30) and (9.31)-(9.34), respectively.

The following lemma indicates a very important connection between the solvability of the primal and dual problems.

LEMMA 9.2 *If objective function $\psi(y)$ of dual problem (9.31)-(9.34) is unbounded from below on its feasible set, then primal LFPT problem (9.27)-(9.30) is unsolvable because its feasible set S is empty.*

It is obvious that this lemma indicates if the total demand exceeds the total supply, i.e.

$$\sum_{j=1}^{n} a_j \geq \sum_{i=1}^{m} b_i.$$

THEOREM 9.6 (THE STRONG DUALITY THEOREM) *If primal LFPT problem (9.27)-(9.30) is solvable and x^* is its optimal solution, then its dual problem (9.31)-(9.34) is also solvable and for any optimal solutions y^* of dual problem (9.31)-(9.34) the following equality*

$$Q(x^*) = \psi(y^*). \qquad (9.36)$$

holds. Conversely, if dual problem (9.31)-(9.34) is solvable and y^* is its optimal solution, then the primal LFPT problem (9.27)-(9.30) is also solvable and for any its optimal solution x^* equality (9.36) holds.

Suppose that x^* is an optimal solution of primal LFPT problem (9.27)-(9.30) and vector y^* is an optimal solution of dual problem (9.31)-(9.34). Let us choose two indices $1 \leq r \leq m$ and $1 \leq k \leq n$ and replace rth supply b_r and kth demand a_k in LFPT problem (9.27)-(9.30) in the following way

$$b_r \rightarrow b'_r = b_r + \delta, \quad a_k \rightarrow a'_k = a_k + \delta, \qquad (9.37)$$

where δ is small enough. Let x' denote an optimal solution of the modified LFPT problem (with replaced supply b_r and demand a_k). The following theorem indicates an important role of dual variables $y_0,\ u_1, u_2, \dots, u_m$ and $v_1, v_2, \dots, v_n$ in a sensitivity analysis.

THEOREM 9.7 *If x^* is an optimal solution of LPFT problem (9.27)-(9.30), y^* is an optimal solution of dual problem (9.31)-(9.34), x' is an optimal solution of the modified LFPT problem (with replaced supply b_r and demand a_k), and δ is small enough, then the following equality*

$$Q(x') = y_0^* + \frac{\delta(u_r^* - v_k^*)}{D(x')} \qquad (9.38)$$

holds.

Note that we may find the values of the elements for new optimal optimal solution x' as follows:

- If x_{rk} is a basic variable in optimal solution x^*, simply increase x_{rk} by δ;
- If x_{rk} is a non-basic variable in optimal solution x^*, we have to find the circle, which involves cell rk and some of the basic cells. Then we go

around this circle alternately increasing and decreasing basic variables in the circle by δ.

To illustrate formula (9.38) we reconsider the balanced LFPT problem (9.21)-(9.24) given in Section 1.4, page 267. The optimal solution of this balanced LFPT problem is vector

$$x^* = \begin{pmatrix} 0 & 0 & 0 & 150 \\ 0 & 250 & 0 & 0 \\ 150 & 0 & 50 & 0 \end{pmatrix},$$

such that $P(x^*) = 7000$, $D(x^*) = 5370$ and $Q(x^*) = 1.30353818$. Solving dual problem

$$\phi(y) = y_0 \longrightarrow \min \tag{9.39}$$

subject to

$$\left.\begin{array}{rcl} 15\,y_0 + u_1 - v_1 & \geq & 10, \\ 12\,y_0 + u_1 - v_2 & \geq & 14, \\ 16\,y_0 + u_1 - v_3 & \geq & 8, \\ 8\,y_0 + u_1 - v_4 & \geq & 12, \end{array}\right\} i = 1; \tag{9.40}$$

$$\left.\begin{array}{rcl} 10\,y_0 + u_2 - v_1 & \geq & 8, \\ 6\,y_0 + u_2 - v_2 & \geq & 12, \\ 13\,y_0 + u_2 - v_3 & \geq & 14, \\ 12\,y_0 + u_2 - v_4 & \geq & 8, \end{array}\right\} i = 2; \tag{9.41}$$

$$\left.\begin{array}{rcl} 13\,y_0 + u_3 - v_1 & \geq & 9, \\ 15\,y_0 + u_3 - v_2 & \geq & 6, \\ 12\,y_0 + u_3 - v_3 & \geq & 15, \\ 10\,y_0 + u_3 - v_4 & \geq & 9, \end{array}\right\} i = 3; \tag{9.42}$$

$$u_i \geq 0, \;\; i = 1,2,3, \qquad v_j \geq 0, \;\; j = 1,2,3,4, \tag{9.43}$$

we obtain optimal solution

$$\begin{array}{llll} y_0^* = 1.303538, & & & \\ u_1^* = 1.571695, & u_2^* = 4.178771, & u_3^* = 0, & \\ v_1^* = 7.945996, & v_2^* = 0, & v_3^* = 0.642458, & v_4^* = 0, \end{array} \tag{9.44}$$

which allows us to predict the change in optimal value of objective function $Q(x)$ if some change occurs in supply vector $b = (150, 250, 200)^T$ and demand vector $a = (150, 250, 50, 150)^T$. For example, if we increase supply $b_1 = 150$ and demand $a_4 = 150$ by $\delta = 1$ unit, then for new optimal solution x' we have

$$x' = \begin{pmatrix} 0 & 0 & 0 & 150+\delta \\ 0 & 250 & 0 & 0 \\ 150 & 0 & 50 & 0 \end{pmatrix},$$

while $D(x') = 5378$. Thus, in accordance with formula (9.38) for new optimal solution x' we have

$$Q(x') = Q(x^*) + \frac{u_1^* - v_4^*}{D(x')} = 1.30353818 + \frac{1.571695 - 0}{5378} = 1.303830425.$$

Closing this discussion of the LFPT problem we note that all results given above may be easily adapted to the case of the transportation problem of linear programming. To obtain the corresponding formula or statement for a linear case we simply have to keep in mind that

$$d_{ij} = 0, \quad i = 1, 2, \ldots, m, \; j = 1, 2, \ldots, n, \qquad d_0 = 1.$$

2. The Transshipment Problem

When considering a transportation LFP problem in Section 1 we assumed that all supply points are directly connected with all demand points, so all direct shipments are allowed. However, in real-world problems sometimes not all supply points and demand points have direct connections. For example, goods may be shipped by a truck to a sea port, then must be reloaded onto a ship and then transported to another sea port, where it must be reloaded again onto a truck, and finally may be shipped to the demand point. So in more realistic transportation problems there may also be points (called*transshipment points*) through which goods can be transshipped on their journey from a supply point to a demand point. Such transportation problems are called*transshipment problems*. Fortunately, transshipment problems of LFP may be easily transformed to the form of conventional transportation LFP problems and, hence may be solved by the transportation simplex method described in Section 1.2.

Consider a situation when all direct shipments from m supply points SP_i to n demand points DP_j are disabled, so the goods may be delivered only via k transshipment points TP_l, see Figure 9.1. To formulate the corresponding transshipment LFP problem we introduce the following unknown variables: x_{il}, $(i = 1, 2, \ldots, m, \; l = 1, 2, \ldots, k,)$ - unknown amount of the goods to be shipped from supply point SP_i to transshipment point TP_l; y_{lj}, $(l = 1, 2, \ldots, k, \; j = 1, 2, \ldots, n,)$ - unknown amount of the goods to be shipped from transshipment point TP_l to demand point DP_j. So the transshipment LFP problem in canonical form may be formulated as follows.

$$Q(x) = \frac{P(x)}{D(x)} = \frac{\sum_{i=1}^{m}\sum_{l=1}^{k} p'_{ij}\, x_{il} + \sum_{l=1}^{k}\sum_{j=1}^{n} p''_{ij}\, y_{lj} + p_0}{\sum_{i=1}^{m}\sum_{l=1}^{k} d'_{ij}\, x_{il} + \sum_{l=1}^{k}\sum_{j=1}^{n} d''_{ij}\, y_{lj} + d_0} \longrightarrow \max \qquad (9.45)$$

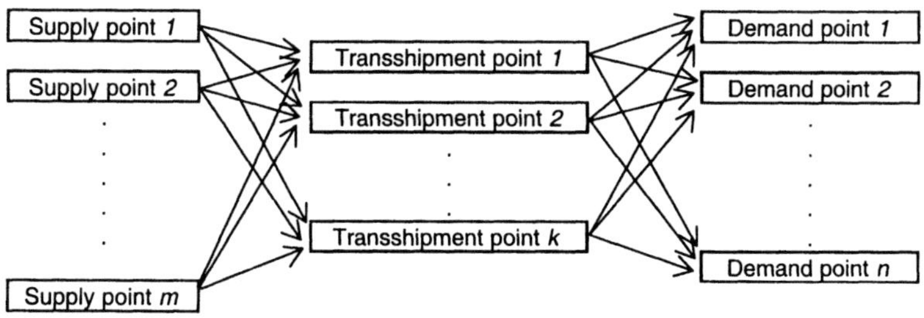

Figure 9.1. Transshipment LFP problem with disabled direct connections.

subject to

$$\sum_{l=1}^{k} x_{il} = b_i, \quad i = 1, 2, \ldots, m, \tag{9.46}$$

$$\sum_{i=1}^{m} x_{il} = \sum_{j=1}^{n} y_{lj}, \quad l = 1, 2, \ldots, k, \tag{9.47}$$

$$\sum_{l=1}^{k} y_{lj} = a_j, \quad j = 1, 2, \ldots, n, \tag{9.48}$$

$$x_{il} \geq 0, \quad i = 1, 2, \ldots, m, \ l = 1, 2, \ldots, k, \tag{9.49}$$

$$y_{lj} \geq 0, \quad l = 1, 2, \ldots, k, \ j = 1, 2, \ldots, n, \tag{9.50}$$

where conditions (9.47) provide a necessary connection between variablesx_{il} and y_{lj} and guarantee that in any transshipment point a total inflow is equal to a total outflow.

We suppose that function $D(x)$ is strictly positive for all points of feasible set S defined by constraints (9.46)-(9.50). Further, we assume also that

$$b_i > 0, \ \ a_j > 0, \ \ i = 1, 2, \ldots, m; \ j = 1, 2, \ldots, n \tag{9.51}$$

and the problem is balanced, i.e.

$$\sum_{i=1}^{m} b_i = \sum_{j=1}^{n} a_j. \tag{9.52}$$

Strictly speaking, problem (9.45)-(9.50) is not a transportation LFP problem since it includes special constraints (9.47). However, the problem may be transformed into the form of a conventional LFPT problem.

	TP_1	TP_2	$\ldots$	TP_k	DP_1	DP_2	$\ldots$	DP_n	
SP_1	p'_{11} d'_{11}	p'_{12} d'_{12}	$\ldots$	p'_{1k} d'_{1k}	0 M	0 M	$\ldots$	0 M	b_1
SP_2	p'_{21} d'_{21}	p'_{22} d'_{22}	$\ldots$	p'_{2k} d'_{2k}	0 M	0 M	$\ldots$	0 M	b_2
$\vdots$	$\vdots$	$\vdots$	$\vdots$	$\vdots$	$\vdots$	$\vdots$	$\vdots$	$\vdots$	$\vdots$
SP_m	p'_{m1} d'_{m1}	p'_{m2} d'_{m2}	$\ldots$	p'_{mk} d'_{mk}	0 M	0 M	$\ldots$	0 M	b_m
TP_1	0 M	0 M	$\ldots$	0 M	p''_{11} d''_{11}	p''_{12} d''_{12}	$\ldots$	p''_{1n} d''_{1n}	K
TP_2	0 M	0 M	$\ldots$	0 M	p''_{21} d''_{21}	p''_{22} d''_{22}	$\ldots$	p''_{2n} d''_{2n}	K
$\vdots$	$\vdots$	$\vdots$	$\vdots$	$\vdots$	$\vdots$	$\vdots$	$\vdots$	$\vdots$	$\vdots$
TP_k	0 M	0 M	$\ldots$	0 M	p''_{k1} d''_{k1}	p''_{k2} d''_{k2}	$\ldots$	p''_{kn} d''_{kn}	K
	K	K	$\ldots$	K	a_1	a_2	$\ldots$	a_n	

Table 9.24. Representation of Transshipment LFP problem as Balanced Transportation LFP problem.

To construct a transportation tableau for this problem we use the tableau associated with a conventional LFPT problem (see Table 9.1) and add to it k additional rows and columns for transshipment points $TP_l,\ l = 1, 2, \ldots, k$. So the total number of rows in the tableau will be $m + k$, while the total number of columns will be $n + k$. Each supply point SP_i will have a supply equal to its original supply b_i, while each transshipment point TP_l will have a supply equal to total available supply, i.e.

$$K = \sum_{i=1}^{m} b_i.$$

Each demand point DP_j will have a demand equal to its original demand a_j, and each transshipment point TP_l will have a demand equal to total demand, i.e. for balanced problem

$$K = \sum_{j=1}^{n} a_j = \sum_{i=1}^{m} b_i.$$

To disable direct shipments from supply points SP_i to demand points DP_j we have to associate with these paths a zero valued profit and high enough cost M which may be set as

$$M = \sum_{i=1}^{m} \sum_{l=1}^{k} d'_{il} + \sum_{l=1}^{k} \sum_{j=1}^{n} d''_{lj}.$$

The result is presented in Table 9.24. In this problem we assume that all shipments between transshipment points are also disabled. This is why in the bottom left part of the tableau all cells contain a profit equal to zero and a cost equal to M.

If some or all of the direct connections between supply points SP_i and demand points DP_j are allowed, the only change we have to make is to replace in the tableau the corresponding zero profits and shipping costs M with suitable coefficients. Similarly, if some or all shipments between transshipment points are allowed we have to use in the corresponding cells the proper coefficients of profit and cost.

Consider the following numerical example. Given are three supply points $SP_1,\ SP_2,\ SP_3,$ with a supply of 150, 250, 200 units respectively. There are four demand points $DP_1,\ DP_2,\ DP_3,\ DP_4,$ with a demand of 100, 150, 200 and 150 units respectively. Also, we have two transshipment points TP_1, and TP_2 with shipments allowed between these two points. The profit and cost coefficients associated with all possible paths are shown in Table 9.25. Constructing the corresponding transportation simplex tableau we obtain the tableau shown in Table 9.26, where cells associated with paths between transshipment points TP_l contain zero profits and zero costs to enable shipments between transshipment points. Obviously, this tableau is of the form of a conventional transportation LFP problem, so it may be solved by the transportation simplex method. If we apply the transportation simplex method to this problem (here we set $p_0 = 0$ and $d_0 = 0$) we obtain the following optimal solution

$$SP_1 \to TP_1 = 150,\quad SP_2 \to TP_2 = 150,\quad SP_3 \to TP_2 = 200,$$

$$SP_2 \to DP_1 = 100,\quad TP_1 \to TP_1 = 450,\quad TP_1 \to TP_2 = 150,$$

$$TP_2 \to TP_2 = 100,\quad TP_2 \to DP_2 = 150,\quad TP_2 \to DP_3 = 200,$$

	TP_1	TP_2	DP_1	DP_2	DP_3	DP_4
SP_1	12	8	15	15	18	16
	10	8	18	20	22	18
SP_2	10	9	12	10	19	20
	12	10	16	15	20	22
SP_3	8	12	14	18	15	12
	9	12	18	22	22	16

	DP_1	DP_2	DP_3	DP_4
TP_1	12	14	12	15
	16	18	22	13
TP_2	12	6	14	8
	16	4	10	4

Table 9.25. Transshipment LFP example – Profits and costs.

$$TP_2 \rightarrow DP_4 = 150.$$

Other cells contain shipments equal to zero, so $P(x^*) = 11650$, $D(x^*) = 10200$, and $Q(x^*) \approx 1,142167$.

When interpreting the solution obtained we simply ignore the shipments from a transshipment point to itself, namely

$$TP_1 \rightarrow TP_1 = 450, \quad \text{and} \quad TP_2 \rightarrow TP_2 = 100.$$

Closing this section we have to note that if shipments between transshipment points TP_l are disabled, we have to replace zero costs in the corresponding cells of Table 9.26 with a large cost M.

3. The Assignment Problem

This problem, also known as the **marriage problem**, was proposed as an application of linear programming to sociology in the early 1950s. Later, in 1970s it was generalized to a form with linear-fractional objective function [152]. The linear-fractional assignment problem (LFAP) is to find such an

	TP_1	TP_2	DP_1	DP_2	DP_3	DP_4	
SP_1	12 10	8 8	15 18	15 20	18 22	16 18	150
SP_2	10 12	9 10	12 16	10 15	19 20	20 22	250
SP_3	8 9	12 12	14 18	18 22	15 22	12 16	200
TP_1	0 0	0 0	12 16	14 18	12 22	15 13	600
TP_2	0 0	0 0	12 16	6 4	14 10	8 4	600
	600	600	100	150	200	150	

Table 9.26. Transshipment LFP example – Initial tableau.

assignment $x = ||x_{ij}||_{n \times n}$ that maximizes (or minimizes) linear-fractional objective function

$$Q(x) = \frac{P(x)}{D(x)} = \frac{\sum_{i=1}^{n}\sum_{j=1}^{n} p_{ij}\, x_{ij} + p_0}{\sum_{i=1}^{n}\sum_{j=1}^{n} d_{ij}\, x_{ij} + d_0} \tag{9.53}$$

subject to the following restrictions

$$\sum_{i=1}^{n} x_{ij} = 1, \quad i = 1, 2, \dots, n, \tag{9.54}$$

$$\sum_{j=1}^{n} x_{ij} = 1, \quad j = 1, 2, \dots, n, \tag{9.55}$$

$$x_{ij} \in \{0, 1\}, \quad i = 1, 2, \dots, n, \;\; j = 1, 2, \dots, n. \tag{9.56}$$

It is conventional to interpret the problem as follows. Given n persons (i.e. $i = 1, 2, \ldots, n$) and n tasks (i.e. $j = 1, 2, \ldots, n$), and for each index pair (ij) a profit p_{ij} and a cost d_{ij} are known. The assignment problem is to assign each person to one and only one task in such a manner that each task gets covered by someone (exactly by one person) and efficiency (calculated as ratio*(total profit)/(total cost)*) of the assignments is maximized. Here

$$x_{ij} = \begin{cases} 1, & \text{if person } i \text{ is assigned task } j, \\ 0, & \text{otherwise.} \end{cases}$$

Constraints (9.54) express the requirement that each person is assigned exactly one task. While constraints (9.55) require that every task be covered exactly by one person.

Obviously, ignoring for the moment the integrality constraints (9.56), we can say that problem (9.53)-(9.56) is a special case of a balanced LFP transportation problem, i.e. it is such a balanced LFP transportation problem in which all supplies and demands are equal to 1. Further, since all supplies and demands are equal to 1, in accordance with Theorem 9.3 we may replace integrality restrictions (9.56) with conventional non-negativity requirements

$$x_{ij} \geq 0, \quad i = 1, 2, \ldots, n, \quad j = 1, 2, \ldots, n. \tag{9.57}$$

and then apply the transportation simplex method to solve problem (9.53)-(9.55), (9.57) instead of original (9.53)-(9.56). Although the transportation simplex method appears to be very efficient, in the case of assignment problems it may often be very inefficient. For the assignment problem of linear programming, many algorithms have been developed. Perhaps the most widespread of them is the so-called **Hungarian method**, developed by H.W.Kuhn in [120] and based on the work of Hungarian mathematician J.Egerváry. For more information see e.g. [188]. In linear-fractional programming, if we have to solve an assignment problem we may use special methods developed especially for this class of LFP problems. Here we just note that one of such methods (proposed by M.Shigeno, Y.Saruwatari and T.Matsui in [169]) is based on Dinkelbach's parametric approach (see Chapter 3, Section 4) and incorporates the Hungarian method for solving a sequence of LP assignment problems.

4. Discussion Questions and Exercises

9.1 An automobile company wishes to transport cars from three supply points W_1, W_2 and W_3 to three different sales locations M_1, M_2 and M_3. The supply points have available 250, 480 and 120 cars, respectively. The demands at the three sales locations are 250, 480 and 120 cars, respectively. The profit and the cost per car to ship from supply to sales locations are given in the following tables

Profit-matrix

	M_1	M_2	M_3
W_1	120	140	180
W_2	140	120	100
W_3	110	120	125

Cost-matrix

	M_1	M_2	M_3
W_1	140	220	285
W_2	160	190	210
W_3	320	195	225

Formulate a maximization LFPT problem with the objective function expressed as the ratio (profit/cost). Then, using Northwest Corner rule find an initial BFS.

9.2 For the LFPT problem from the previous exercise use Vogel's method to find an initial BFS. Compare the BFS obtained with the one determined in the previous exercise.

9.3 For the LFPT problem given in exercise 9.1 find an optimal solution which maximizes the efficiency of shipment calculated as*profit/cost* if $p_0 = 250$ and $d_0 = 1500$.

9.4 For the transshipment LFP problem given in Section 2, Table 9.26, apply Vogel's method to determine an initial BFS.

9.5 For the maximization transshipment LFP problem given in previous exercise find optimal solution if $p_0 = 1250$ and $d_0 = 1500$.

9.6 A company producing a single product has three plants and four customers. These plants P_1, P_2, and P_3, can produce monthly 1000, 2000, and 3000 units of the product, respectively. The company has made a commitment to sell 1500 units to customer C_1, 1000 units to customer C_2, 1500 units to customer C_3, and at least 1000 units to customer C_4. Customer C_4 wants to buy as many of the remaining units as possible. Some direct shipments are impossible and may be resolved via two transshipment pointsTP_1 and TP_2. The profit and cost associated with shipping a unit of the product is given in the following tableaus

	TP_1	TP_2	C_1	C_2	C_3	C_4
P_1	10 12	8 8	10 16	–	12 20	–
P_2	8 11	9 10	–	10 15	19 20	20 22
P_3	15 14	10 12	–	10 12	–	12 16

	C_1	C_2	C_3	C_4
TP_1	4 6	4 8	6 8	4 5
TP_2	10 10	8 6	4 8	6 8

where "-" indicates that a shipment is impossible. Formulate a transshipment LFP problem that can be used to maximize the company's efficiency calculated as *profit/cost.*

9.7 Five employees are available to perform five jobs. The time it takes each person to perform each job, and the cost associated with each possible assignment (person↔job) are given in the following tables

Time

	Job 1	*Job* 2	*Job* 3	*Job* 4	*Job* 5
Person 1	9	8	9	16	12
Person 2	6	14	9	12	12
Person 3	12	9	15	15	14
Person 4	12	12	10	12	11
Person 5	8	12	15	10	15

Cost

	Job 1	*Job* 2	*Job* 3	*Job* 4	*Job* 5
Person 1	6	2	6	7	6
Person 2	4	9	5	3	8
Person 3	8	4	8	1	5
Person 4	7	8	6	5	2
Person 5	7	5	4	10	8

Assume that some preparation must be made before performing the jobs, so it takes 5 units of time and costs 7 units of money. Using the transportation simplex method determine the assignment of employees to jobs that minimizes the specific cost calculated as (total cost)/(total time).

Chapter 10

ADVANCED METHODS AND ALGORITHMS IN LFP

In this chapter, we describe the state of the art in LFP methods. We start by presenting some special variants of the simplex method (including the so-called *Dual Simplex Method* and the *Criss-Cross Method*), and then we go on to discuss one of the so-called *Interior-Point Methods (IPM)* of linear-fractional programming, namely the *Method of Analytic Centers* proposed by A.S.Nemirovskii [139], [140].

1. The Dual Simplex Method in LFP

In this section, we study what happens if we apply the simplex method to the dual problem of LFP. The basic idea of this approach is that when we use the simplex method to solve a maximization LFP problem (in this case we will refer to the simplex method as a **primal** one) we begin with an initial basic solution which is feasible with respect to the primal problem (**primal feasible**) but not feasible in terms of a dual problem (**dual infeasible**). Through a sequence of simplex iterations the method maintains the feasibility of all basic solutions traversed through and obtains an optimal basic solution when dual feasibility is attained. The **dual simplex method** does just the opposite. It starts with an initial basic solution which is feasible with respect to the dual problem (**dual feasible**) but is not feasible for the primal one (**primal infeasible**). And then, at each iteration it traverses through dual feasible basic solutions. If a primal feasible basis is obtained, the dual simplex method terminates since it is an optimal basis.

Consider the following maximization LFP problem in canonical form

$$Q(x) = \frac{P(x)}{D(x)} = \frac{\sum_{j=1}^{n} p_j x_j + p_0}{\sum_{j=1}^{n} d_j x_j + d_0} \longrightarrow \max, \tag{10.1}$$

subject to

$$\sum_{j=1}^{n} a_{ij} x_j = b_i, \quad i = 1, 2, \ldots, m, \tag{10.2}$$

$$x_j \geq 0, \quad j = 1, 2, \ldots, n, \tag{10.3}$$

where $D(x) > 0$ for all $x = (x_1, x_2, \cdots, x_n)^T$, which satisfy constraints (10.2)-(10.3). We assume that feasible set S is a regular set, i.e. is non-empty and bounded.

Let vector x be a basic solution of linear-fractional programming problem (10.1)-(10.3) associated with basis

$$B = (A_{s_1}, A_{s_2}, \ldots, A_{s_m}), \quad \text{where} \quad A_j = (a_{1j}, a_{2j}, \ldots, a_{mj})^T,$$

and $x_{s_1}, x_{s_2}, \ldots, x_{s_m}$ denote the corresponding basic variables with values $x^*_{s_1}, x^*_{s_2}, \ldots, x^*_{s_m}$, respectively.

The following optimality criterion is used in the dual simplex method.

THEOREM 10.1 (DUAL CRITERIA OF OPTIMALITY) *The current basic solution x is an optimal solution of linear-fractional programming problem (10.1)-(10.3) if and only if all basic variables x_{s_i} have non-negative values, i.e.*

$$x^*_{s_i} \geq 0, \quad i = 1, 2, \ldots, m.$$

When using the dual simplex method we have to proceed as follows.

Step 1 (Initial basis). Start with a dual feasible basis and create a corresponding simplex tableau. Find an initial basic but not feasible (i.e. containing negative basic variables) solution x. Go to Step 2.

Step 2 (Termination test). If all basic variables just obtained are non-negative, the process must be terminated since the current basic vector x is an optimal solution. Otherwise Calculate all Δ'_j, Δ''_j, and $\Delta_j(x)$ and go to Step 3.

Step 3 (Pivot row). Pick in the simplex tableau just obtained the row containing the most negative basic variable. Let it be variable x_{s_r}, so rth row is the pivot row and variable x_{s_r} leaves the basis. Go to Step 4.

Step 4 (Pivot column). To select the variable that enters the basis, we calculate the following ratio for each variable x_j that has a negative coefficient in the pivot row

$$\frac{\Delta_j(x)}{x_{rj}}, \quad j \in J^- = \{j \in J | \, x_{rj} < 0\}, \quad J = \{1, 2, \ldots, n\},$$

and then choose the ratio with the largest value. Column k for which this ratio occurred is the pivot column and variable x_k must enter the basis. Go to Step 5.

If all entries in the pivot row are non-negative, the original LFP problem has no feasible solutions. Stop.

Step 5 (Iteration). Perform simplex iteration as for the primal simplex method and go to Step 2.

To illustrate how the method works we consider the following numerical example.

$$Q(x) = \frac{P(x)}{D(x)} = \frac{-4x_1 - 35x_2 - 20x_3 - 8}{5x_1 + 40x_2 + 35x_3 + 30} \longrightarrow \max \tag{10.4}$$

subject to

$$\left. \begin{array}{rcl} 1x_1 - 1x_2 - 1x_3 & \leq & -2, \\ -1x_1 - 3x_2 - 1x_3 & \leq & -3, \end{array} \right\} \tag{10.5}$$

$$x_j \geq 0, \quad j = 1, 2, 3. \tag{10.6}$$

First of all, adding slack variables x_4 and x_5 we convert the problem to canonical form. Observe that slack variables are associated with unit vectors A_4 and A_5, respectively, and these vectors give us a primal infeasible but dual feasible initial basis. The initial tableau is shown in Table 10.1. Since the current basic solution contains negative basic variables, it is not an optimal one. We choose row 2 as a pivot row since variable x_5 has the most negative value -3. The ratio test picks vector A_1 as a pivot column. After performing a simplex iteration we obtain a new tableau shown in Table 10.2. The termination test at this step fails because the current basis contains negative basic variable $x_4 = -5$. So in Table 10.2 we choose row 1 as a pivot row and then after performing a ratio test we choose vector A_3 as a pivot column. It leads to the tableau shown in Table 10.3. Since the simplex tableau shown in Table 10.3 contains only non-negative basic variables, it means that the optimal solution for original problem (10.4)-(10.6) has been found

$$x^* = (0.5, \, 0, \, 2.5)^T, \quad Q(x^*) = -60/120 = -0.5\,.$$

B	p_B	d_B	x_B	A_1	A_2	A_3	A_4	A_5	
A_4	0	0	-2	1	-1	-1	1	0	
A_5	0	0	-3	-1	-3	-1	0	1	$\Rightarrow$
$P(x) = -8$				4	35	20	0	0	
$D(x) = 30$				-5	-40	-35	0	0	
$Q(x) = -4/15$				$8/3$	$73/3$	$32/3$	0	0	
			Ratio	$-8/3$	$-73/9$	$-32/3$	N/A	N/A	
				$\Uparrow$					

Table 10.1. The Dual Simplex Method – Initial tableau.

B	p_B	d_B	x_B	A_1	A_2	A_3	A_4	A_5	
A_4	0	0	-5	0	-4	-2	1	1	$\Rightarrow$
A_1	-4	5	3	1	3	1	0	-1	
$P(x) = -20$				0	23	16	0	4	
$D(x) = 45$				-0	-25	-30	0	-5	
$Q(x) = -4/9$				0	$107/9$	$8/3$	0	$16/9$	
			Ratio	N/A	$-107/36$	$-4/3$	N/A	N/A	
						$\Uparrow$			

Table 10.2. The Dual Simplex Method – After first iteration.

B	p_B	d_B	x_B	A_1	A_2	A_3	A_4	A_5
A_3	-20	35	$5/2$	0	2	1	$-1/2$	$-1/2$
A_1	-4	5	$1/2$	1	1	0	$1/2$	$-1/2$
$P(x) = -60$				0	-9	0	8	12
$D(x) = 120$				-0	35	0	-15	-20
$Q(x) = -1/2$				0	$17/2$	0	$1/2$	2

Table 10.3. The Dual Simplex Method – Optimal tableau.

The dual simplex method is especially useful in the following cases:

Case 1. If an initial dual feasible basic solution is easily available (since it allows to avoid the I phase of the primal simplex method).

Case 2. If we have to re-optimize the solution after a constraint has been added, and hence, the current optimal solution may be no longer feasible (since it allows to avoid solving the new problem from scratch).

We now discuss these two cases. The first case has just been illustrated in the numerical example above. Indeed, after entering slack variables x_4 and x_5 (to convert original maximization LFP problem (10.4)-(10.6) to canonical form) we immediately obtained a unit sub-matrix

$$\left.\begin{array}{rcrcr|ccc|cr} 1x_1 & - & 1x_2 & - & 1x_3 & + & x_4 & & = & -2, \\ -1x_1 & - & 3x_2 & - & 1x_3 & & & + \; x_5 & = & -3, \end{array}\right\} \quad (10.7)$$

which may serve as initial basis $B = (A_4, A_5)$. Observe that given basis B is primal infeasible. Thus, to apply the primal simplex method to this problem, in addition to slack variables x_4 and x_5 we would have to enter two more artificial variables x_6 and x_7, and then perform the first phase of the primal simplex method to determine an initial basic feasible solution for system (10.7).

Case 2 usually occurs in the integer programming problems when we use the branch-and-bound method or the cutting plane method of Gomory to maintain an integrality restriction. Suppose we have to add to system (10.7) constraint $x_1 \geq 1$. Since the current optimal solution $x^* = (0.5,\ 0,\ 2.5)^T$ has $x_1^* = 0.5$, it is no longer feasible and hence, cannot be optimal. So we have to re-optimize the simplex tableau shown in Table 10.3. First, we introduce new artificial variable x_6 and then convert the constraint to be added to the following form

$$x_1 - x_6 = 1\ ,$$

or

$$-x_1 + x_6 = -1\ . \quad (10.8)$$

Let constraint (10.8) be appended to the original constraints as they appear in the optimal tableau (Table 10.3). We have

$$\left.\begin{array}{rrrrrcr} & 2x_2 & +1x_3 & -1/2x_4 & -1/2x_5 & & = \;\; 5/2\,, \\ 1x_1 & +1x_2 & & +1/2x_4 & -1/2x_5 & & = \;\; 1/2\,, \\ -1x_1 & & & & & +1x_6 & = \;\; -1\,. \end{array}\right\} \quad (10.9)$$

Since variable x_1 appears in Table 10.3 as a basic variable and it is associated in the optimal simplex tableau with a unit vector, we cannot append restriction (10.8) to the optimal tableau in the form as it is, since otherwise x_1 will no longer be a basic variable. To avoid this problem, we replace the third constraint in (10.9) by another one obtained as the sum of row 2 and row 3. Thus, instead of (10.9) we have

$$\left.\begin{array}{rrrrrrr} & 2x_2 & +1x_3 & -1/2x_4 & -1/2x_5 & & = & 5/2 \\ 1x_1 & +1x_2 & & +1/2x_4 & -1/2x_5 & & = & 1/2 \\ & 1x_2 & & +1/2x_4 & -1/2x_5 & +1x_6 & = & -1/2 \end{array}\right\} \quad (10.10)$$

The system obtained contains three unit columns A_1, A_3, and A_6, which may be used to construct initial basis $B = (A_3, A_1, A_6)$. So, to begin re-optimization we can use the initial tableau shown in Table 10.4. Obviously, as we can see from

B	p_B	d_B	x_B	A_1	A_2	A_3	A_4	A_5	A_6	
A_3	-20	35	$5/2$	0	2	1	$-1/2$	$-1/2$	0	
A_1	-4	5	$1/2$	1	1	0	$1/2$	$-1/2$	0	
A_6	0	0	$-1/2$	0	1	0	$1/2$	$-1/2$	1	$\Rightarrow$
$P(x) = -60$				0	-9	0	8	11.5	0	
$D(x) = 120$				0	35	0	-15	-20	0	
$Q(x) = -1/2$				0	$17/2$	0	$1/2$	$3/2$	0	
			Ratio	N/A	N/A	N/A	N/A	-3	N/A	
								$\Uparrow$		

Table 10.4. The Dual Simplex Method – With a new constraint.

Table 10.4, the current dual feasible basis $B = (A_3, A_1, A_6)$ is neither optimal nor primal feasible. Further, variable x_6 is the only and the most negative basic variable. Hence, it must be removed from the current basis. Meanwhile, the ratio test gives us vector A_5 as a pivot column. The new basic solution is shown in Table 10.5. In Table 10.5 all basic variables are non-negative, hence, we have obtained an optimal solution. So, after re-optimization we have

$$x^* = (1,\ 0,\ 3)^T, \quad Q(x^*) = -72/140. \qquad (10.11)$$

Thus, if the constraint $x_1 \geq 1$ is added to original problem (10.4)-(10.6), the optimal solution becomes as shown in (10.11).

In bringing this discussion of the dual simplex method to a close, we have to note that adding a new constraint to a problem may lead to the infeasibility of the new problem. In this case, the dual simplex method will indicate (see Step 4) that the given problem has no feasible solutions.

B	p_B	d_B	x_B	A_1	A_2	A_3	A_4	A_5	A_6
A_3	-20	35	3	0	1	1	-1	0	-1
A_1	-4	5	1	1	0	0	0	0	-1
A_5	0	0	1	0	-2	0	-1	1	-2
$P(x) = -72$				0	15	0	20	0	24
$D(x) = 140$				0	-5	0	-35	0	-40
$Q(x) = -18/35$				0	$87/7$	0	2	0	$24/7$

Table 10.5. The Dual Simplex Method – After re-optimization.

2. The Criss-Cross Method

The main uniting inherent properties of various modifications of the simplex method are as follows

- during iterations the method preserves feasibility (primal or dual) of the basic solutions inspected;
- the method forces monotonicity of the objective value, i.e for a maximization problem an objective value in the next iteration will be not less than the current value;
- the new basis differs from the previous one exactly by one element (vector), i.e the new vertex is a neighbor of the previous one.

Similar to the simplex method, the *Criss-Cross Method* (CCM) is based on pivoting, and in a finite number of iterations it either solves the problem or indicates that the problem is unsolvable (infeasible or unbounded). Contrary to the simplex method, the criss-cross method traverses through different (not necessarily feasible) vertices (not necessarily neighbors) of a feasible set, and does not preserve the monotonicity of the objective value.

The criss-cross method was first proposed for linear programming problems by T.Terlaky in [179], [180] and was referred to as a *finite* (i.e. convergent) *criss-cross method.* Later, the method was generalized by T.Illés, Á.Szirmai and T.Terlaky in [99] for the class of linear-fractional programming problems.

The main (and most attractive) features of the method are

- it can be started from any initial, not necessarily feasible, basic solution;

- since the initial basic solution may be infeasible, the method does not require artificial variables, and hence, solves the problem in one phase;
- the method can solve linear-fractional programming problems both with bounded and unbounded feasible sets.

The aim of this section is to describe the CCM and to show how it can be used to solve LFP problems.

Consider the following maximization LFP problem in canonical form

$$Q(x) = \frac{P(x)}{D(x)} = \frac{\sum_{j=1}^{n} p_j x_j + p_0}{\sum_{j=1}^{n} d_j x_j + d_0} \longrightarrow \max, \tag{10.12}$$

subject to

$$\sum_{j=1}^{n} a_{ij} x_j = b_i, \quad i = 1, 2, \dots, m, \tag{10.13}$$

$$x_j \geq 0, \quad j = 1, 2, \dots, n, \tag{10.14}$$

where

$$D(x) > 0, \ \ \forall x \in S \tag{10.15}$$

and S denotes a feasible set determined by constraints (10.13)-(10.14).

The method we are going to describe is based on the following idea: we try to solve the original LFP problem as it is (i.e. in its original fractional form) but during performing iterations we use a piece of information related to the linear analogue (see Chapter 3, Section 3) of the problem.

Let B denote a basis (not necessary feasible), i.e.

$$B = (A_{s_1}, A_{s_2}, \dots, A_{s_m}),$$

where $A_j = (a_{1j}, a_{2j}, \dots, a_{mj})^T$, $j = 1, 2, \dots, n$, while J_B and J_N denote a set of basic and non-basic indices j, respectively, such that $J = \{1, 2, \dots, n\} = J_B \cup J_N$ and $J_B = \{s_1, s_2, \dots, s_m\}$. Vector x be the basic solution of problem (10.12)-(10.14) associated with the current basis B. Further, similar to the conventional simplex method, we introduce the following notations:

$$\Delta'_j = \sum_{i=1}^{m} p_{s_i} x_{ij} - p_j, \ \ \Delta''_j = \sum_{i=1}^{m} d_{s_i} x_{ij} - d_j, \ \ j = 1, 2, \dots, n,$$

$$\Delta_j(x) = \Delta'_j - Q(x)\Delta''_j, \ \ j = 1, 2, \dots, n,$$

where coefficients x_{ij} are determined from the following linear combinations of basic vectors $A_{s_i},\ i = 1, 2, \ldots, m,$

$$\sum_{i=1}^{m} A_{s_i} x_{ij} = A_j, \quad j = 1, 2, \ldots, n.$$

In addition, in accordance with the rules of Charnes&Cooper's transformation (see Chapter 3, Section 3) we introduce the following notation:

$$t_j = \frac{x_j}{D(x)}, \quad j = 1, 2, \ldots, n.$$

Finally, let

$$u_{ij} = x_{ij} - \frac{x_{s_i} \Delta''_j}{D(x)}, \quad i = 1, 2, \ldots, m, \quad j = 1, 2, \ldots, n,$$

where u_{ij} are coefficients of the simplex tableau (associated with the current basis B) constructed for linear analogue of LFP problem (10.12)-(10.14).

Now we formulate the following statements that provide theoretical foundations for the method (see [99] for proofs).

THEOREM 10.2 (OPTIMALITY CRITERIA) *If for all indices $j = 1, 2, \ldots, n$*

$$x_j \geq 0, \quad \textit{and} \quad \Delta_j(x) \geq 0, \quad j = 1, 2, \ldots, n,$$

then vector x is an optimal solution for LFP problem (10.12)-(10.14), while vector $y = (y_0, y_1, y_2, \ldots, y_m)$ is an optimal solution of the dual problem, where

$$y_0 = \frac{P(x)}{D(x)}, \quad y_i = \Delta'_{s_i} - y_0 \Delta''_{s_i}, \quad i = 1, 2, \ldots, m.$$

THEOREM 10.3 (INFEASIBILITY CRITERIA) *If there exists index $s_i \in J_B$ such that*

$$x_{s_i} < 0, \quad \textit{and} \quad u_{ij} \geq 0, \quad j = 1, 2, \ldots, n,$$

then LFP problem (10.12)-(10.14) is infeasible, i.e. its feasible set S is empty.

THEOREM 10.4 (UNBOUNDEDNESS CRITERIA) *If there exists index $j_0 \in J_N$ such that*

$$\Delta_{j_0}(x) < 0, \quad \textit{and} \quad u_{ij_0} < 0, \quad i = 1, 2, \ldots, m,$$

then LFP problem (10.12)-(10.14) is unbounded, i.e. its objective function $Q(x)$ has no finite upper bound over feasible set S.

DEFINITION 10.1 *The simplex tableau transformation shown in Table 10.6 and Table 10.7 is called* ***external transformation****, where*

$$t' = \frac{P(x)}{D(x)}, \quad t_0 = \frac{1}{D(x)}, \quad t_{s_i} = \frac{t_{s_i}}{D(x)}, \quad i = 1, 2, \dots, m,$$

and

$$\tilde{\Delta}'_j = \Delta'_j - Q(x)\Delta''_j, \quad \tilde{\Delta}''_j = \frac{\Delta''_j}{D(x)}, \quad j = 1, 2, \dots, n.$$

	p_1 d_1	$\dots$ $\dots$	p_k d_k	$\dots$ $\dots$	p_n d_n	
x_B	A_1	$\dots$	A_k	$\dots$	A_n	t_B
x_{s_1}	x_{11}	$\dots$	x_{1k}	$\dots$	x_{1n}	0
$\vdots$	$\vdots$	$\dots$	$\vdots$	$\vdots$	$\vdots$	$\vdots$
x_{s_r}	x_{r1}	$\dots$	x_{rk}	$\dots$	x_{rn}	0
$\vdots$	$\vdots$	$\dots$	$\vdots$	$\vdots$	$\vdots$	$\vdots$
x_{s_m}	x_{m1}	$\dots$	x_{mk}	$\dots$	x_{mn}	0
$P(x)$	Δ'_1	$\dots$	Δ'_k	$\dots$	Δ'_n	0
$D(x)$	Δ''_1	$\dots$	Δ''_k	$\dots$	Δ''_n	1
$Q(x)$	$\Delta_1(x)$	$\dots$	$\Delta_k(x)$	$\dots$	$\Delta_n(x)$	

Table 10.6. External transformation – Original tableau.

When using the criss-cross method we proceed as follows.

Step 0 (Initial basis). Determine an initial basic solution x with corresponding basic B such that $D(x) \neq 0$ and assumption (10.15) holds. Go to Step 1.

Step 1 (Optimality test). If

$$x_j \geq 0, \quad \text{and} \quad \Delta_j(x) \geq 0, \quad \forall j \in J = \{1, 2, \dots, n\},$$

the current basic solution x is feasible and optimal. Stop.
Otherwise, we choose index

$$k := \min\{j |\ j \in J^0\},$$

where

$$J^0 = \{j \in J_B |\ x_j < 0\} \cup \{j \in J_N |\ \Delta_j(x) < 0\},$$

	p_1 d_1	$\cdots$	p_k d_k	$\cdots$	p_n d_n	
x_B	A_1	$\cdots$	A_k	$\cdots$	A_n	t_B
0	x_{11}	$\cdots$	x_{1k}	$\cdots$	x_{1n}	t_{s_1}
$\vdots$	$\vdots$	$\cdots$	$\vdots$	$\vdots$	$\vdots$	$\vdots$
0	x_{r1}	$\cdots$	x_{rk}	$\cdots$	x_{rn}	t_{s_r}
$\vdots$	$\vdots$	$\cdots$	$\vdots$	$\vdots$	$\vdots$	$\vdots$
0	x_{m1}	$\cdots$	x_{mk}	$\cdots$	x_{mn}	t_{s_m}
0	$\tilde{\Delta}'_1$	$\cdots$	$\tilde{\Delta}'_k$	$\cdots$	$\tilde{\Delta}'_n$	t'
1	$\tilde{\Delta}''_1$	$\cdots$	$\tilde{\Delta}''_k$	$\cdots$	$\tilde{\Delta}''_n$	t_0
$Q(x)$	$\Delta_1(x)$	$\cdots$	$\Delta_k(x)$	$\cdots$	$\Delta_n(x)$	

Table 10.7. External transformation – Resulting tableau.

and then go to Step 2 if $k \in J_B$, or go to Step 3 if $k \in J_N$.

Step 2 (Dual iteration). First, we construct the following index set

$$J^- := \{j \in J_N | \ u_{kj} < 0\}.$$

If set $J^- = \emptyset$ then the problem is primal infeasible, that is the feasible set of LFP problem (10.12)-(10.14) is empty. Stop.
Otherwise, let $r := \min\{j | \ j \in J^-\}$ and go to Step 4.

Step 3 (Primal iteration). Let

$$I^+ := \{i | \ 1 \le i \le m, \ u_{ik} > 0\}.$$

If set $I^+ = \emptyset$ then the problem is dual infeasible, that is objective function (10.12) has no upper bound. Stop.
Otherwise, let $r := \min\{i | \ i \in I^+\}$ and go to Step 4. go to Step 4.

Step 4 (Pivot transformation). If $x_{rk} = 0$ perform double pivot transformation: first, external transformation, then the standard simplex transformation at position r, k.
Otherwise, perform standard simplex transformation at positionr, k.
Go to Step 1.

To illustrate this method we consider the following maximization LFP problem with unbounded feasible set.

$$Q(x) = \frac{P(x)}{D(x)} = \frac{x_1 - 2x_2 + 2}{x_1 + x_2 + 0.5} \longrightarrow \max \tag{10.16}$$

subject to

$$\left.\begin{array}{rcr} -1x_1 + 1x_2 & \leq & 1, \\ 1x_1 - 4x_2 & \leq & 2, \\ -1x_1 - 2x_2 & \leq & -2, \end{array}\right\} \tag{10.17}$$

$$x_1 \geq 0, \quad x_2 \geq 0. \tag{10.18}$$

First of all, adding slack variables $x_3,\ x_4$ and x_5 we convert the problem to canonical form. Observe that slack variables are associated with unit vectors $A_3,\ A_4$ and A_5, respectively. Obviously, these vectors may serve as an initial primal infeasible basis B.

Starting with initial primal infeasible basic solution $x = (0, 0, 1, 2, -2)^T$, $Q(x) = 2/0.5 = 4$, and the simplex tableau shown in Table 10.8, and then performing in the given order the following pivot transformations

Iteration 1 $(x_2 \rightarrow x_3)$. Gives basis $B = (A_2, A_4, A_5)$ and associated feasible basic solution $x = (0, 1, 0, 6, 0)^T$ with objective value $Q(x) = 0/1.5 = 0$;

Iteration 2 $(x_3 \rightarrow x_5)$. Results in basis $B = (A_2, A_4, A_3)$ and the same feasible basic solution $x = (0, 1, 0, 6, 0)^T$ with $Q(x) = 0/1.5 = 0$;

Iteration 3 $(x_1 \rightarrow x_2)$. Leads to optimal basis $B = (A_1, A_4, A_3)$, optimal basic solution $x = (2, 0, 3, 0, 0)^T$ with $Q(x) = 4/2.5 = 1.6$,

for the original LFP problem we obtain primal feasible and optimal basic solution $x^* = (2, 0)^T$ and $Q(x^*) = 4/2.5 = 1.6$.

In conclusion, we note that replacing an LFP problem with its linear analogue and then applying the LP simplex method to the latter will lead to the same results.

3. The Interior-Point Methods

All forms of the simplex method reach the optimum by traversing a series of basic (feasible or infeasible) solutions. Since each basic feasible solution of an LFP problem represents an extreme point of the feasible set, the track followed by the simplex algorithm moves around the boundary of the primal or dual feasible region. In the worst case, it may be necessary to examine most (if not all) of the vertices of the feasible set. This may be disgracefully inefficient

				1	−2	0	0	0	
				1	1	0	0	0	
B	p_B	d_B	x_B	A_1	A_2	A_3	A_4	A_5	t_B
A_3	0	0	1	−1	1	1	0	0	0
A_4	0	0	2	1	−4	0	1	0	0
A_5	0	0	−2	−1	−2	0	0	1	0
$P(x) = 2$				−1	2	0	0	0	0
$D(x) = 0.5$				−1	−1	0	0	0	1
$Q(x) = 4$				3	6	0	0	0	

Table 10.8. The Criss-Cross Method Example – Initial tableau.

given that the number of extreme point grows very fast (exponentially) with the size of problem n and m.

The running time of an algorithm as a function of the problem size is known as its *computational complexity*. In practice, the simplex method works surprisingly well, often exhibiting linear complexity, i.e., proportional to the expression $n + m$. However, a lot of researchers have long tried to develop methods for LP and LFP whose worst-case running times are a polynomial function of the problem size. The first success was attributed to the Soviet mathematician, Leonid Khachian, who proposed the *Ellipsoid Method* for linear programming problems, which has a running time proportional ton^6 (see L.G.Khachian [111], N.Z.Shor [170] for a full discussion of the approach). Though theoretically efficient, the software tools developers were never able to realize an implementation that matched the performance of concurrent simplex method codes.

Just about the time when interest in the ellipsoid method was waning, a new technique to solve linear programming problems was proposed by N.Karmarkar in [108]. His idea was to approach the optimal solution from the strict interior of the feasible region. This led to the series of*Interior Point Methods* (IPM) that combined the advantages of the simplex method with the geometry of the ellipsoid algorithm. IPMs are of interest from the theoretical point of view because they have produced solutions to many real-world industrial problems that were hitherto intractable.

There are at least three major types of IPMs: (1) the potential reduction algorithms which most closely embody the idea of Karmarkar, (2) the affine scaling algorithms which may be considered to be the simplest to implement, and (3) path following algorithms which arguably combine excellent behavior in theory and practice.

The landmark paper of Karmarkar initiated investigation activity in fractional programming as well as in linear-fractional programming.

In the 90's, a Karmarkar-like algorithm was proposed by R.W.Freund and F.Jarre in [65] and [66] for a special class of fractional programming problems with convex constraints. They showed that a so-called *short-step* version of their algorithm converges at a polynomial time.

The further improvement and expansion of the algorithm was made by A.Nemirovskii and Y.Nesterov in [138], where the authors adapted the algorithm to a generalized linear-fractional problem (see Chapter 11, Section 1) and proved its polynomiality. Later, in [139] and [140] the so-called *Method of Analytic Centers* (which may be classified as a path following method) and its *long-step* algorithm were proposed for a class of optimization problems formulated as follows

$$\phi(t,x) = t \longrightarrow \min \tag{10.19}$$

subject to

$$tB(x) - A(x) \in K, \quad x \in G, \tag{10.20}$$

where $G \subset R^n$ and $K \subset R^m$ are closed convex sets, while $A(x)$ and $B(x)$ are linear functions. Set G is assumed to be bounded.

Strictly speaking, problem (10.19)-(10.20) is not a linear-fractional problem, since its objective function $\phi(t,x)$ is linear. Actually, if $m = 1$ and linear function $B(x) > 0$ for all $x \in G$, from problem (10.19)-(10.20) as a special case we can obtain the LP problem considered by W.Dinkelbach in [54] and used to solve an LFP problem in conventional form (see Chapter 3, Section 4).

So the method of analytic centers is beyond the scope of the book and we restrict our consideration of the method to the brief description of its steps (for detailed information on interior-point methods in linear programming see e.g. [141], [153]). The method may proceed as follows: first, we have to associate with sets G and K the appropriate *barriers* (special interior penalty functions) $\Phi_G(x)$ and $\Phi_K(y)$, respectively, and then, trace the path given by the following rule

$$\begin{aligned} x^*(t) &= \arg\min \Phi_t(x), \\ \Phi_t(x) &\equiv \Phi_G(x) + \Omega_K \Phi_K(tB(x) - A(x)) + \Omega_K \Phi_K(B(x)), \end{aligned}$$

where Ω_K denotes a special positive constant.

In concluding this discussion of the interior-point methods in LFP, we just note that most of the known IPM algorithms without any adaptations may be applied to the linear analogue of an LFP problem obtained from the original LFP problem after applying Charnes&Cooper's transformation (see Chapter 3, Section 3).

4. Discussion Questions and Exercises

10.1 Reconsider numerical example (10.4)-(10.6) given in Section 1 (page 289). Its optimal simplex tableau is shown in Table 10.3. Suppose we have to add to system (10.7) a new constraint $x_1 \leq 0$. Using the dual simplex method re-optimize the optimal simplex tableau shown in Table 10.3.

10.2 In the numerical example given in the previous exercise replace restriction $x_1 \leq 0$ with the following one

$$x_2 + x_3 \leq 1$$

and perform the re-optimization.

10.3 For numerical example (10.16)-(10.18) given in Section 2 (see page 298) using Charnes&Cooper transformation construct the associated linear analogue and solve it by the primal simplex method. Compare the sequence of pivots traversed by the primal simplex method with the sequence traversed by the criss-cross method as shown in Section 2.

Chapter 11

ADVANCED TOPICS IN LFP

In this chapter, we briefly indicate several new directions of investigations in fractional programming made in the last decades. We discuss here the following extensions of linear-fractional programming: generalized LFP and LFP problems with multiple objective functions.

1. Generalized LFP

A *generalized linear-fractional programming* problem is specified as a non-linear problem

$$\left.\begin{array}{rl} \lambda = & \max\limits_{1\le l\le q}\{\dfrac{P_l(x)}{D_l(x)}\} \longrightarrow \min \\ \text{subject to} & \\ & x \in S\,, \end{array}\right\} \tag{11.1}$$

where

$$x = (x_1, x_2, \dots, x_n)^T, \quad P_l(x) = \sum_{j=1}^{n} p_{lj}x_j + p_{l0}, \quad D_l(x) = \sum_{j=1}^{n} d_{lj}x_j + d_{l0},$$

S denotes a non-empty and possibly unbounded feasible set given by constraints

$$\sum_{j=1}^{n} a_{ij}x_j \le b_i, \quad i = 1, 2, \dots, m;$$
$$x_j \ge 0, \quad j = 1, 2, \dots, n;$$

and

$$D_l(x) > 0 \quad \forall x \in S, \quad l = 1, 2, \dots, k.$$

It is obvious that this problem is a generalization of a linear-fractional programming problem ($k = 1$) which has been investigated in previous chapters. Problem (11.1) has a wide-range application area in science, economics and industry. One of the first applications of generalized linear-fractional programming problem ($k > 1$) is a model of expanding economy developed in the 1940's by John von Neumann, [142]. Problems of type (11.1) appear in so-called*goal programming* if the aim of the decision maker is to bring several ratio-type objective functions as close as possible to some predefined values, [114]. Another instance of applications for problems of generalized liner-fractional programming is multi-objective programming where several fractional objective functions must be optimized simultaneously and the main aim is maximization (minimization) of the smallest (largest) of these ratios, [7]. Also, some allocation models considered by A.I.Barros [18] lead to the problem of generalized fractional programming with non-linear functions $P_l(x)$ and $D_l(x)$.

In this section we briefly overview the algorithmic aspects of the generalized linear-fractional programming problem formulated in the form of (11.1). More detailed information on this topic (including duality for generalized linear-fractional programming) may be found in [18], [19], [32], [44], [45], [46], [104],[138]. Other forms of generalization for LFP problems available in the literature, see for example [110], are beyond the scope of this book.

One of the most popular algorithmic procedures for solving problem (11.1) available in the literature is based on the well-known Dinkelbach's algorithm developed for LFP problems (see Chapter 3, Section 4). This method corresponds to solving a sequence of the following parametric problems

$$F(\lambda) = \min_{x \in S}\{\max_{1 \le l \le q}\{P_l(x) - \lambda D_l(x)\}\}, \tag{11.2}$$

Before discussing the method we have to recall that

DEFINITION 11.1 *Function $f(x)$ defined over set X is said to be lower (upper)* ***semi-continuous*** *at point $x' \in X$ if*

$$\lim_{x \to x'-0} f(x) = f(x') \quad (\lim_{x \to x'+0} f(x) = f(x'))$$

The following statements establish relations between the original problem (11.1) of generalized linear-fractional programming and the problem (11.2) with parametric objective function $F(\lambda)$.

LEMMA 11.1 *([46], Proposition 2.1) Let* $\bar{\lambda} = \min_{x \in S}\{\max_{1 \le l \le q}\{\frac{P_l(x)}{D_l(x)}\}\}$ *then*

1 Parametric function $F(\lambda) < +\infty$. Moreover, $F(\lambda)$ is upper semi-continuous and non-increasing;

2 *$F(\lambda) < 0$ if and only if $\lambda > \overline{\lambda}$;*

3 *$F(\overline{\lambda}) \geq 0$;*

4 *If problem (11.1) is solvable then $F(\overline{\lambda}) = 0$;*

5 *If $F(\overline{\lambda}) = 0$ then problem (11.1) and problem (11.2) have the same set of optimal solutions.*

LEMMA 11.2 *([46], Theorem 4.1) If feasible set S is compact then*

1 *Parametric function $F(\lambda) < +\infty$. Moreover, $F(\lambda)$ is continuous and strictly decreasing;*

2 *Problems (11.1) and (11.2) always have optimal solutions.*

3 *$\overline{\lambda}$ is finite and $F(\overline{\lambda}) = 0$;*

4 *$F(\lambda) = 0$ implies $\lambda = \overline{\lambda}$;*

These two lemmas provide the necessary theoretical basis for a generalization of Dinkelbach's algorithm as shown in Figure 11.1.

Generalization of Dinkelbach's Algorithm

Step 0. Take arbitrary $x^{(0)} \in S$,
compute $\lambda^{(1)} := \max\limits_{1\leq l\leq q}\{\dfrac{P_l(x^{(0)})}{D_l(x^{(0)})}\}$, and
let $k := 1$;

Step 1. Determine an optimal solution $x^{(k)}$ of problem
$F(\lambda^{(k)}) = \min\limits_{x\in S}\{\max\limits_{1\leq l\leq q}\{P_l(x) - \lambda^{(k)}D_l(x)\}\}$;

Step 2. If $F(\lambda^{(k)}) = 0$ then $x^* = x^{(k)}$ is an optimal solution of (11.1),
$\lambda^{(k)}$ is the optimal value,
Stop;

Step 3. Let $\lambda^{(k+1)} := \max\limits_{1\leq l\leq q}\{\dfrac{P(x^{(k)})}{D(x^{(k)})}\}$,
let $k := k + 1$,
go to Step 1;

Figure 11.1. Algorithm – Generalized Dinkelbach's Algorithm.

The convergence of sequence $\{\lambda^{(k)}\}$ generated by the algorithm is guaranteed by the following properties of the sequence:

- For all $k \geq 0, \quad \lambda^{(k+1)} = \max_{1 \leq l \leq q} \{ \frac{P(x^{(k)})}{D(x^{(k)})} \} \geq \overline{\lambda};$
- The sequence $\{\lambda^{(k)}\}$ is monotone decreasing.

We have to note that some special variation of the Dinkelbach-type algorithm may be derived if we apply Charnes&Cooper's transformation (see Chapter 3, Section 3) to problem (11.1), [18]. For further information connected with solving generalized fractional programming problems see, e.g., [19].

Before closing this discussion of generalized linear-fractional programming, we will consider the following numeric example.

$$\lambda = \max_{l=1,2} \{ \frac{P_l(x)}{D_l(x)} \} \longrightarrow \min \tag{11.3}$$

subject to

$$\left. \begin{array}{r} x_1 \leq 5 \,, \\ x_1 \geq 0 \,, \end{array} \right\} \tag{11.4}$$

where

$$P_1(x) = 4x_1 + 5 \,, \quad D_1(x) = 2x_1 + 1 \,,$$

and

$$P_2(x) = 1x_1 + 8 \,, \quad D_2(x) = 2x_1 + 1 \,.$$

To apply the Dinkelbach-type algorithm for solving this problem we associate with (11.3)-(11.4) a sequence of parametric problems (11.2). The algorithm proceeds as follows:

Step 0. Let $x_1^{(0)} = 4 \in S = [0; 5]$, then

$$\lambda^{(1)} = \max\{ \frac{4 \times 4 + 5}{2 \times 4 + 1} ; \frac{1 \times 4 + 8}{2 \times 4 + 1} \} = \max\{2.3333;\ 1.3333\} = 2.3333 \,.$$

Let $k := 1$.

Step 1 ($k = 1$). Now, for $\lambda^{(1)} = 2.3333$ we construct problem

$$F(\lambda^{(1)}) = \min_{x \in S} \{ \max\{ P_1(x) - \lambda^{(1)} D_1(x) \,, P_2(x) - \lambda^{(1)} D_2(x) \} \}.$$

Solving this problem we obtain $x_1^{(1)} = 5.0$,
hence $F(\lambda^{(1)}) = -0.6667$.

Step 2 ($k = 1$). Since $F(\lambda^{(1)}) \neq 0$ go to Step 3.

Step 3 ($k = 1$). Let

$$\lambda^{(2)} = \max\{\frac{4 \times 5 + 5}{2 \times 5 + 1}, \frac{1 \times 5 + 8}{2 \times 5 + 1}\} = \max\{2.2727;\ 1.1818\} = 2.2727\ .$$

Let $k := k + 1 = 2$.

Step 1 ($k = 2$). For $\lambda^{(2)} = 2.2727$ we construct problem

$$F(\lambda^{(2)}) = \min_{x \in S}\{\max\{P_1(x) - \lambda^{(2)} D_1(x)\ , P_2(x) - \lambda^{(2)} D_2(x)\}\}.$$

Solving this problem we obtain $x_1^{(2)} = 5.0,\ \ F(\lambda^{(2)}) = 0.0$.

Step 2 ($k = 2$). Since $F(\lambda^{(2)}) = 0$, Stop.
$x_1^{(2)} = 5.0$ is an optimal solution of problem (11.3)-(11.4). The optimal value of objective function (11.3) is $\overline{\lambda} = \lambda^{(2)} = 2.2727$.

2. Multi-objective LFP

The branch of mathematical programming where the problem has several objective functions is well developed and is referred to as*multi-objective programming* or *vector optimization.* In recent decades, a number of researchers (see e.g., [26], [41], [85], [86], [105], [106], [116], [117], [134], [143] [190], etc.) extended the theory of multi-objective programming to the case of linear-fractional programming when the problem contains several linear-fractional objective functions. Such problems arise in corporate planning, marine transportation, health care, educational planning, network flows, etc., when there are several (generally speaking, conflicting) objectives that cannot be optimized simultaneously, and a decision maker has to find a most preferred solution.

Consider the following multi-objective LFP (MOLFP) problem

$$Q(x) = (Q_1(x),\ Q_2(x),\ \ldots,\ Q_K(x)) \longrightarrow \max \tag{11.5}$$

subject to

$$\sum_{j=1}^{n} a_{ij} x_j = b_i, \quad i = 1, 2, \ldots, m, \tag{11.6}$$

$$x_j \geq 0, \quad j = 1, 2, \ldots, n, \tag{11.7}$$

where

$$Q_k(x) = \frac{P_k(x)}{D_k(x)} = \frac{\sum_{j=1}^{n} p_{kj}x_j + p_{k0}}{\sum_{j=1}^{n} d_{kj}x_j + d_{k0}}, \qquad k = 1, 2, \dots, K,$$

$$D_k(x) > 0, \quad \forall x \in S, \quad \forall k = 1, 2, \dots, K,$$

and S denotes the feasible set of MOLFP problem (11.5)-(11.7).

DEFINITION 11.2 *A point $x^* \in S$ is said to be an **efficient solution** (or **Pareto optimal**) of MOLFP problem (11.5)-(11.7) if there does not exist another feasible point $x' \in S$ such that*

a. $Q_k(x') \geq Q_k(x^*), \quad k = 1, 2, \dots, K,$ *and*

b. *there is at least one index $k_0 \in \{1, 2, \dots, K\}$ such that*

$$Q_{k_0}(x') > Q_{k_0}(x^*).$$

There are at least the following two general approaches to solving mathematical programming problems with multiple objective functions:

1. (Weighted sum) Original objective function (11.5) must be replaced with the following, usually referred to as a *composite objective function*:

$$Q'(x) = \sum_{k=1}^{K} w_k Q_k(x) \longrightarrow \max$$

   where vector of weights $w = (w_1, w_2, \dots, w_K)$ consists of positive weights $w_k > 0$ which are the subject of the preferences of the decision maker.

2. (Lexicographic) When using this approach we have to fix in advance a *lexicographical order* for functions $Q_k, \ k = 1, 2, \dots, K,$

$$Q_{k_1}(x) \succeq Q_{k_2}(x) \succeq \dots \succeq Q_{k_K}(x)$$

   and then perform successively the following simple objective optimizations

$$\begin{array}{ll} (k_1): & \max\limits_{x \in S_1} Q_{k_1}(x), \\ (k_2): & \max\limits_{x \in S_2} Q_{k_2}(x), \\ \vdots & \vdots \\ (k_K): & \max\limits_{x \in S_K} Q_{k_K}(x), \end{array}$$

where

$$\begin{aligned} S_1 &= S, \\ S_2 &= \{x \in S_1 |\, Q_{k_1}(x) = Q^*_{k_1}\}, \\ S_3 &= \{x \in S_2 |\, Q_{k_2}(x) = Q^*_{k_2}\}, \\ &\vdots \\ S_K &= \{x \in S_{K-1} |\, Q_{k_{K-1}}(x) = Q^*_{k_{K-1}}\}, \end{aligned}$$

and $Q^*_{k_i}$ is the optimal objective value of problem(k_i), $i = 1, 2, \ldots, K-1$.

Both approaches result in an efficient solution (if it does exist) and under certain assumptions can be used to generate the set of all efficient points usually referred to as an *efficient frontier.*

The approach based on the use of the weighted sum is closely connected with investigations in the domain of fractional programming problems with such special objective functions as a sum and product of two or more linear-fractional functions, see e.g. [3], [33], [49], [60], [97], [113], [151], [164]. When solving a multi-objective LFP problem by the weighted sum approach, the weights represent the value of relative importance associated with the single objective functions $Q_k(x)$, $k = 1, 2, \ldots, K$. Obviously these values usually are imprecise and affect the efficient solution very dramatically. This is why it is important to analyze the sensitivity of the solution with respect to the deviation of weights. In this case the so-called *tolerance approach* (see, e.g., [6]) may provide the necessary tools for such analysis.

Owing to its simplicity, the lexicographical approach does not require any further investigations. Obviously, it may be used only in the case if feasible set S_2 consists of more than one point.

The main approach proposed by several researchers, especially for linear-fractional problems with multiple objective functions is based on the reduction of the original MOLFP problem to a special multi-objective LP problem.

For example, I.Nykowski and Z.Zolkiewski in [143] developed an approach which instead of the original objective function (11.5) uses the following linear multi-objective function

$$F(x) = (P_1(x),\, P_2(x),\, \ldots,\, P_K(x),\, D_1(x),\, D_2(x),\, \ldots,\, D_K(x))$$

or

$$F(x) = (P_1(x),\, P_2(x),\, \ldots,\, P_K(x),\, -D_1(x),\, -D_2(x),\, \ldots,\, -D_K(x))$$

on the same feasible set S. An approach based on Charnes&Cooper's transformation is used in [58] and [128].

In [173] the following linear programming problem with multiple objective functions is associated with the original MOLFP problem (11.5)-(11.7).

Consider point $x^{(0)} \in S$. With this point $x^{(0)}$ we associate coefficients

$$h_{kj} = p_{kj}D(x^{(0)}) - d_{kj}P(x^{(0)}), \quad k = 1, 2, \ldots, K, \quad j = 1, 2, \ldots, n,$$

and the following multi-objective LP problem

$$G(x) = (G_1(x),\ G_2(x),\ \ldots,\ G_K(x)) \longrightarrow \max \tag{11.8}$$

subject to (11.6)-(11.7), where

$$G_k(x) = \sum_{j=1}^{n} h_{kj}x_j, \quad k = 1, 2, \ldots, K.$$

The following theorem establishes the main relation between the original MOLFP problem (11.5)-(11.7) and multi-objective LP problem (11.8),(11.6)-(11.7).

THEOREM 11.1 ([173]) *Vector $x^{(0)} \in S$ is an efficient solution of original MOLFP problem (11.5)-(11.7) if and only if $x^{(0)}$ is an efficient solution of multi-objective LP problem (11.8),(11.6)-(11.7).*

In closing this discussion we just note that readers interested in such advanced topics of multi-objective LFP as duality theory in MOLFP or integer MOLFP can find detailed information on these topics in [105] and [86].

Chapter 12

COMPUTATIONAL ASPECTS

Let us consider the following LFP problem in a canonical form:

$$Q(x) = \frac{P(x)}{D(x)} = \frac{\sum_{j=1}^{n} p_j x_j + p_0}{\sum_{j=1}^{n} d_j x_j + d_0} \longrightarrow \max \tag{12.1}$$

subject to

$$\sum_{j=1}^{n} a_{ij} x_j = b_i, \quad i = 1, 2, \ldots, m; \tag{12.2}$$

$$x_j \geq 0, \quad j = 1, 2, \ldots, n, \tag{12.3}$$

where $D(x) > 0, \ \ \forall x = (x_1, x_2, \ldots, x_n)^T \in S, \ \ S$ - is a feasible set defined by constraints (12.2) and (12.3).

It is a well-known fact that large linear-programming (LP) models

$$P(x) = \sum_{j=1}^{n} p_j x_j \longrightarrow \max \tag{12.4}$$

subject to constraints (12.2)-(12.3);

require hundreds of thousands to millions of floating-point arithmetic calculations to solve. Because of the finite precision inherent in computer arithmetic, small numerical errors occur in these calculations. These errors typically have a cumulative effect, leading to a numerically unstable problem and possibly large errors in the "solution" obtained. The same computational problems occur in large-scale LFP problems too.

To avoid such problems, all well-made industrial LP solvers include special sophisticated techniques that dramatically reduce a cumulative effect of rounding and often lead to considerable improvement in the solvers' performance.

One of the most easy, relatively effective and widespread techniques of this type is *scaling*[1]. This technique means that those rows and/or columns of matrix $A = \|a_{ij}\|_{m\times n}$ in the original optimization problem, which are *poorly (or badly) scaled*, that is have a wide range of entries, must be divided (or multiplied) with their own scaling factors ρ_i^r, $i = 1, 2, \dots, m$ and/or ρ_j^c, $j = 1, 2, \dots, n$, respectively. In most real-world LP and LFP applications, the model originally is very poorly scaled — for example, with dollar amounts in millions for some constraints and return figures in percentages for others. This is why before beginning the simplex or other method the program package must re-scale columns, rows, and right-hand sides to a common magnitude[2].

Such scaling may include or may not the coefficients of an objective function. In the case of LP problems, scaling matrix A, right-hand-side vector b and objective function $P(x)$ does not lead to any difficulties because of the linearity of the constraints and the objective function. In most cases scaling improves the numerical properties of the problem to be solved so it is justified to use it. Moreover, sometimes it can dramatically reduce the number of iterations in simplex method. Most professionally developed LP solvers automatically use scaling methods to maintain numerical stability. Normally, you can choose among "No Scaling", "Row Scaling Only", "Column Scaling", or "Row and Column Scaling" with or without scaling the objective function.

In the case of LFP problem (12.1)-(12.3), when scaling we should keep in mind the main difference between LP and LFP problems - the non-linear objective function $Q(x)$.

Another widespread way to reduce the cumulative effect of rounding is the so-called *re-initialization* (or *re-factorization*) of the basis matrix. This technique means recalculating coefficients x_{ij} of the simplex tableau using some direct methods of linear algebra. Most well-made public-domain and commercial solvers usually use *LU-decomposition* (or *LU-factorization*) or some other special methods (for example, *Cholesky* factorization, requires a symmetric

[1] The pre-solution transformation of the data of a problem that attempts to make the magnitudes of all the data as close as possible

[2] A frequently used scaling algorithm is to divide each row by the largest absolute element in it, and then divide each resulting column by the largest absolute element in it. This ensures that the largest absolute value in the matrix is 1.0 ant that each column and row has at least one element equal to 1.0.

matrix) of the basic matrix and apply this updating periodically (at most every 100 iterations) during performing the simplex method. We should note here that this technique allows dramatically to improve the numerical stability of the algorithm, but on the other hand, the re-initialization of the simplex tableau is a very expensive operation, especially for problems with a high aspect ratio n/m.

In this chapter we consider the theoretical backgrounds of the techniques that are usually used to make solvers more stable and can help to improve their performance.

1. Scaling LFP Problems

We begin this section considering a numeric example which shows that in finite precision computations bad scaled or non-scaled equations and rounding can cause serious problems.

The following system of linear equations

$$\begin{pmatrix} 0.003 & 59.140 \\ 5.291 & -6.130 \end{pmatrix} \cdot \begin{pmatrix} x_1 \\ x_2 \end{pmatrix} = \begin{pmatrix} 59.17 \\ 46.78 \end{pmatrix}$$

has the exact solution

$$x_1 = 10.000; \quad x_2 = 1.000\ .$$

Let us solve this system with Gaussian elimination and using 4 decimal digit rounding. Choose the entry $a_{11} = 0.003$ as a pivot and calculate the multiplier $\lambda = a_{21}/a_{11} = 5.291/0.003 = 1763.(6)$ which rounds to 1763.6667 (we use only 4 decimal places rounding!). After performing elementary row operations (row 2) - λ(row 1) $\to$ (row 2) with $\lambda = 1763.6667$ we obtain

$$\begin{pmatrix} 0.0030 & 59.1400 \\ 0.0000 & -104309.3786 \end{pmatrix} \cdot \begin{pmatrix} x_1 \\ x_2 \end{pmatrix} = \begin{pmatrix} 59.1700 \\ -104309.3786 \end{pmatrix}$$

instead of correct values

$$\begin{pmatrix} 0.0030 & 59.1400 \\ 0.0000 & -104309.37(6) \end{pmatrix} \cdot \begin{pmatrix} x_1 \\ x_2 \end{pmatrix} = \begin{pmatrix} 59.1700 \\ -104309.37(6) \end{pmatrix} .$$

Backward substitution yields: $x_2 = 1.00000000001$, which is very close to the correct result $x_2 = 1.000$. However, using relatively correct value $x_2 = 1.00000000001$ to calculate unknown x_1, we obtain

$$x_1 \approx \frac{59.17 - (59.140)(1.00000000001)}{0.003} = -9.71(3)$$

instead of the exact value $x_1 = 10.000$. It is clear what is happening: the value of x_2 contains the small error of 0.00000000001 but it is multiplied by $59.121/0.003$.

In this example we used 4 decimal rounding. Of course, most computers in the world are significantly more precise but they work with **finite precision**! For most of them the IEEE standard relative precision means only 16 digits after the decimal point:

$$\epsilon = 2.23 \times 10^{-16}.$$

It means that

$$1 + \epsilon = 1 + \epsilon$$

but

$$1 + \frac{1}{2}\,\epsilon = 1.$$

A relatively simple way to avoid such problems with precision in linear algebra when solving systems of linear equations is accomplished through scaling. This approach may be fruitfully used in linear-fractional programming too. Scaling in LFP problems affects the accuracy of a computed solution and may lead in the simplex method to a change in the selection of pivots.

When scaling an LFP problem, we have to distinguish the following possible cases:

1 scaling constraints:

- right-hand-side vector $b = (b_1, b_2, \ldots, b_m)^T$;
- columns $A_j,\ \ j = 1, 2, \ldots, n$; of matrix A;
- rows of matrix A;

2 scaling the objective function:

- only vector $p = (p_0, p_1, p_2, \ldots, p_n)$ of numerator $P(x)$
- only vector $d = (d_0, d_1, d_2, \ldots, d_n)$ of denominator $D(x)$
- both vector p and d of objective function $Q(x)$

Below, we investigate all these possible cases.

1.1 RHS Vector $b \rightarrow \rho b$

Suppose that vector x^* is an optimal solution for LFP problem (4.1)-(4.3), so

$$\sum_{j=1}^{n} A_j x_j^* = b \quad \text{and} \quad x^* \geq 0,$$

and matrix $B = (A_{s_1}, A_{s_2}, \ldots, A_{s_m})$ is its basis.

Let us replace RHS vector b with some other vector $b' = \rho b$, where $\rho > 0$. Consider the new vector $x' = \rho x^*$. It is obvious that this vector x' satisfies constraints

$$\sum_{j=1}^{n} A_j(\rho x_j^*) = \rho b \quad \text{and} \quad x' = \rho x^* \geq 0,$$

so vector x' is a feasible solution of LFP problem

$$Q(x) = \frac{P(x)}{D(x)} = \frac{\sum_{j=1}^{n} p_j x_j + p_0'}{\sum_{j=1}^{n} d_j x_j + d_0'} \longrightarrow \max \tag{12.5}$$

subject to

$$\sum_{j=1}^{n} a_{ij} x_j = \rho b_i, \quad i = 1, 2, \ldots, m; \tag{12.6}$$

$$x_j \geq 0, \quad j = 1, 2, \ldots, n. \tag{12.7}$$

Now we have to check this vector x' if it is an optimal solution of problem (12.5)-(12.7).

Since vector x^* is an optimal solution of the original LFP problem (4.1)-(4.3), we have

$$\Delta_j(x^*) \geq 0, \quad j = 1, 2, \ldots, n, \tag{12.8}$$

where

$$\begin{aligned} \Delta_j(x^*) &= D(x^*)\Delta_j' - P(x^*)\Delta_j'', \quad j = 1, 2, \ldots, n, \\ \Delta_j' &= \sum_{i=1}^{m} p_{s_i} x_{ij} - p_j, \quad j = 1, 2, \ldots, n, \\ \Delta_j'' &= \sum_{i=1}^{m} d_{s_i} x_{ij} - d_j, \quad j = 1, 2, \ldots, n, \end{aligned}$$

coefficient x_{ij} are defined from the systems

$$\sum_{i=1}^{m} A_{s_i} x_{ij} = A_j, \quad j = 1, 2, \ldots, n. \tag{12.9}$$

and A_j denotes the column-vectors

$$A_j = (a_{1j}, a_{2j}, \ldots, a_{mj})^T, \quad j = 1, 2, \ldots, n$$

of matrix $A = \|a_{ij}\|_{m \times n}$.

Observe that reduced costs Δ'_j and Δ''_j do not depend on RHS vector b, so substitution $b \rightarrow \rho b$ does not affect values of Δ'_j and Δ''_j. But values of functions $P(x)$ and $D(x)$ depend on RHS vector b, so we have to consider the new reduced costs $\Delta_j(x')$, where $x' = \rho x^*$, for LFP problem (12.5)-(12.7). We have

$$
\begin{aligned}
\Delta_j(\rho x^*) &= D(\rho x^*)\,\Delta'_j - P(\rho x^*)\,\Delta''_j = \\
&= (\sum_{j=1}^{n} d_j(\rho x_j^*) + d'_0)\,\Delta'_j - (\sum_{j=1}^{n} p_j(\rho x_j^*) + p'_0)\,\Delta''_j = \\
&= (\sum_{j=1}^{n} d_j(\rho x_j^*) + d'_0 + \rho d_0 - \rho d_0)\,\Delta'_j - \\
&\quad -(\sum_{j=1}^{n} p_j(\rho x_j^*) + p'_0 + \rho p_0 - \rho p_0)\,\Delta''_j = \\
&= \rho D(x^*)\,\Delta'_j + (d'_0 - \rho d_0)\,\Delta'_j - \\
&\quad -\rho P(x^*)\,\Delta''_j - (p'_0 - \rho p_0)\,\Delta''_j = \\
&= \rho\Delta_j(x^*) + (d'_0 - \rho d_0)\,\Delta'_j - (p'_0 - \rho p_0)\,\Delta''_j = \\
&= \rho\Delta_j(x^*) - G_j,
\end{aligned}
\qquad (12.10)
$$

where

$$G_j = \begin{vmatrix} p'_0 - \rho p_0 & \Delta'_j \\ d'_0 - \rho d_0 & \Delta''_j \end{vmatrix}.$$

The (12.10) means that if p'_0 and d'_0 are such that

$$\rho\Delta_j(x^*) - G_j \geq 0, \quad j = 1, 2, \dots, n,$$

or, in particular, if

$$p'_0 = \rho p_0 \quad \text{and} \quad d'_0 = \rho d_0,$$

then

$$\Delta_j(\rho x^*) \overset{(12.10)}{=} \rho\Delta_j(x^*) \overset{(12.8)}{\geq} 0, \quad \forall j = 1, 2, \dots, n,$$

and hence, vector x' is an optimal solution of LFP problem (12.5)-(12.7).

So, if we substitute RHS vector b with some other vector $b' = \rho b, \ \rho > 0,$ we have simultaneously to replace coefficients p_0 and d_0 in the original objective function $Q(x)$ with $p'_0 = \rho p_0$ and $d'_0 = \rho d_0$, respectively. These two substitutions will guarantee the equivalence between the original problem (4.1)-(4.3) and the new scaled LFP problem (12.5)-(12.7).

It is obvious that if vector x' is an optimal solution of the new (scaled) LFP problem (12.5)-(12.7), then vector $x^* = x'/\rho$ is an optimal solution of the original LFP problem (4.1)-(4.3).

1.2 Column $A_j \longrightarrow \rho A_j$

In this section we consider the scaling columns $A_j,\ j = 1, 2, \dots, n$, of matrix $A = \|a_{ij}\|_{m\times n}$.

We suppose that vector x^* is an optimal solution for the original LFP problem (4.1)-(4.3), so

$$\sum_{j=1}^{n} A_j\, x_j^* = b \quad \text{and} \quad x_j^* \geq 0,\ j = 1, 2, \dots, n,$$

and matrix $B = (A_{s_1}, A_{s_2}, \dots, A_{s_m})$ is its basis.

Let us replace some vector $A_r,\ r \in J = \{1, 2, \dots, n\}$, with some other vector $A'_r = \rho A_r$, where $\rho > 0$.

It is obvious that new vector

$$x' = (x_1^*,\ x_2^*,\ \dots,\ x_{r-1}^*,\ \frac{x_r^*}{\rho},\ x_{r+1}^*,\ \dots,\ x_n^*)$$

will satisfy constraints

$$\sum_{\substack{j=1\\ j\neq r}}^{n} A_j x_j^* + \rho A_r \frac{x_r^*}{\rho} = b,$$

$$x'_j \geq 0,\ \ j = 1, 2, \dots, n,$$

and, hence, vector x' is a feasible solution of the new scaled LFP problem

$$Q(x) = \frac{P'(x)}{D'(x)} = \frac{\sum\limits_{\substack{j=1\\ j\neq r}}^{n} p_j x_j + p'_r x_r + p_0}{\sum\limits_{\substack{j=1\\ j\neq r}}^{n} d_j x_j + d'_r x_r + d_0} \longrightarrow \max \tag{12.11}$$

subject to

$$\sum_{\substack{j=1\\ j\neq r}}^{n} A_j x_j + A'_r x_r = b, \tag{12.12}$$

$$x_j \geq 0, \quad j = 1, 2, \ldots, n. \tag{12.13}$$

Our aim now is to examine if vector x' is an optimal solution of the scaled LFP problem (12.11)-(12.13)?

Since vector x^* is an optimal solution of the original problem (4.1)-(4.3), we have that

$$\Delta_j(x^*) \;=\; D(x^*)\Delta'_j - P(x^*)\Delta''_j \geq 0, \quad j = 1, 2, \ldots, n. \tag{12.14}$$

Let us suppose that A_r is a basic vector, i.e. $r \in J_B = \{s_1, s_2, \ldots, s_m\}$. In this case, for the new scaled problem (12.11)-(12.13) we have

$$\begin{aligned}
\Delta_j(x') &= D'(x')\Delta'_j - P'(x')\Delta''_j = \\
&= (\sum_{\substack{j=1\\ j\neq r}}^{n} d_j x_j^* + d'_r\frac{x_r^*}{\rho} + d_0)(\sum_{\substack{i=1\\ s_i\neq r}}^{m} p_{s_i}x_{ij} + p'_r\frac{x_{rj}}{\rho} - p_j) - \\
&- (\sum_{\substack{j=1\\ j\neq r}}^{n} p_j x_j^* + p'_r\frac{x_r^*}{\rho} + p_0)(\sum_{\substack{i=1\\ s_i\neq r}}^{m} d_{s_i}x_{ij} + d'_r\frac{x_{rj}}{\rho} - d_j) = \\
&= (\sum_{\substack{j=1\\ j\neq r}}^{n} d_j x_j^* + d'_r\frac{x_r^*}{\rho} + d_0 + d_r x_r^* - d_r x_r^*) \times \\
&\times (\sum_{\substack{i=1\\ s_i\neq r}}^{m} p_{s_i}x_{ij} + p'_r\frac{x_{rj}}{\rho} - p_j + p_r x_{rj} - p_r x_{rj}) - \\
&- (\sum_{\substack{j=1\\ j\neq r}}^{n} p_j x_j^* + p'_r\frac{x_r^*}{\rho} + p_0 + p_r x_r^* - p_r x_r^*) \times \\
&\times (\sum_{\substack{i=1\\ s_i\neq r}}^{m} d_{s_i}x_{ij} + d'_r\frac{x_{rj}}{\rho} - d_j + d_r x_{rj} - d_r x_{rj}) = \\
&= (D(x^*) - d_r x_r^* + d'_r\frac{x_r^*}{\rho})(\Delta'_j - p_r x_{rj} + p'_r\frac{x_{rj}}{\rho}) - \\
&- (P(x^*) - p_r x_r^* + p'_r\frac{x_r^*}{\rho})(\Delta''_j - d_r x_{rj} + d'_r\frac{x_{rj}}{\rho})
\end{aligned} \tag{12.15}$$

Expression (12.15) makes it obvious that if $p'_r = p_r\rho$ and $d'_r = d_r\rho$, then

$$\Delta_j(x') \overset{(12.15)}{=} \Delta_j(x^*) \overset{(12.14)}{\geq} 0, \; j = 1, 2, \ldots, n.$$

The latter means that in this case vector x' is an optimal solution of the scaled LFP problem (12.11)-(12.13).

So, if we substitute some basic vector A_r with some other vector $A'_r = \rho A_r$, $\rho > 0$, we have simultaneously to replace coefficients p_r and d_r in the original objective function $Q(x)$ with $p'_r = \rho p_r$ and $d'_r = \rho d_r$, respectively. These two substitutions will guarantee the equivalence between the original problem (4.1)-(4.3) and the new scaled LFP problem (12.11)-(12.13).

It is obvious that if vector x' is an optimal solution of the new (scaled) LFP problem (12.11)-(12.13), then vector

$$x^* = (x'_1,\, x'_2,\, \dots,\, x'_{r-1},\, x'_r\rho,\, x'_{r+1},\, \dots,\, x'_n)$$

will be an optimal solution of the original LFP problem (4.1)-(4.3).

Now, we have to consider the case when substituted vector A_r is a non-basic vector, i.e. $r \in J_N = J \setminus J_B$.

As in the previous case, we simultaneously replace original coefficients p_r and d_r with ρp_r and ρd_r, respectively. Since index r is non-basic and $x^*_r = 0$, it is obvious that

$$x' = x^*, \quad P'(x') = P(x^*), \quad D'(x') = D(x^*) \quad \text{and, hence,} \quad Q'(x') = Q(x^*).$$

So replacement $A_r \rightarrow \rho A_r$, $r \in J_N$, affects only values of Δ'_r, Δ''_r, and $\Delta_r(x')$.

Indeed, if in the original LFP problem (4.1)-(4.3) for non-basic vector A_r we had (see (12.9)), that

$$\sum_{i=1}^{m} A_{s_i} x_{ir} = A_r, \quad j = 1, 2, \dots, n,$$

then after replacement $A_r \rightarrow A'_r$, where $A'_r = \rho A_r$, we obtain the following representation of the new vector A'_r in the same basis B:

$$\sum_{i=1}^{m} A_{s_i} (\rho x_{ir}) = \rho A_r, \quad j = 1, 2, \dots, n.$$

If when replacing $A_r \rightarrow \rho A_r$, we simultaneously substitute $p_r \rightarrow p'_r$, where $p'_r = \rho p_r$, and $d_r \rightarrow d'_r$, where $d'_r = \rho d_r$, then for new $\tilde{\Delta}'_r$, $\tilde{\Delta}''_r$, and $\tilde{\Delta}_r(x')$ we have

$$\tilde{\Delta}'_r \;=\; \sum_{i=1}^{m} p_{s_i}\,(\rho x_{ir}) - (\rho p_r) = \rho \Delta'_r,$$

$$\begin{aligned}
\tilde{\Delta}''_r &= \sum_{i=1}^{m} d_{s_i}\,(\rho x_{ir}) - (\rho d_r) = \rho\Delta''_r, \\
\tilde{\Delta}_r(x') &= D(x^*)\,\tilde{\Delta}'_r - P(x^*)\,\tilde{\Delta}''_r = \\
&= D(x^*)\,(\rho\Delta'_r) - P(x^*)\,(\rho\Delta''_r) \;=\; \rho\Delta_r(x^*) \overset{(12.14)}{\geq} 0.
\end{aligned}$$

The latter means that in this case vector x^* is an optimal solution of the scaled LFP problem (12.11)-(12.13).

So, if we substitute some non-basic vector A_r with some other vector $A'_r = \rho\, A_r,\ \rho > 0,$ we have simultaneously to replace coefficients p_r and d_r in the original objective function $Q(x)$ with $p'_r = \rho p_r$ and $d'_r = \rho d_r,$ respectively. These two substitutions will guarantee the equivalency between the original problem (4.1)-(4.3) and the new scaled LFP problem (12.11)-(12.13). Moreover, it will guarantee that $x^*_r = x'_r = 0$.

1.3 Row $a_i \;\longrightarrow\; \rho a_i$

Let us replace row-vector $a_r = (a_{r1}, a_{r2}, \ldots, a_{rn})$ of matrix $A = ||a_{ij}||_{m\times n}$ in LFP problem (4.1)-(4.3) with some other row-vector $a'_r = \rho a_r$. In this case we have to distinguish the following two cases:

1 simultaneously with replacement $a_r \longrightarrow \rho a_i$ we substitute the r-th element of RHS vector b, that is $b_r \longrightarrow b'_r = \rho b_r$.

2 we do not modify any element in RHS column-vector, so scaling must be performed only in matrix A.

In case 1 we have:
instead of original constraint in the rth row

$$\sum_{j=1}^{n} a_{rj}x_j = b_r,$$

we have

$$\sum_{j=1}^{n}(\rho a_{rj})x_j = (\rho b_r).$$

It is well-known that such scaling does not affect the structure of feasible set S. So the new scaled problem is absolutely equivalent with the original one.

In case 2 we do not modify RHS vector b. Such scaling leads to unpredictable deformations in feasible set S, so we cannot provide any guarantee that the optimal basis of the scaled problem will be the same as in the original one.

So, the only negotiable method of scaling rows in matrix A is the following

$$\tilde{a}_r \longrightarrow \tilde{a}'_r$$

where

$$\tilde{a}_r = (a_{r1},\ a_{r2},\ \ldots,\ a_{rn},\ b_r), \qquad \tilde{a}'_r = (\rho a_{r1},\ \rho a_{r2},\ \ldots,\ \rho a_{rn},\ \rho b_r).$$

Obviously, the optimal solutions x' and x^* of the scaled problem and the original problem, respectively, are exactly the same. So we need not any "unscaling" in this case.

Note that in the simplex method only elements of the pivotal column are compared. Hence, the choice of pivotal row depends on the row scaling. Since a bad choice of pivots can lead to large errors in the computed solution, it means that a proper row scaling is very important.

1.4 Numerator Vector $p \longrightarrow \rho p$

Let us replace vector $p = (p_0, p_1, \ldots, p_n)$ in the numerator $P(x)$ of the objective function $Q(x)$ with some other vector $p' = (p'_0, p'_1, \ldots, p'_n)$, where $p'_j = \rho p_j,\ j = 0, 1, 2, \ldots, n$.

It is clear that such replacement does not affect either the optimal value of denominator $D(x)$ or the values of reduced costs $\Delta''_j,\ j = 1, 2, \ldots, n$, but changes the optimal values of functions $P(x)$ and $Q(x)$ and the values of reduced costs Δ'_j and $\Delta_j(x),\ j = 1, 2, \ldots, n$.

So, for the new values $\tilde{\Delta}'_j,\ P'(x^*),\ Q'(x^*)$, and $\tilde{\Delta}_j(x^*),\ j = 1, 2, \ldots, n$, we have:

$$\begin{aligned}
\tilde{\Delta}'_j &= \sum_{i=1}^{m} p'_{s_i} x_{ij} - p'_j = \\
&= \sum_{i=1}^{m} (\rho\, p_{s_i}) x_{ij} - (\rho\, p_j) = \rho\, \Delta'_j, \quad j = 1, 2, \ldots, n, \\
P'(x^*) &= \sum_{j=1}^{n} p'_j x^*_j + p'_0 = \sum_{j=1}^{n} (\rho\, p_j) x^*_j + (\rho\, p_0) = \rho\, P(x^*), \\
Q'(x^*) &= P'(x^*)/D(x^*) = \rho\, P(x^*)/D(x^*) = \rho\, Q(x^*),
\end{aligned}$$

and hence,

$$\begin{aligned}
\tilde{\Delta}_j(x^*) &= D(x^*)\tilde{\Delta}'_j - P'(x^*)\Delta''_j = \\
&= D(x^*)(\rho\, \Delta'_j) - (\rho\, P(x^*))\Delta''_j = \rho\, \Delta_j(x^*), \quad j = 1, 2, \ldots, n.
\end{aligned}$$

From the latter equation we obtain

$$\tilde{\Delta}_j(x^*) = \rho\,\Delta_j(x^*) \overset{(12.14)}{\geq} 0, \quad j = 1, 2, \dots, n.$$

Finally, we have to note that replacement $p \;\rightarrow\; \rho p$ does not lead to any changes in the optimal basis or in optimal solution x^*. So, if we have solved the scaled LFP problem, in order to "un-scale" the optimal solution obtained we have to use the following formula $Q(x^*) = \dfrac{1}{\rho}Q'(x^*)$, because the optimal solution x' of the scaled problem is exactly the same as optimal solution x^* of the original problem.

1.5 Denominator Vector $d \;\rightarrow\; \rho d$

Let us replace vector $d = (d_0, d_1, \dots, d_n)$ in the denominator $D(x)$ of the objective function $Q(x)$ with some other vector $d' = (d'_0, d'_1, \dots, d'_n)$, where $d'_j = \rho d_j$, $j = 0, 1, \dots, n$.

It is obvious that such replacement leads to some changes in the optimal values of denominator $D(x)$, objective function $Q(x)$ and values Δ''_j, $\Delta_j(x)$, $j = 1, 2, \dots, n$; but does not affect the optimal value of numerator $P(x)$ or the values of reduced costs Δ'_j, $j = 1, 2, \dots, n$.

So for new values $\tilde{\Delta}''_j$, $D'(x^*)$, $Q'(x^*)$, and $\tilde{\Delta}_j(x^*)$, $j = 1, 2, \dots, n$, we have

$$\begin{aligned}
\tilde{\Delta}''_j &= \sum_{i=1}^{m} d'_{s_i} x_{ij} - d'_j = \\
&= \sum_{i=1}^{m} (\rho\, d_{s_i}) x_{ij} - (\rho\, d_j) = \rho\,\Delta''_j, \quad j = 1, 2, \dots, n, \\
D'(x^*) &= \sum_{j=1}^{n} d'_j x^*_j + d'_0 = \sum_{j=1}^{n} (\rho\, d_j) x^*_j + (\rho\, d_0) = \rho\, D(x^*), \\
Q'(x^*) &= P(x^*)/D'(x^*) = P(x^*)/(\rho\, D(x^*)) = Q(x^*)/\rho,
\end{aligned}$$

and hence,

$$\begin{aligned}
\tilde{\Delta}_j(x^*) &= D'(x^*)\Delta'_j - P(x^*)\tilde{\Delta}''_j = \\
&= (\rho\, D(x^*))\Delta'_j - P(x^*)(\rho\,\Delta''_j) = \rho\,\Delta_j(x^*), \quad j = 1, 2, \dots, n.
\end{aligned}$$

From the latter formula we obtain

$$\tilde{\Delta}_j(x^*) = \rho\,\Delta_j(x^*) \overset{(12.14)}{\geq} 0, \quad j = 1, 2, \dots, n.$$

Finally, we have to note that replacement $d \longrightarrow \rho d$ does not lead to any changes in optimal basis B or in optimal solution x^*. So once we have solved the scaled LFP problem, in order to "un-scale" the optimal solution obtained we have to use the following formula $Q(x^*) = \rho\, Q'(x^*)$, because the optimal solution x' of the scaled problem is exactly the same as optimal solution x^* of the original problem.

1.6 Scaling Factors

In this section we briefly overview two rules for calculating scaling factors ρ. Both techniques have been implemented in several commercial and freely usable LP codes, and bring a compromise between provided stability and computational efficiency. For more information on scaling rules with detailed theoretical backgrounds see, for instance [47], [57], [146], [171], etc.

Consider a matrix $A = ||a_{ij}||_{m\times n}$ and an RHS vector $b = (b_1, b_2, \dots, b_m)^T$. A measure of *ill-scaling* of the system $Ax = b$ is

$$\sigma(A) \;=\; \frac{\max\limits_{i,j\in J_+} (|a_{ij}|)}{\min\limits_{i,j\in J_+} (|a_{ij}|)},$$

where $J_+ = \{i, j \mid a_{ij} \neq 0\}$. The larger is the magnitude between the largest and the smallest absolute values of non-zero entries a_{ij}, the worse scaled is the system.

DEFINITION 12.1 *We will say that given matrix A is* ***poorly scaled*** *or* ***badly scaled****, if* $\sigma(A) >= 1E+5$.

The aim of scaling is to make measure $\sigma(A)$ as small as possible. To reach this aim we can scale columns and rows as many times as we need.

1.6.1 Hall-rule

In accordance with this rule we define the following column-vector ρ^r of scaling factors for rows

$$\rho^r \;=\; (\rho_1^r, \rho_2^r, \dots, \rho_m^r)^T, \tag{12.16}$$

where

$$\rho_i^r \;=\; (\prod_{j\in J_i^+} a_{ij})^{1/K_i^r}, \quad i = 1, 2, \dots, m;$$

$J_i^+ = \{j : a_{ij} \neq 0\}, \quad i = 1, 2, \dots, m$, is a row related set of indices j of non-zero entries a_{ij} in row i, and K_i^r denotes the number of non-zero entries a_{ij} in row i.

Analogically, to scale columns, we use the following factors organized in a row-vector ρ^c

$$\rho^c = (\rho_1^c, \rho_2^c, \ldots, \rho_n^c), \tag{12.17}$$

where

$$\rho_j^c = \Big(\prod_{i \in I_j^+} a_{ij}\Big)^{1/K_j^c}, \quad j = 1, 2, \ldots, n;$$

$I_j^+ = \{i : a_{ij} \neq 0\}, \quad j = 1, 2, \ldots, n$, is a column related set of indices i of non-zero entries a_{ij} in column j, and K_j^c denotes the number of non-zero entries a_{ij} in column j.

1.6.2 Gondzio-rule

As an alternative to the scaling factors calculated in accordance with Hall-rule, we can define the following column-vector ρ^r of scaling factors for rows

$$\rho^r = (\rho_1^r, \rho_2^r, \ldots, \rho_m^r)^T, \tag{12.18}$$

where

$$\rho_i^r = \sqrt{r_i' \, r_i''}, \quad i = 1, 2, \ldots, m;$$

and

$$r_i' = \max_{j:\, a_{ij} \neq 0} |a_{ij}|, \quad r_i'' = \min_{j:\, a_{ij} \neq 0} |a_{ij}|, \quad i = 1, 2, \ldots, m.$$

Analogically to row scaling factors, for columns we have to define the following row-vector ρ^c of scaling factors

$$\rho^c = (\rho_1^c, \rho_2^c, \ldots, \rho_n^c), \tag{12.19}$$

where

$$\rho_j^c = \sqrt{c_j' \, c_j''}, \quad j = 1, 2, \ldots, n;$$

and

$$c_j' = \max_{i:\, a_{ij} \neq 0} |a_{ij}|, \quad c_j'' = \min_{i:\, a_{ij} \neq 0} |a_{ij}|, \quad j = 1, 2, \ldots, n.$$

1.6.3 Implementation Issues

To scale an LFP problem we have to calculate and then to store scaling factors for rows, columns and the objective function (separately for numerator and denominator). One of the possible ways to store factors is to expand the matrix of the problem as follows

$$\tilde{A} = \left(\begin{array}{c|cccc|c} \rho_1^r & a_{11} & a_{12} & \dots & a_{1n} & b_1 \\ \rho_2^r & a_{21} & a_{22} & \dots & a_{2n} & b_2 \\ \vdots & \vdots & \vdots & \dots & \vdots & \vdots \\ \rho_m^r & a_{m1} & a_{m2} & \dots & a_{mn} & b_n \\ \hline \rho_{m+1}^r & p_1 & p_2 & \dots & p_n & p_0 \\ \rho_{m+2}^r & d_1 & d_2 & \dots & d_n & d_0 \\ \hline & \rho_1^c & \rho_2^c & \dots & \rho_n^c & \rho_{n+1}^c \end{array} \right).$$

If we scale rows and columns multiple times we have to accumulate scaling factors for post-optimization un-scaling as shown in the algorithm presented in Figure 12.1.

Scaling an LFP Problem

{Initialization}
For $i := 1$ **To** $m+2$ **Do** $\rho_i^r := 1.0$;
For $j := 1$ **To** $n+1$ **Do** $\rho_j^c := 1.0$;
{Scaling block}
Repeat {Repeat scaling several times}
  Begin
  {Scaling rows}
  For $i := 1$ **To** $m+2$ **Do** {Loop over all rows}
    Begin
    temp :=*get_row_factor(i)*; {Calculate row factor}
    $\rho_i^r := \rho_i^r * temp$; {Update factor}
    scale_row(i, temp); {Row scaling with factor *temp*}
    End
  {Scaling columns}
  For $j := 1$ **To** $n+1$ **Do** {Loop over all columns}
    Begin
    temp :=*get_col_factor(j)*; {Calculate column factor}
    $\rho_j^c := \rho_j^c * temp$; {Update factor}
    scale_col(j, temp); {Column scaling with factor *temp*}
    End
  End

Figure 12.1. Algorithm – Scaling an LFP Problem.

Before closing this section, we just note that instead of precisely calculated values of scaling factors ρ several linear programming codes usually use (depends on the options selected by users) the nearest powers of two as a "binary approximation" of these values. The reason is that for computers based on the binary system, it may dramatically improve performance of scaling since in this case the relatively expensive operation of multiplication may be implemented as very fast shifting of data to the left or right, depending on the power of2 used for such "approximation".

1.7 Numeric examples

In the previous section we considered two rules for calculating scaling factors[3]. Both of them are suitable to be used for automatic scaling in programming packages and allow relatively easy achievement of a well-scaled problem.

To illustrate how these scaling factors work, we consider the following rectangular matrix of order 7×5

$$A = \begin{pmatrix} 0.0005 & 3.000 & 0.340 & 234.000 & 34.000 \\ 2.0000 & 4.000 & 345.000 & 1234.000 & 234.000 \\ 30000.0000 & 5.000 & 4565643.000 & 34.000 & 234.000 \\ 9.0000 & 6.000 & 0.001 & 567.000 & 4.000 \\ 567.0000 & 7.000 & 234.000 & 24.000 & 234.000 \\ 56.0000 & 8.000 & 345.000 & 0.001 & 3.000 \\ 45000.0000 & 9.000 & 4.000 & 3.000 & 123.000 \end{pmatrix}. \tag{12.20}$$

This matrix may be said to be **badly scaled** since

$$\max_{i,j \in J_+} (|a_{ij}|) = a_{33} = 4565643.000 = 4.565643E + 06,$$

$$\min_{i,j \in J_+} (|a_{ij}|) = a_{11} = 0.0005 = 5.000000E - 04;$$

[3]These rules have been implemented in the linear programming codes developed at Edinburgh University, Department of Mathematics and Statistics, Scotland. Gondzio-rule is used in the package developed by J.Gondzio for sparse and dense large-scale linear programming; the package implements some special algorithm of the method of interior point. Hall-rule is implemented in the package of J.A.J.Hall for very sparse large-scale linear programming problems; the package is based on the revised simplex method.

and

$$\sigma(A) = \frac{\max\limits_{i,j\in J_+}(|a_{ij}|)}{\min\limits_{i,j\in J_+}(|a_{ij}|)} = \frac{4.565643E+06}{5.00E-04} = 9.13E+09,$$

i.e. the magnitude between the largest and the smallest absolute values of non-zero entries a_{ij} is of order 10 ($\sigma(A) = 9.13E+09 \approx 1.0E+10$).

First, we apply successively Gondzio-factors for rows and columns to scale matrix A. The results of scaling are as follows

Original matrix: In accordance with rule (12.18), vector ρ^r of row scaling factors for original matrix A is

$$\rho^r = (0.3421, 49.6790, 4777.8881, 0.7530, 63.0000, 0.5874, 367.4235)^T.$$

Perform row scaling.

After 1st row scaling: For modified matrix we calculate measure $\sigma(A)$ of ill-scaling:

$$\max_{i,j\in J_+}(|a_{ij}|) = 9.56E+02; \quad \min_{i,j\in J_+}(|a_{ij}|) = 1.05E-03;$$

$$\sigma(A) = \frac{9.56E+02}{1.05E-03} = 9.13E+05.$$

We use rule (12.19) to calculate vector ρ^c of column scaling factors:

$$\rho^c = (0.4231, 0.1194, 1.1265, 1.1322, 2.2064).$$

Perform column scaling.

After 1st column scaling: For the modified matrix we calculate measure $\sigma(A)$ of ill-scaling:

$$\max_{i,j\in J_+}(|a_{ij}|) = 8.48E+02; \quad \min_{i,j\in J_+}(|a_{ij}|) = 1.18E-03;$$

$$\sigma(A) = \frac{8.48E+02}{1.18E-03} = 7.20E+05.$$

Vector ρ^r of row scaling factors:

$$\rho^r = (1.4448, 1.4448, 2.3090, 0.8854, 2.6752, 0.8854, 1.4448)^T.$$

Perform row scaling.

After 2nd row scaling: For the modified matrix we calculate measure $\sigma(A)$ of ill-scaling:

$$\max_{i,j\in J_+} (|a_{ij}|) = 7.51E+02; \quad \min_{i,j\in J_+} (|a_{ij}|) = 1.33E-03;$$

$$\sigma(A) = \frac{7.51E+02}{1.33E-03} = 5.64E+05.$$

Vector ρ^c of column scaling factors:

$$\rho^c = (0.7801, 0.6994, 0.8854, 1.1294, 0.5475).$$

Perform column scaling.

After 2nd column scaling: For the modified matrix we calculate measure $\sigma(A)$ of ill-scaling:

$$\max_{i,j\in J_+} (|a_{ij}|) = 6.65E+02; \quad \min_{i,j\in J_+} (|a_{ij}|) = 1.50E-03;$$

$$\sigma(A) = \frac{6.65E+02}{1.50E-03} = 4.42E+05.$$

Vector ρ^r of row scaling factors:

$$\rho^r = (1.0654, 1.0654, 1.0000, 1.0000, 1.0654, 1.0000, 1.0654)^T.$$

Perform row scaling.

After 3rd row scaling: For the modified matrix we calculate measure $\sigma(A)$ of ill-scaling:

$$\max_{i,j\in J_+} (|a_{ij}|) = 6.65E+02; \quad \min_{i,j\in J_+} (|a_{ij}|) = 1.50E-03;$$

$$\sigma(A) = \frac{6.65E+02}{1.50E-03} = 4.42E+05.$$

Vector ρ^c of column scaling factors:

$$\rho^c = (0.9688, 1.0000, 1.0000, 1.0000, 0.9688).$$

After performing multiple successive scaling operations for rows and columns, we obtain scaling factors both for rows and columns with values close to 1. Hence, there is no reason to continue this process, since the further improvement of ill-scaling measure $\sigma(A)$ for matrix A becomes more and more expensive.

So, starting from the original matrix A with $\sigma(A) = 9.13E+09$ we obtained its scaled modification with $\sigma(A) = 4.42E+05$. As we can see, the improvement of magnitude achieved is of order$5 = 10 - 5$.

Now, let us apply Hall-rule factors to scale the same matrix A given in (12.20). We have the following results

Original matrix: In accordance with rule (12.16) we calculate vector ρ^r of row scaling factors

$$\rho^r = (1.3233, 60.2959, 1403.6460, 2.6158, 87.7940, 3.4138, 56.9257)^T.$$

Perform row scaling.

After 1st row scaling: For the modified matrix we calculate measure $\sigma(A)$ of ill-scaling:

$$\max_{i,j \in J_+} (|a_{ij}|) = 3.25E + 03; \quad \min_{i,j \in J_+} (|a_{ij}|) = 2.93E - 04;$$

$$\sigma(A) = \frac{3.25E + 03}{2.93E - 04} = 1.11E + 07.$$

Row-vector ρ^c of column scaling factors calculated in accordance with rule (12.17)

$$\rho^c = (1.8606, 0.2321, 1.6591, 0.6973, 2.0014).$$

Perform column scaling.

After 1st column scaling: For the modified matrix we calculate measure $\sigma(A)$ of ill-scaling:

$$\max_{i,j \in J_+} (|a_{ij}|) = 1.96E + 03; \quad \min_{i,j \in J_+} (|a_{ij}|) = 2.03E - 04;$$

$$\sigma(A) = \frac{1.96E + 03}{2.03E - 04} = 9.65E + 06.$$

Column-vector ρ^r of row scaling factors will be

$$\rho^r = (1.0000, 1.0000, 1.0000, 1.0000, 1.0000, 1.0000, 1.0000)^T.$$

Moreover, rule (12.17) used to calculate new row-vector ρ^c of column scaling factors gives

$$\rho^c = (1.0000, 1.0000, 1.0000, 1.0000, 1.0000).$$

After performing two successive scaling operations for rows and columns, we obtain scaling factors both for rows and columns with values exactly equal to 1. Hence, there is no reason to continue this process, since the further improvement of ill-scaling measure $\sigma(A)$ for matrix A using this rule is impossible.

So, starting from the original matrix A with $\sigma(A) = 9.13E + 09$ we obtained its scaled modification with $\sigma(A) = 9.65E + 06$. As we can see, the improvement of magnitude achieved is of order$4 = 10 - 6$.

2. Factorization of Basis Matrix

Most real-world applications of LP and LFP have hundreds of thousands of unknown variables and main constraints. Usually, such large-scale problems require thousands of iterations in the simplex method and millions of floating-point operations. Since precision of computer arithmetic is finite, small numerical errors occur in these calculations. Because of the iterative nature of the simplex method, this inaccuracy typically has a cumulative effect and leads to large errors in calculations.

The main problem of this type occurs in the simplex method during pivot transformations (see Section 5, Table 4.2) in iterations when calculating elements

$$x'_{ij} = \begin{cases} x_{ij} - \dfrac{x_{rj}x_{ik}}{x_{rk}}, & i = 1, 2, \dots, m, \ i \neq r, \\ \dfrac{x_{rj}}{x_{rk}}, & i = r, \end{cases} \qquad j \in J_N = J \setminus J_B,$$

where $J = \{1, 2, \dots, n\}$ is an index-set of all vectors $A_j, \ j = 1, 2, \dots, n$, $J_B = \{s_1, s_2, \dots, s_m\}$ is an index-set of the basic vectors, r is index of the vector leaving the current basis, and k is index of the vector that should be entered into basis.

While performing the simplex method, this formula provides a connection between "old" and "new" coefficients x_{ij} in iterations when interchanging vectors in the simplex tableau and re-calculating "new" elements x'_{ij} on the basis of "old" entries x_{ij}. Small rounding inaccuracy caused by finite precision of the computer repeated many times in the iterations may result in big problems and even in incorrect "solutions". This is why when performing simplex iterations from time to time we have to recalculate the main elements x_{ij} of the simplex tableau directly from the original matrix A and the current basis $B = (A_{s_1}, A_{s_2}, \dots, A_{s_m})$. So we have to solve the following systems of linear equations

$$\sum_{i=1}^{m} A_{s_i} x_{ij} = A_j, \quad \text{for each} \quad j \in J_N = J \setminus J_B.$$

that is, in other words, a set of systems of linear equations $Ax = b$ with the same left-hand side Ax and multiple right-hand side vectors b. Using Gaussian elimination or the Gauss-Jordan method to solve such systems would not be the very best decision, because both methods share the disadvantage that all right-hand side vectors must be known in advance since they are used during calculations. Another reason is that both methods are very expensive and require about $m^3/2+m^2$ floating point operations (*flops*) to reduce the original matrix A

to triangular form and then perform the backward (or forward) substitution. So these methods are $O(m^3)$ expensive. The method considered in the next section does not share that deficiency and is more efficient in providing a solution with any number of arbitrary right-hand sides.

2.1 *LU*-factorization

In this section, we discuss solving systems of linear equations given in the form

$$Ax = b, \tag{12.21}$$

where A is an invertible $m \times m$ square matrix and b is an arbitrary column-vector with m elements $b_1, b_2, \ldots, b_m$.

It is a conventional convenience to denote by A^{-1} the inverse of the matrix A so that the solution to system (12.21) is given by $A^{-1}b$. However, there is almost no occasion when it is appropriate to compute the inverse in order to solve a set of linear equations. There are usually far more computationally efficient methods (direct as well as iterative) of doing this than to compute the inverse.

The most common direct methods use a factorization of the coefficient matrix to facilitate the solution. One of the most wide-spread and well-known factorizations for nonsymmetric systems is *LU-factorization* (or *LU-decomposition*), where matrix A (or rather a permutation of it) is expressed as the product of the lower triangle matrix L and the upper triangle matrix U. Thus

$$P_r A P_c = LU, \tag{12.22}$$

where

$$L = \begin{pmatrix} l_{11} & 0 & \ldots & 0 \\ l_{21} & l_{22} & \ldots & 0 \\ \ldots & \ldots & \ldots & \ldots \\ l_{m1} & l_{m2} & \ldots & l_{mm} \end{pmatrix}, \quad U = \begin{pmatrix} u_{11} & u_{12} & \ldots & u_{1m} \\ 0 & u_{22} & \ldots & u_{2m} \\ \ldots & \ldots & \ldots & \ldots \\ 0 & 0 & \ldots & u_{mm} \end{pmatrix}$$

and, P_r and P_c are permutation matrices used to interchange rows and columns, respectively.

This factorization can then be used to solve the system (12.21) through the following two steps:

$$Ly = P_r b, \tag{12.23}$$

and then

$$Uz = y, \tag{12.24}$$

hence solution x is just a permutation of vector z, i.e.,

$$x = P_c z. \tag{12.25}$$

This decomposition is very useful, because the solution of triangular matrices is easily accomplished by successive backward or forward substitution in the corresponding linear system.

The algorithm for solving system $Ax = b$ with LU-decomposition may be formulated as follows:

1 Factorize original matrix A into form $A = LU$;

2 In the system $LUx = b$ denote Ux by y and solve system $Ly = b$ for y by forward substitution, so $y = L^{-1}b$;

3 With known vector y solve system $Ux = y$ for unknown x, so $x = U^{-1}(L^{-1}b)$;

Observe that not every matrix can be LU-decomposed. For example

$$A = \begin{pmatrix} 0 & 4 \\ 4 & 2 \end{pmatrix} = \begin{pmatrix} 1 & 0 \\ l_{21} & 1 \end{pmatrix} \begin{pmatrix} u_{11} & u_{12} \\ 0 & u_{22} \end{pmatrix},$$

so we have that

$$u_{11} = 0; \quad \text{and} \quad l_{21}u_{11} = 4.$$

The latter system cannot be satisfied. At the same time,

$$B = \begin{pmatrix} 4 & 2 \\ 0 & 4 \end{pmatrix} = \begin{pmatrix} 1 & 0 \\ 0 & 1 \end{pmatrix} \begin{pmatrix} 4 & 2 \\ 0 & 4 \end{pmatrix},$$

where matrix B is obtained from A by interchanging rows. Matrix B can be LU-decomposed because all its diagonal entries are nonzero. Such re-arrangement of rows (or columns) is always possible if matrix A is non-singular, that is its determinant is nonzero, and hence, system $Ax = b$ has a unique solution. Note that non-singularity is not the necessary condition for the existence of LU-decomposition. For example, the following singular matrix

$$A = \begin{pmatrix} 1 & 1 \\ 1 & 1 \end{pmatrix}$$

has the LU-decomposition

$$A = \begin{pmatrix} 1 & 1 \\ 1 & 1 \end{pmatrix} = \begin{pmatrix} 1 & 0 \\ 1 & 1 \end{pmatrix} \begin{pmatrix} 1 & 1 \\ 0 & 0 \end{pmatrix} = LU.$$

As we saw in section 1 such rearrangement of rows or columns may be performed with permutation matrices. This is why in formula (12.22) we applied the LU-decomposition in general form using permutation matrices P_r and P_c.

The following statement summarizes facts mentioned above:

THEOREM 12.1 (EXISTENCE OF LU-DECOMPOSITION) *If square matrix A is non-singular, then there exist such permutations P_r and P_c, a unit lower triangular matrix L, and a non-singular upper triangular matrix U, that*

$$P_r A P_c = LU.$$

In fact, only one of permutation matrices P_r and P_c is necessary.

Another useful statement gives an answer to the question how to determine the necessary permutation matrix (or matrices).

THEOREM 12.2 *If for given non-singular square matrix A the Gaussian elimination can be performed in $Ax = b$ without row interchange, then decomposition $A = LU$ is possible.*

This statement means that before performing LU-decomposition we have to check if all diagonal entries in the original matrix A are nonzero. If not, we have to perform necessary row (or column) interchanges to produce the main diagonal without zero entries.

Note that decomposition $A = LU$ is not unique. For example,

$$A = \begin{pmatrix} 4 & 1 \\ 1 & 2 \end{pmatrix} = \begin{pmatrix} 4 & 0 \\ 1 & 7/4 \end{pmatrix}\begin{pmatrix} 1 & 1/4 \\ 0 & 1 \end{pmatrix} = \begin{pmatrix} 1 & 0 \\ 1/4 & 1 \end{pmatrix}\begin{pmatrix} 4 & 1 \\ 0 & 7/4 \end{pmatrix}.$$

Thus, in general we can move the diagonal entries around.

Then how can we perform LU-decomposition for given matrix A ? First, we rewrite expression $A = LU$ in the following form

$$\begin{aligned} A &= \begin{pmatrix} a_{11} & a_{12} & \dots & a_{1m} \\ a_{21} & a_{22} & \dots & a_{2m} \\ \dots & \dots & \dots & \dots \\ a_{m1} & a_{m2} & \dots & a_{mm} \end{pmatrix} = \\ &= \begin{pmatrix} l_{11} & 0 & \dots & 0 \\ l_{21} & l_{22} & \dots & 0 \\ \dots & \dots & \dots & \dots \\ l_{m1} & l_{m2} & \dots & l_{mm} \end{pmatrix}\begin{pmatrix} u_{11} & u_{12} & \dots & u_{1m} \\ 0 & u_{22} & \dots & u_{2m} \\ \dots & \dots & \dots & \dots \\ 0 & 0 & \dots & u_{mm} \end{pmatrix} = LU. \end{aligned}$$

This system allows us to write out all necessary operations. Indeed, for every index i and j we can write out

$$a_{ij} = l_{i1}u_{1j} + \dots , \quad i = 1, 2, \dots, m; \quad j = 1, 2, \dots, m.$$

If $i = 1$, then

$$a_{1j} = l_{i1}u_{1j}, \quad j = 1, 2, \ldots, m.$$

For $i = 2$ we have

$$\begin{aligned} a_{21} &= l_{21}u_{11}; \\ a_{22} &= l_{21}u_{12} + l_{22}u_{22}. \end{aligned}$$

If $i = 3$ we obtain

$$\begin{aligned} a_{31} &= l_{31}u_{11}; \\ a_{32} &= l_{31}u_{12} + l_{32}u_{22}; \\ a_{32} &= l_{31}u_{13} + l_{32}u_{23} + l_{33}u_{33}. \end{aligned}$$

Actually, the number of terms in the sum depends, however, on whether index i or j is a smaller number. We have, in fact, the following three cases

$$l_{i1}u_{1j} + l_{i2}u_{2j} + \ldots + l_{ii}u_{ij} = a_{ij}, \quad \text{for } i < j; \tag{12.26}$$

$$l_{i1}u_{1j} + l_{i2}u_{2j} + \ldots + l_{ii}u_{jj} = a_{ij}, \quad \text{for } i = j; \tag{12.27}$$

$$l_{i1}u_{1j} + l_{i2}u_{2j} + \ldots + l_{ij}u_{jj} = a_{ij}, \quad \text{for } i > j; \tag{12.28}$$

Observe that the system of equations (12.26)-(12.28) includes totally m^2 equations and $m^2 + m$ unknowns l_{ij} and u_{ij} (the elements on the main diagonal are represented twice). Since the number of unknowns is greater than the number of equations, we may fix any m unknowns l_{ij} and u_{ij} to arbitrarily determined value(s) and then solve the system for other non-fixed unknowns. In fact, it is always possible to take

$$l_{ii} = 1, \quad \text{for all} \quad i = 1, 2, \ldots, m.$$

The following procedure used to solve system (12.26)-(12.28) is usually referred to as *Crout's algorithm*:

1 Set $l_{ii} = 1, \; i = 1, 2, \ldots, m$

2 For each index $j = 1, 2, \ldots, m$ perform the following two steps:

(a) First, for $i = 1, 2, \ldots, j$ use system (12.26)-(12.28) to determine

$$u_{ij} = \begin{cases} a_{ij}, & \text{if } i = 1; \\ a_{ij} - \sum_{k=1}^{i-1} l_{ik}u_{kj}, & \text{if } i > 1; \end{cases} \tag{12.29}$$

(b) Second, for $i = j+1, j+2, \ldots, m$ use equations (12.28) to determine

$$l_{ij} = (a_{ij} - \sum_{k=1}^{j-1} l_{ik}u_{kj})/u_{jj}. \qquad (12.30)$$

To illustrate how the method works, we consider the following example. Let

$$A = \begin{pmatrix} 1 & 1 & 1 \\ 3 & 1 & 2 \\ 4 & 2 & 1 \end{pmatrix}.$$

In accordance with step 1 we have to set all l_{ii} to 1

$$l_{11} = 1, \quad l_{22} = 1, \quad l_{33} = 1.$$

In the second phase we have

for $j = 1$:

$$\begin{aligned} u_{11} &= a_{11} = 1; \\ l_{21} &= \frac{1}{u_{11}}\, a_{21} = \frac{1}{1} \times 3 = 3; \\ l_{31} &= \frac{1}{u_{11}}\, a_{31} = \frac{1}{1} \times 4 = 4; \end{aligned}$$

for $j = 2$:

$$\begin{aligned} u_{12} &= a_{12} = 1; \\ u_{22} &= a_{22} - l_{21}u_{12} = 1 - 3 \times 1 = 1 - 3 = -2; \\ l_{32} &= \frac{1}{u_{22}}\,(a_{32} - l_{31}u_{12}) = \frac{1}{-2}\,(2 - 4 \times 1) = 1; \end{aligned}$$

finally, for $j = 3$:

$$\begin{aligned} u_{13} &= a_{13} = 1; \\ u_{23} &= a_{23} - l_{21}u_{13} = 2 - 3 \times 1 = 2 - 3 = -1; \\ u_{33} &= a_{33} - l_{31}u_{13} - l_{32}u_{23} = \\ &= 1 - 4 \times 1 - 1 \times (-1) = 1 - 4 + 1 = -2\,. \end{aligned}$$

So we obtain

$$A = \begin{pmatrix} 1 & 1 & 1 \\ 3 & 1 & 2 \\ 4 & 2 & 1 \end{pmatrix} = \begin{pmatrix} 1 & 0 & 0 \\ 3 & 1 & 0 \\ 4 & 1 & 1 \end{pmatrix} \begin{pmatrix} 1 & 1 & 1 \\ 0 & -2 & -1 \\ 0 & 0 & -2 \end{pmatrix} = LU.$$

Working through these iterations of Crout's algorithm, we saw that elements l_{ij} and u_{ij} that appear on the right-hand side of equations (12.29) and (12.30) are already determined by the time they are needed. We saw also that every entry a_{ij} of the original matrix A is used only once and never again. This means that the corresponding elements l_{ij} or u_{ij} can be stored in the same place (computer memory) that the original element a_{ij} used to occupy. So, the LU-decomposition may be performed "in place". Observe that the main diagonal unity elements l_{ij} need not be stored at all. In other words, Crout's method allows us to transform the original square matrix A to the LU form 'in place', i.e. without requiring any extra memory and hence, using computer memory very effectively.

Since equations (12.30) contain a division operation, it is obvious that pivoting is an absolutely essential and generally speaking unavoidable part of Crout's algorithm. Being more simple for efficient implementation, only partial pivoting (i.e. interchange of rows) is necessary to make the method stable ([147]).

The algorithm shown in Figure 12.2 replaces given matrix $A = ||a_{ij}||_{m\times m}$ with its LU-decomposition without storing diagonal elements $l_{jj} = 1,\ j = 1,2,\dots,m$.

The LU-decomposition performed with Crout's method requires about $m^3/3$ executions of multiplications and the same number of additions. The forward and backward substitutions for systems $Ly = b$ and $Ux = y$ are both of order m^2. Thus, the operation count for solving system $Ax = b$ with one (or a few different) right-hand side vector b is better than in the case of the Gaussian elimination or Gauss-Jordan method (see section 7 and section 8, respectively). Moreover, if we have to solve a set of systems $Ax = b$ with multiple right-hand side vectors b it may be more preferable to use the LU-decomposition, since in this case we have to perform this factorization only once and then it may be used as many times as right-hand side vectors b we have. Since the LU-factorization is of order $O(m^3)$ and backward (forward) substitution requires only $O(m^2)$ operations it is obvious that backward and forward substitution can be done much faster than LU-factorization. The greater is the size of the system to be solved, the greater is the part of computational expenses the LU-factorization phase takes in the total amount of computations. Indeed, for $m = 100$ we have

$$\frac{LU\text{-phase}}{LU\text{-phase} + \text{Substitution}} = \frac{m^3}{m^3 + m^2} = \frac{1000000}{1010000} \approx 0.990\,,$$

but if $m = 1000$, we obtain 0.999. If the total number k of the right-hand side vectors b the system $Ax = b$ must be solved for is, say 50, then for $m = 100$ we have

$$k(m^3 + m^2) = 50500000\,.$$

Crout's Algorithm

For $j := 1$ **To** m **Do** {Loop over all columns}
  Begin
  {Loop for equation (12.29) except $i = j$}
  For $i := 1$ **To** $j - 1$ **Do**
    Begin
    $sum := a_{ij};$
    For $k := 1$ **To** i **Do** $sum := sum - a_{ik}a_{kj};$
    $a_{ij} := sum;$
    End
  {Equation (12.29) for $i = j$ and}
  {loop for equation (12.30) $i = j + 1, \ldots, m$}
  For $i := j$ **To** m **Do**
    Begin
    $sum := a_{ij};$
    For $k := 1$ **To** j **Do** $sum := sum - a_{ik}a_{kj};$
    $a_{ij} := sum;$
    {Here we have to interchange rows, if necessary}
    {Finally, divide by the pivot element}
    $\mu := 1.0/a_{jj};$
    For $i := j + 1$ **To** m **Do** $a_{ij} := \mu a_{ij};$
    End
  End

Figure 12.2. Algorithm – Crout's method.

If we LU-decompose matrix A and then use it to solve all k systems, the total computational cost is

$$m^3 + km^2 = 1000000 + 500000 = 1500000\,,$$

i.e. the reduction is $1 - 1500000/50500000 = 1 - 0.02970297 = 0.97029703$ (that is 97%!).

Therefore, it is very important to perform as few LU-factorizations as possible. Re-using and updating LU-factorization (and other types of factorization) is the subject of the next sections.

Before closing this section, we just note that finding an LU-factorization of a matrix is equivalent to Gaussian elimination in the sense that multiplying matrix

A on the left by L^{-1} has the effect of applying elementary row operations to A to put it into upper triangular form U. This is a topic of the next section.

2.2 LU-factorization and Gaussian Elimination

Analytically, the LU-factorization of a matrix has a very close connection with Gaussian elimination. In this section we discuss this connection and show that these two processes can be combined with each another.

First of all, recall that when reducing a given matrix

$$A = \begin{pmatrix} a_{11} & a_{12} & \dots & a_{1m} \\ a_{21} & a_{22} & \dots & a_{2m} \\ \vdots & \vdots & \vdots & \vdots \\ a_{m1} & a_{m2} & \dots & a_{mm} \end{pmatrix}$$

to the upper triangular form in the Gaussian elimination, on the first step we subsequently replace row i in the augmented matrix $(A|b)$ with expression (row i - (row 1)$*\mu_{i1}$), where $\mu_{i1} = a_{i1}/a_{11}$. Using matrix notation this operation may be written as follows

$$A^{(2)} = M^{(1)} A, \tag{12.31}$$

where

$$A^{(2)} = \begin{pmatrix} a_{11} & a_{12} & a_{13} & \dots & a_{1m} \\ 0 & a_{22}^{(2)} & a_{23}^{(2)} & \dots & a_{2m}^{(2)} \\ 0 & a_{32}^{(2)} & a_{33}^{(2)} & \dots & a_{3m}^{(2)} \\ \vdots & \vdots & \vdots & \vdots & \vdots \\ 0 & a_{m2}^{(2)} & a_{m3}^{(2)} & \dots & a_{mm}^{(2)} \end{pmatrix}$$

and

$$M^{(1)} = \begin{pmatrix} 1 & 0 & 0 & \dots & 0 \\ -\mu_{21} & 1 & 0 & \dots & 0 \\ -\mu_{31} & 0 & 1 & \dots & 0 \\ \vdots & \vdots & \vdots & \vdots & \vdots \\ -\mu_{m1} & 0 & 0 & \dots & 1 \end{pmatrix}.$$

On step 2 we construct matrix

$$M^{(2)} = \begin{pmatrix} 1 & 0 & 0 & \dots & 0 \\ 0 & 1 & 0 & \dots & 0 \\ 0 & -\mu_{32} & 1 & \dots & 0 \\ \vdots & \vdots & \vdots & \vdots & \vdots \\ 0 & -\mu_{m2} & 0 & \dots & 1 \end{pmatrix},$$

where $\mu_{i2} = a_{i2}^{(2)}/a_{22}^{(2)}$, $i = 3, 4, \ldots, m$. Then we can express matrix $A^{(3)}$ as the following product

$$A^{(3)} = M^{(2)} M^{(1)} A \,. \tag{12.32}$$

If pivot entries a_{11} or a_{22} are zero before performing these steps we have to interchange rows, i.e apply a row permutation to the corresponding matrix. So instead of equations (12.31) and (12.32) we have to write

$$A^{(2)} = M^{(1)} P^{(1)} A \,, \tag{12.33}$$

and

$$A^{(3)} = M^{(2)} P^{(2)} M^{(1)} P^{(1)} A \,, \tag{12.34}$$

where $P^{(1)}$ is a suitable permutation matrix of order m applied to interchange rows in matrix A, and permutation matrix $P^{(2)}$ is applied to interchange rows in the product matrix $(M^{(1)} P^{(1)} A)$.

Generalizing this process we obtain

$$A^{(k+1)} = M^{(k)} P^{(k)} M^{(k-1)} P^{(k-1)} \ldots M^{(1)} P^{(1)} A \,.$$

The process ends with

$$A^{(m)} = M^{(m-1)} P^{(m-1)} M^{(m-2)} P^{(m-2)} \ldots M^{(1)} P^{(1)} A \,, \tag{12.35}$$

where the matrix $A^{(m)}$ is an upper triangular one

$$A^{(m)} = \begin{pmatrix} a_{11}^{(1)} & a_{12}^{(1)} & a_{13}^{(1)} & \ldots & a_{1m}^{(1)} \\ 0 & a_{22}^{(2)} & a_{23}^{(2)} & \ldots & a_{2m}^{(2)} \\ 0 & 0 & a_{33}^{(3)} & \ldots & a_{3m}^{(3)} \\ \vdots & \vdots & \vdots & \vdots & \vdots \\ 0 & 0 & 0 & \ldots & a_{mm}^{(m)} \end{pmatrix} = U \,, \tag{12.36}$$

and $a_{1j}^{(1)} = a_{1j}$, $j = 1, 2, \ldots, m$.

Further, let us multiply expression (12.35) from the left by inverse$(M^{(m-1)})^{-1}$ and then by $(P^{(m-1)})^{-1}$. We have

$$\begin{aligned} &(P^{(m-1)})^{-1}(M^{(m-1)})^{-1} A^{(m)} = \\ &= (P^{(m-1)})^{-1}(M^{(m-1)})^{-1} M^{(m-1)} P^{(m-1)} \ldots M^{(1)} P^{(1)} A \,. \end{aligned}$$

Since

$$(M^{(m-1)})^{-1} M^{(m-1)} = I \quad \text{and} \quad (P^{(m-1)})^{-1} P^{(m-1)} = I$$

we obtain

$$(P^{(m-1)})^{-1}(M^{(m-1)})^{-1}A^{(m)} = M^{(m-2)}\,P^{(m-2)}\dots M^{(1)}\,P^{(1)}A\,.$$

Then we repeat this step with $(M^{(m-2)})^{-1}$ and $(P^{(m-2)})^{-1}$, and so on. Finally, we have

$$(P^{(1)})^{-1}(M^{(1)})^{-1}\dots(P^{(m-1)})^{-1}(M^{(m-1)})^{-1}A^{(m)} = A$$

or, using (12.36)

$$(P^{(1)})^{-1}(M^{(1)})^{-1}\dots(P^{(m-1)})^{-1}(M^{(m-1)})^{-1}\,U = A\,. \qquad (12.37)$$

Now we tackle the matrix

$$M^{(k)} = \begin{pmatrix} 1 & \dots & 0 & 0 & 0 & \dots & 0 & 0 \\ \vdots & \vdots & \vdots & \vdots & \vdots & \vdots & \vdots & \vdots \\ 0 & \dots & 1 & 0 & 0 & \dots & 0 & 0 \\ 0 & \dots & 0 & 1 & 0 & \dots & 0 & 0 \\ 0 & \dots & 0 & -\mu_{k+1,k} & 1 & \dots & 0 & 0 \\ 0 & \dots & 0 & -\mu_{k+2,k} & 0 & \dots & 0 & 0 \\ \vdots & \vdots & \vdots & \vdots & & \vdots & \vdots & \vdots \\ 0 & \dots & 0 & -\mu_{m,k} & 0 & \dots & 0 & 1 \end{pmatrix}, \qquad (12.38)$$

which corresponds to the kth step of the Gaussian elimination and means the subtracting product μ_{ki}(row $k-1$) from (row i), $i = k, k+1, \dots, m$. Intuitively, we can see that the inverse operation is obtained by adding μ_{ki} times (row $k-1$) to (row i), $i = k, k+1, \dots, m$. Thus for matrix $M^{(k)}$ we have

$$(M^{(k)})^{-1} = \begin{pmatrix} 1 & \dots & 0 & 0 & 0 & \dots & 0 & 0 \\ \vdots & \vdots & \vdots & \vdots & \vdots & \vdots & \vdots & \vdots \\ 0 & \dots & 1 & 0 & 0 & \dots & 0 & 0 \\ 0 & \dots & 0 & 1 & 0 & \dots & 0 & 0 \\ 0 & \dots & 0 & \mu_{k+1,k} & 1 & \dots & 0 & 0 \\ 0 & \dots & 0 & \mu_{k+2,k} & 0 & \dots & 0 & 0 \\ \vdots & \vdots & \vdots & \vdots & & \vdots & \vdots & \vdots \\ 0 & \dots & 0 & \mu_{m,k} & 0 & \dots & 0 & 1 \end{pmatrix} \qquad (12.39)$$

and we can easily check that $(M^{(k)})^{-1}M^{(k)} = M^{(k)}(M^{(k)})^{-1} = I$. Hence, we can determine the inverse for each matrix $M^{(k)}$ merely by changing the signs of the off-diagonal elements, and each $(M^{(k)})^{-1}$ is lower triangular.

We know that the product of two lower triangular matrices is also lower triangular, so

$$(M^{(1)})^{-1}\,(M^{(2)})^{-1}\ldots(M^{(m-1)})^{-1} \;=\; L\,. \tag{12.40}$$

Consider the product matrix

$$\tilde{L} \;=\; (P^{(1)})^{-1}(M^{(1)})^{-1}\ldots(P^{(m-1)})^{-1}(M^{(m-1)})^{-1} \tag{12.41}$$

from equation (12.37). Since for permutation matrices P we have that $P^{-1} = P$, we can rewrite (12.41) in the following form

$$\tilde{L} \;=\; P^{(1)}(M^{(1)})^{-1}\ldots P^{(m-1)}(M^{(m-1)})^{-1}. \tag{12.42}$$

Unfortunately, even so we cannot say that matrix $\tilde{L}$ is necessarily a lower triangular matrix. Rather it is a permutation of the lower triangular matrix L defined in (12.40). However, if we keep track of the row interchanges performed by the permutation matrices $P^{(1)}$, $P^{(2)}$, …, $P^{(m-1)}$, we can get a matrix $\tilde{A}$, which is just A with its rows permuted. This permuted matrix $\tilde{A}$ does have the decomposition

$$\tilde{A} \;=\; LU\,,$$

where lower triangular matrix L is determined in (12.40) and upper triangular matrix U is determined by (12.36).

To illustrate this method we consider the following example:

$$A = \begin{pmatrix} 1 & 1 & 1 \\ 2 & 1 & 1 \\ 1 & 2 & 3 \end{pmatrix}.$$

Let us find the LU-decomposition of this matrix and then use it to solve the system $Ax = b$, where $b = (1,2,2)^T$. Note that the first step of the Gaussian elimination does not require any interchange of rows. Eliminating the second and the third elements in column 1 and recording the multipliers gives

$$A^{(2)} \;=\; M^{(1)}A \;=\; \begin{pmatrix} 1 & 1 & 1 \\ 0 & -1 & -1 \\ 0 & 1 & 2 \end{pmatrix}, \quad \text{where } M^{(1)} \;=\; \begin{pmatrix} 1 & 0 & 0 \\ \mathbf{-2} & 1 & 0 \\ \mathbf{-1} & 0 & 1 \end{pmatrix},$$

where $\mu_{21} = 2,\ \mu_{31} = 1$. Eliminating the third element in column 2 gives

$$A^{(3)} \;=\; M^{(2)}M^{(1)}A \;=\; U \;=\; \begin{pmatrix} 1 & 1 & 1 \\ 0 & -1 & -1 \\ 0 & 0 & 1 \end{pmatrix},$$

where

$$M^{(2)} = \begin{pmatrix} 1 & 0 & 0 \\ 0 & 1 & 0 \\ 0 & \mathbf{1} & 1 \end{pmatrix},$$

with $\mu_{32} = -1$.

Therefore

$$\begin{aligned} A &= (M^{(1)})^{-1}(M^{(2)})^{-1}U = LU = \\ &= \begin{pmatrix} 1 & 0 & 0 \\ 2 & 1 & 0 \\ 1 & -1 & 1 \end{pmatrix} \begin{pmatrix} 1 & 1 & 1 \\ 0 & -1 & -1 \\ 0 & 0 & 1 \end{pmatrix}, \end{aligned}$$

where

$$(M^{(1)})^{-1} = \begin{pmatrix} 1 & 0 & 0 \\ 2 & 1 & 0 \\ 1 & 0 & 1 \end{pmatrix} \quad \text{and} \quad (M^{(2)})^{-1} = \begin{pmatrix} 1 & 0 & 0 \\ 0 & 1 & 0 \\ 0 & -1 & 1 \end{pmatrix}.$$

To solve the given system $Ax = b$, we should first to solve the system $Ly = b$ for unknown vector y, i.e.

$$\begin{pmatrix} 1 & 0 & 0 \\ 2 & 1 & 0 \\ 1 & -1 & 1 \end{pmatrix} \begin{pmatrix} y_1 \\ y_2 \\ y_3 \end{pmatrix} = \begin{pmatrix} 1 \\ 2 \\ 2 \end{pmatrix}.$$

Using forward substitution we obtain

$$y_1 = 1, \; y_2 = 0, \; y_3 = 1 .$$

In the next phase we have to solve the system $Ux = y$ for unknown vector x with known y, i.e.

$$\begin{pmatrix} 1 & 1 & 1 \\ 0 & -1 & -1 \\ 0 & 0 & 1 \end{pmatrix} \begin{pmatrix} x_1 \\ x_2 \\ x_3 \end{pmatrix} = \begin{pmatrix} 1 \\ 0 \\ 1 \end{pmatrix}.$$

After performing backward substitution we have

$$x_3 = 1, \; x_2 = -1, \; x_1 = 1 .$$

Observe that for simplicity we considered a numeric example that does not require any row interchanges, since all used pivots were nonzero. We have to note that permutation matrices may be used not only to avoid a zero pivot, but also to improve the accuracy of computations. Because of finite precision of computers it is often worth choosing at step k such a pivot row that ensures the pivot entry with a maximal absolute value, i.e.

$$|a_{kk}^{(k)}| = \max_i\{|a_{ik}^{(k)}| : i = k, k+1, \dots, m\}.$$

2.3 Updating LU-factorization

In the previous section we showed how to use the LU-factorization of square $A = ||a_{ij}||_{m\times m}$ matrix to solve a system of linear equations given in the form

$$Ax = b.$$

Since the basis matrix in the simplex method does not change much from one iteration to the next (columns of the basis matrix get replaced by new ones one at a time), it is obvious that we could improve the performance of computations, if it were possible to avoid the repetition of the LU decomposition from scratch for the new basis but instead of it to re-use the existing LU representation (obtained in earlier iterations) of the matrix A somewhere later in the next iterations of the simplex method. There are several such update methods. These update methods can be applied when matrix A is only slightly modified at each subsequent step.

2.3.1 Fundamentals

First of all, let us introduce the necessary notations. Let B denote the current basis (for which an LU-factorization has already been computed) and let $\tilde{B}$ denote the basis of the next iteration. So the new basis $\tilde{B}$ differs from B in only one column, say jth, which holds in the basis B column-vector $A_r = (a_{1r}, a_{2r}, \dots, a_{mr})^T$, associated with the leaving variable x_r, but in the new basis $\tilde{B}$ this vector A_r is replaced with another one, say $A_k = (a_{1k}, a_{2k}, \dots, a_{mk})^T$, associated with the new basic variable x_k. Using matrix notation, this fact may be expressed as follows:

$$\tilde{B} = B + (A_k - A_r)\, e_j^T, \tag{12.43}$$

where e_j denotes the jth column of the unit matrix I of order m. To see why this formula is correct, consider the following example. Let

$$B = \begin{pmatrix} a_{12} & a_{11} & a_{13} \\ a_{22} & a_{21} & a_{23} \\ a_{32} & a_{31} & a_{33} \end{pmatrix}$$

If we replace vector $a_2 = (a_{12}, a_{22}, a_{32})^T$ in column 1 with some other column $c = (c_1, c_2, c_3)^T$, we have to write

$$\begin{aligned}\tilde{B} &= \begin{pmatrix} a_{12} & a_{11} & a_{13} \\ a_{22} & a_{21} & a_{23} \\ a_{32} & a_{31} & a_{33} \end{pmatrix} + \left(\begin{pmatrix} c_1 \\ c_2 \\ c_3 \end{pmatrix} - \begin{pmatrix} a_{12} \\ a_{22} \\ a_{32} \end{pmatrix} \right) (1, 0, 0) = \\ &= \begin{pmatrix} a_{12} & a_{11} & a_{13} \\ a_{22} & a_{21} & a_{23} \\ a_{32} & a_{31} & a_{33} \end{pmatrix} + \begin{pmatrix} c_1 - a_{12} & 0 & 0 \\ c_2 - a_{22} & 0 & 0 \\ c_3 - a_{32} & 0 & 0 \end{pmatrix} = \\ &= \begin{pmatrix} c_1 & a_{11} & a_{13} \\ c_2 & a_{21} & a_{23} \\ c_3 & a_{31} & a_{33} \end{pmatrix}.\end{aligned}$$

In a more general case, any addition or substraction of an $m \times m$ matrix that is an outer product uv^T of two nonzero column-vectors $u = (u_1, u_2, \ldots, u_m)^T$ and $v = (v_1, v_2, \ldots, v_m)^T$ to the original matrix $A = ||a_{ij}||_{m \times m}$ in matrix notation may be expressed as follows:

$$\tilde{A} = A \pm uv^T.$$

This is called a *rank-one modification* because the outer product matrix

$$uv^T = \begin{pmatrix} u_1 \\ u_2 \\ \vdots \\ u_m \end{pmatrix} (v_1, v_2, \ldots, v_m) = \begin{pmatrix} u_1 v_1 & u_1 v_2 & \ldots & u_1 v_m \\ u_2 v_1 & u_2 v_2 & \ldots & u_2 v_m \\ \vdots & \vdots & & \vdots \\ u_m v_1 & u_m v_2 & \ldots & u_m v_m \end{pmatrix}$$

has rank one (i.e. only one linearly independent row or column). If a single entry of the matrix A changes, say $a_{ij} \rightarrow \tilde{a}_{ij} = a_{ij} + \alpha$, then the new matrix

$$\tilde{A} = A + \alpha e_i e_j^T, \tag{12.44}$$

where e_i and e_j are the ith and jth columns of the identity matrix, respectively.

Let us suppose now that the current basis B has been changed in accordance with (12.43) and our aim is to update its LU-decomposition.

The first efficient and numerically stable implementation of update methods was given by R.H.Bartels and G.H.Golub in [20]. Because of its advantages - simplicity and efficiency - this method is probably the most widely used in practical applications. There exist many efficient variations and implementations of this method (for more information see, for example, [70]), including extensions of P.E.Gill et al. [71], improvements made by J.K.Reid [149] and modifications of U.H.Suhl et al. [174]. Later, J.J.H.Forrest and J.A.Tomlin [64] developed a method especially adjusted for solving large LP problems with a sparse matrix. Nowadays this method is used in most commercial codes for very large-scale sparse LP problems. However, this method is not the most stable and requires monitoring and full re-factorization from scratch. More stable methods were developed by M.A.Saunders [158] (a variation of the Bartels-Golub method) and R.Fletcher and S.P.J.Matthews [61], [62] (almost ideal for dense matrices). J.Gondzio developed a stable method for updating dense LU-factorization [83] and another (*parallelisable*) method for computing and updating an inverse representation of large and sparse nonsymmetric matrices [83] for different types of changes in matrix A (row and column exchange, row and column addition or deletion). For further information on this topic, see e.g. [42], [63], [79], etc.

We now give a brief overview of the main ideas of these methods.

2.3.2 The Bartels-Golub Updating

Here we consider a variant of the Bartels-Golub method based on ideas by Suhl et al. [174]. The basic idea is reordering and exploiting the structure of existing LU-decomposition.

Let the corresponding LU-decomposition of the current basis B, which consists of column-vectors $A_1,\ A_2,\ \ldots,\ A_m,$ where

$$A_j \;=\; (a_{1j},\ a_{2j},\ \ldots,\ a_{mj})^T,\ \ j = 1,2,\ldots,m,$$

be

$$M^{(m-1)}\,P^{(m-1)}\,M^{(m-2)}\,P^{(m-2)}\ldots M^{(1)}\,P^{(1)}\,B \;=\; U,$$

where $M^{(i)},\ i\ =\ 1,2,\ldots,m\ -\ 1,$ are the Gauss transformation matrices defined in section 2.2, formula (12.38), $P^{(i)},\ i = 1,2,\ldots,m-1,$ are permutation matrices ensuring a maximum pivot before the application of the corresponding Gaussian transformation $M^{(i)}$, and U is the resulting upper triangular matrix

$$U \;=\; \begin{pmatrix} a_{11}^{(1)} & a_{12}^{(1)} & a_{13}^{(1)} & \ldots & a_{1m}^{(1)} \\ 0 & a_{22}^{(2)} & a_{23}^{(2)} & \ldots & a_{2m}^{(2)} \\ 0 & 0 & a_{33}^{(3)} & \ldots & a_{3m}^{(3)} \\ \vdots & \vdots & \vdots & \vdots & \vdots \\ 0 & 0 & 0 & \ldots & a_{mm}^{(m)} \end{pmatrix}. \tag{12.45}$$

To avoid the unnecessary superscripts in the upper triangular matrix U we rewrite it as follows

$$U = (U_1, U_2, \ldots, U_m) = \begin{pmatrix} u_{11} & u_{12} & u_{13} & \ldots & u_{1m} \\ 0 & u_{22} & u_{23} & \ldots & u_{2m} \\ 0 & 0 & u_{33} & \ldots & u_{3m} \\ \vdots & \vdots & \vdots & \vdots & \vdots \\ 0 & 0 & 0 & \ldots & u_{mm} \end{pmatrix}. \qquad (12.46)$$

Now, let us suppose that one of the columns of B must be changed. For example, let vector A_r in column r of matrix B be the leaving column-vector and vector A_k be the entering column-vector which must replace vector A_r. So basis

$$B = (A_1, A_2, \ldots, A_{r-1}, A_r, A_{r+1}, \ldots, A_m)$$

transforms to

$$\tilde{B} = (A_1, A_2, \ldots, A_{r-1}, A_k, A_{r+1}, \ldots, A_m)$$

To preserve the nice triangular structure of the existing LU-factorization, we reorder the new basis $\tilde{B}$ so that all the columns to the right of the column-vector A_k are shifted by one position to the left and column-vector A_k moves to the rightmost column. So the new basis $\tilde{B}$ is reordered as follows

$$\tilde{B}P^{(R)} = (A_1, A_2 \ldots, A_{r-1}, A_{r+1}, A_{r+2}, \ldots, A_m, A_k), \qquad (12.47)$$

where $P^{(R)}$ is a suitable permutation matrix.

Let f denote the following product

$$f = M^{(m-1)} P^{(m-1)} M^{(m-2)} P^{(m-2)} \ldots M^{(1)} P^{(1)} A_k. \qquad (12.48)$$

If we multiply the new permuted basis $\tilde{B}$ from the left by

$$M^{(m-1)} P^{(m-1)} M^{(m-2)} P^{(m-2)} \ldots M^{(1)} P^{(1)}$$

we obtain

$$M^{(m-1)} P^{(m-1)} M^{(m-2)} P^{(m-2)} \ldots M^{(1)} P^{(1)} \tilde{B}P^{(R)} = U', \qquad (12.49)$$

where

$$U' = (U_1, U_2, \ldots, U_{r-1}, U_{r+1}, \ldots, U_m, f) \qquad (12.50)$$

or

$$U' = \begin{pmatrix} u_{1,1} & \dots & u_{1,r-1} & u_{1,r+1} & u_{1,r+2} & \dots & u_{1,m} & f_1 \\ 0 & \dots & u_{2,r-1} & u_{2,r+1} & u_{2,r+2} & \dots & u_{2,m} & f_2 \\ \vdots & \vdots & \vdots & \vdots & \vdots & \vdots & \vdots & \vdots \\ 0 & \vdots & u_{r-1,r-1} & u_{r-1,r+1} & u_{r-1,r+2} & \dots & u_{r-1,m} & f_{r-1} \\ 0 & \vdots & 0 & u_{r,r+1} & u_{r,r+2} & \dots & u_{r,m} & f_r \\ 0 & \vdots & 0 & u_{r+1,r+1} & u_{r+1,r+2} & \dots & u_{r+1,m} & f_{r+1} \\ 0 & \vdots & 0 & 0 & u_{r+2,r+2} & \dots & u_{r+2,m} & f_{r+2} \\ \vdots & \vdots & \vdots & \vdots & \vdots & \dots & \vdots & \vdots \\ 0 & 0 & 0 & 0 & 0 & \dots & u_{m,m} & f_m \end{pmatrix}.$$

This matrix U' has a special structure – it is almost upper triangular but with sub-diagonal elements $u_{r+1,r+1}, u_{r+2,r+2}, \dots, u_{m,m}$ – and can be relatively easily reduced back to the upper triangular form by performing several suitable Gaussian transformations. This may be accomplished by repeating the following operations for $j = r, r+1, \dots, m-1$:

Step 1. Permute row j and row $j+1$ by permutation matrix $\tilde{P}_j$ so that

$$\tilde{u}_{j,j} \geq \tilde{u}_{j+1,j}. \tag{12.51}$$

We have to note here that permutation matrix $\tilde{P}_j$ is a unit matrix of order m if no interchange is necessary, i.e. $u_{j,j} \geq u_{j+1,j}$.

Step 2. Perform the Gaussian transformation $\tilde{M}^{(j)}$ to zero the sub-diagonal entry $\tilde{u}_{j+1,j}$, where

$$\begin{aligned} \tilde{M}^{(j)} &= I - (\underbrace{0, 0, \dots, 0}_{j}, \frac{\tilde{u}_{j+1,j}}{\tilde{u}_{j,j}}, 0, \dots, 0)^T e_j^T = \\ &= \begin{pmatrix} 1 & 0 & \dots & 0 & \dots & 0 \\ 0 & 1 & \dots & 0 & \dots & 0 \\ \vdots & \vdots & \dots & \vdots & \dots & \vdots \\ 0 & 0 & \dots & 1 & \dots & 0 \\ 0 & 0 & \dots & -\frac{\tilde{u}_{j+1,j}}{\tilde{u}_{j,j}} & \dots & 0 \\ \vdots & \vdots & \vdots & \vdots & \dots & \vdots \\ 0 & 0 & \dots & 0 & \dots & 1 \end{pmatrix}. \end{aligned}$$

After $m - r$ permutations $\tilde{P}_j$ and the Gaussian transformations $\tilde{M}^{(j)}$, the new matrix U' has been transformed back to the upper triangular form

$$U'' = \tilde{M}^{(m-1)} \tilde{P}^{(m-1)} \tilde{M}^{(m-2)} \tilde{P}^{(m-2)} \dots \tilde{M}^{(r)} \tilde{P}^{(r)} U'$$

or, using (12.49) we obtain

$$U'' = \tilde{M}^{(m-1)} \tilde{P}^{(m-1)} \dots \tilde{M}^{(r)} \tilde{P}^{(r)} M^{(m-1)} P^{(m-1)} \dots M^{(1)} P^{(1)} \tilde{B} P^{(R)} .$$

Note that each of the permutation matrices $\tilde{P}^{(j)}$ permutes only two rows. Also, each of the Gaussian transformation matrices $\tilde{M}^{(j)}$ has only one nonzero off-diagonal entry. All these mean that the procedure of the LU update described above may be performed very fast. However, the method is not absolutely free of disadvantages. The main problem with this procedure is that the queue of the LU factors

$$\tilde{M}^{(m-1)} \tilde{P}^{(m-1)} \dots \tilde{M}^{(r)} \tilde{P}^{(r)} M^{(m-1)} P^{(m-1)} \dots M^{(1)} P^{(1)}$$

gets longer for each update of the basis matrix B. Obviously, the greater is the size of the problem to be solved, i.e. m, the faster the queue of the LU factors gets longer. So, for better numerical accuracy and more effective utilization of memory LU-factorization must be periodically re-evaluated from scratch.

Let us consider the general scheme of applying updated LU-decomposition when solving the system of linear equations in form $\tilde{B}x = b$. Recall that we had an LU-decomposition of the original basis B, i.e. $B = LU$ or $L^{-1}B = U$. So after modifying basis B we have $L^{-1}\tilde{B} = \tilde{U}$, where matrix $\tilde{U}$ has a so-called *spike*

$$\tilde{U} = \begin{pmatrix} * & * & * & * & * & * & * \\ & * & * & * & * & * & * \\ & & * & * & * & * & * \\ & & * & * & * & * & * \\ & & * & & * & * & * \\ & & * & & & * & * \\ & & * & & & & * \end{pmatrix} .$$

containing vector f_k. After applying permutation $P^{(R)}$ from the right to matrix $\tilde{U}$ we obtain matrix U' with sub-diagonal entries

$$L^{-1}\tilde{B}P^{(R)} = \tilde{U}P^{(R)} = \begin{pmatrix} * & * & * & * & * & * & * \\ & * & * & * & * & * & * \\ & & * & * & * & * & * \\ & & * & * & * & * & * \\ & & & * & * & * & * \\ & & & & * & * & * \\ & & & & & * & * \end{pmatrix} = U' .$$

Then, to convert matrix U' to upper-triangular form we perform permutations $\tilde{P}^{(j)}$ and Gauss transformations $\tilde{M}^{(j)}$, $j = r, r+1, \ldots, m-1$. So multiplying from the left both sides of equality

$$L^{-1}\tilde{B}P^{(R)} = \tilde{U}P^{(R)}$$

by

$$\tilde{M}^{(m-1)}\tilde{P}^{(m-1)} \ldots \tilde{M}^{(r)}\tilde{P}^{(r)}$$

we obtain

$$QL^{-1}\tilde{B}P^{(R)} = Q\tilde{U}P^{(R)}, \tag{12.52}$$

where Q denotes product

$$Q = \tilde{M}^{(m-1)}\tilde{P}^{(m-1)} \ldots \tilde{M}^{(r)}\tilde{P}^{(r)}, \tag{12.53}$$

and matrix $Q\tilde{U}P^{(R)}$ is updated upper-triangular matrix U''. So we can re-write (12.52) as follows

$$QL^{-1}\tilde{B}P^{(R)} = U'', \tag{12.54}$$

Finally, from equality (12.54) we obtain LU-decomposition of new basis $\tilde{B}$

$$\tilde{B} = LQ^{-1}U''(P^{(R)})^{-1}. \tag{12.55}$$

Formula (12.55) allows us to solve system $\tilde{B}x = b$ using the following steps.

Step 1. First of all, using formula (12.55) we re-write system $\tilde{B}x = b$ as follows

$$LQ^{-1}U''(P^{(R)})^{-1}x = b\,, \tag{12.56}$$

Step 2. Then using the new variable

$$y = Q^{-1}U''(P^{(R)})^{-1}x$$

we solve system $Ly = b$ to obtain vector $y = L^{-1}b$. We have to note here that generally speaking this system $Ly = b$ may have been solved in the previous step when we had to solve the original system $Bx = b$. So actually we may omit this step.

Step 3. Multiplying by Q from the left both sides of equality

$$y = Q^{-1}U''(P^{(R)})^{-1}x$$

we obtain

$$Qy = QQ^{-1}U''(P^{(R)})^{-1}x$$

or

$$U''(P^{(R)})^{-1}x = Qy\,.$$

Then, using the new variable $z = (P^{(R)})^{-1}x$ and applying the backward substitution method we solve system

$$U''z = Qy$$

Step 4. Finally, we obtain solution $x = P^{(R)}z$.

To illustrate the Bartels-Golub method, we consider matrix

$$B = (A_1, A_2, A_3) = \begin{pmatrix} 1 & 1 & 1 \\ 2 & 1 & 1 \\ 1 & 2 & 3 \end{pmatrix}$$

from the example of Section 2.2, page 341. When transforming this matrix to upper triangular form we established that

$$M^{(2)}M^{(1)}B = \begin{pmatrix} 1 & 0 & 0 \\ 0 & 1 & 0 \\ 0 & 1 & 1 \end{pmatrix}\begin{pmatrix} 1 & 0 & 0 \\ -2 & 1 & 0 \\ -1 & 0 & 1 \end{pmatrix} B =$$

$$= \begin{pmatrix} 1 & 1 & 1 \\ 0 & -1 & -1 \\ 0 & 0 & 1 \end{pmatrix} = U\,.$$

Let us suppose that we have to replace column-vector $A_1 = (1,2,1)^T$ (that is for our example $r = 1$) in the basis $B = (A_1, A_2, A_3)$ with some other column-vector $A_4 = (5,1,3)^T$. It means that in accordance with the general approach to matrix update we have to remove column A_1 from basis B, shift columns A_2 and A_3 to the left by one position, and then put new vector A_4 into the rightmost column. The correspondent permutation matrix $P^{(R)}$ that shifts columns A_2 and A_3 to the left by one position and moves column A_1 to the rightmost position is

$$P^{(R)} = \begin{pmatrix} 0 & 0 & 1 \\ 1 & 0 & 0 \\ 0 & 1 & 0 \end{pmatrix}.$$

So, we obtain a new basis

$$BP^{(R)} = (A_1, A_2, A_3)\,P^{(R)} = (A_2, A_3, A_1) \stackrel{A_4 \to A_1}{\longrightarrow} (A_2, A_3, A_4) =$$

$$= \begin{pmatrix} 1 & 1 & 5 \\ 1 & 1 & 1 \\ 2 & 3 & 3 \end{pmatrix} = \tilde{B}P^{(R)}\,.$$

Applying transformation $M^{(2)}M^{(1)}$ from the left to the new basis $\tilde{B}P^{(R)}$ is equivalent to calculating vector (see (12.48))

$$f = M^{(2)}M^{(1)}A_4 = \begin{pmatrix} 1 & 0 & 0 \\ 0 & 1 & 0 \\ 0 & 1 & 1 \end{pmatrix}\begin{pmatrix} 1 & 0 & 0 \\ -2 & 1 & 0 \\ -1 & 0 & 1 \end{pmatrix}\begin{pmatrix} 5 \\ 1 \\ 3 \end{pmatrix} = \begin{pmatrix} 5 \\ -9 \\ -11 \end{pmatrix},$$

then shifting columns 2 and 3 in the upper triangular matrix U to the left by one position and putting vector f into the rightmost column. In any case we obtain matrix

$$U' = \begin{pmatrix} u_{12} & u_{13} & f_1 \\ u_{22} & u_{23} & f_2 \\ 0 & u_{33} & f_3 \end{pmatrix} = \begin{pmatrix} 1 & 1 & 5 \\ -1 & -1 & -9 \\ 0 & 1 & -11 \end{pmatrix},$$

with sub-diagonal elements u_{22} and u_{33} (see (12.49) and (12.50)). This matrix U' must be reduced to the upper triangular form. To achieve this aim, we perform the following row perturbations and Gaussian transformations:

$$\text{for } j = 1: \quad \tilde{P}^{(1)} = \begin{pmatrix} 1 & 0 & 0 \\ 0 & 1 & 0 \\ 0 & 0 & 1 \end{pmatrix}, \quad \tilde{M}^{(1)} = \begin{pmatrix} 1 & 0 & 0 \\ -\mu_1 & 1 & 0 \\ 0 & 0 & 1 \end{pmatrix},$$

and

$$\text{for } j = 2: \quad \tilde{P}^{(2)} = \begin{pmatrix} 1 & 0 & 0 \\ 0 & 0 & 1 \\ 0 & 1 & 0 \end{pmatrix}, \quad \tilde{M}^{(2)} = \begin{pmatrix} 1 & 0 & 0 \\ 0 & 1 & 0 \\ 0 & -\mu_2 & 1 \end{pmatrix}.$$

where $\mu_1 = -1$ and $\mu_2 = 0$.

Since $u_{12} \geq u_{22}$ we do not need any permutation at the first step. This is why $\tilde{P}^{(1)}$ is a unit matrix and may be omitted. So, after performing transformation for $j = 1$ we have

$$\tilde{M}^{(1)}\tilde{P}^{(1)}U' = \tilde{M}^{(1)}U' = \begin{pmatrix} 1 & 0 & 0 \\ 1 & 1 & 0 \\ 0 & 0 & 1 \end{pmatrix}\begin{pmatrix} 1 & 1 & 5 \\ -1 & -1 & -9 \\ 0 & 1 & -11 \end{pmatrix} =$$

$$= \begin{pmatrix} 1 & 1 & 5 \\ 0 & 0 & -4 \\ 0 & 1 & -11 \end{pmatrix}.$$

Then, because of condition (12.51), we have to interchange row 2 and row 3, i.e.

$$\tilde{P}^{(2)}\tilde{M}^{(1)}U' = \begin{pmatrix} 1 & 0 & 0 \\ 0 & 0 & 1 \\ 0 & 1 & 0 \end{pmatrix}\begin{pmatrix} 1 & 1 & 5 \\ 0 & 0 & -4 \\ 0 & 1 & -11 \end{pmatrix} =$$

$$= \begin{pmatrix} 1 & 1 & 5 \\ 0 & 1 & -11 \\ 0 & 0 & -4 \end{pmatrix} = U'' .$$

Occasionally, after permutation $\tilde{P}^{(2)}$ we obtained an upper triangular matrix. So this is why $\tilde{M}^{(2)}$ is a unit matrix and hence, may be omitted. Summarizing, we can say that to reduce the "almost" upper triangular matrixU' to the "pure" upper triangular form we only have to perform transformation$\tilde{M}^{(1)}$ and then permutation $\tilde{P}^{(2)}$. Finally, we have

$$\tilde{P}^{(2)}\tilde{M}^{(1)}U' = U'' .$$

Once, we have determined all right-hand side components of (12.55), we can use these matrices L, Q^{-1}, U'', and $(P^{(R)})^{-1}$ to calculate solution for system $\tilde{B}x = b$. In accordance with Step 2, we use new variable $y = Q^{-1}U''(P^{(R)})^{-1}\, x$ and solve system $Ly = b$, i.e.

$$\begin{pmatrix} 1 & 0 & 0 \\ 2 & 1 & 0 \\ 1 & -1 & 1 \end{pmatrix} y = \begin{pmatrix} 1 \\ 2 \\ 2 \end{pmatrix},$$

which gives us vector $y = (1, 0, 1)^T$. Further, following prescriptions of Step 3 we solve system$U''z = Qy$, i.e.

$$\begin{pmatrix} 1 & 1 & 5 \\ 0 & 1 & -11 \\ 0 & 0 & -4 \end{pmatrix} z = \begin{pmatrix} 1 & 0 & 0 \\ 0 & 0 & 1 \\ 1 & 1 & 0 \end{pmatrix} \begin{pmatrix} 1 \\ 0 \\ 1 \end{pmatrix}$$

or

$$\begin{pmatrix} 1 & 1 & 5 \\ 0 & 1 & -11 \\ 0 & 0 & -4 \end{pmatrix} z = \begin{pmatrix} 1 \\ 1 \\ 1 \end{pmatrix}$$

and obtain vector $z = (4, -7/4, -1/4)^T$. Finally, using permutation matrix $P^{(R)}$ (see Step 4) we determine vector$x$ as follows

$$x = P^{(R)}z = \begin{pmatrix} 0 & 0 & 1 \\ 1 & 0 & 0 \\ 0 & 1 & 0 \end{pmatrix} \begin{pmatrix} 4 \\ -7/4 \\ -1/4 \end{pmatrix} = \begin{pmatrix} -1/4 \\ 4 \\ -7/4 \end{pmatrix}.$$

Another way of using formula (12.55) for solving system$\tilde{B}x = b$ consists of the following steps.

Step 1. First, we re-write system $\tilde{B}x = b$ in the form of (12.56).

Step 2. Then we multiply from the left both sides of formula (12.56) by L^{-1}

$$L^{-1}LQ^{-1}U''(P^{(R)})^{-1}x = L^{-1}b \tag{12.57}$$

Step 3. Multiply from the left both sides of (12.57) by Q

$$QQ^{-1}U''(P^{(R)})^{-1}x = QL^{-1}b\,. \tag{12.58}$$

Step 4. Further, we calculate inverse matrix of U'' and then multiply equality (12.58) from the left by $(U'')^{-1}$

$$(U'')^{-1}U''(P^{(R)})^{-1}x = (U'')^{-1}QL^{-1}b\,, \tag{12.59}$$

Step 5. Finally, we multiply (12.59) by $P^{(R)}$

$$P^{(R)}(P^{(R)})^{-1}x = P^{(R)}(U'')^{-1}QL^{-1}b\,,$$

and obtain

$$x = P^{(R)}(U'')^{-1}QL^{-1}b\,. \tag{12.60}$$

Before illustrating this approach we have to note that the procedure described above requires extra calculations only for inverse matrices L^{-1} and $(U'')^{-1}$, since matrices Q and P^R are given (they were determined when calculating U'', see (12.47) and (12.53) respectively). Moreover, when changing vectors in basis B all necessary updates of the corresponding LU-factors are carried out in the upper triangular factor U without any changes in the lower triangular factor L. The latter means that matrix L^{-1} should be calculated only once, and then it may be used without any changes as many times as required. Thus, when using the scheme described above, the only extra calculations required are in Step 4 for determining inverse matrix $(U'')^{-1}$. Note also that U'' is an upper triangular matrix, so calculation of its inverse is relatively cheap operation.

To illustrate this approach we reconsider numerical example described above, see page 350. So, we have shown earlier that

$$B = \begin{pmatrix} 1 & 1 & 1 \\ 2 & 1 & 1 \\ 1 & 2 & 3 \end{pmatrix} = LU = \begin{pmatrix} 1 & 0 & 0 \\ 2 & 1 & 0 \\ 1 & -1 & 1 \end{pmatrix} \begin{pmatrix} 1 & 1 & 1 \\ 0 & -1 & -1 \\ 0 & 0 & 1 \end{pmatrix}$$

and

$$U'' = \begin{pmatrix} 1 & 1 & 5 \\ 0 & 1 & -11 \\ 0 & 0 & -4 \end{pmatrix}, \quad P^{(R)} = \begin{pmatrix} 0 & 0 & 1 \\ 1 & 0 & 0 \\ 0 & 1 & 0 \end{pmatrix}, \quad Q = \begin{pmatrix} 1 & 0 & 0 \\ 0 & 0 & 1 \\ 1 & 1 & 0 \end{pmatrix}.$$

Calculating inverse matrices $(U'')^{-1}$ and L^{-1} we obtain

$$(U'')^{-1} = \begin{pmatrix} 1 & -1 & 4 \\ 0 & 1 & -11/4 \\ 0 & 0 & -1/4 \end{pmatrix}, \quad L^{-1} = \begin{pmatrix} 1 & 0 & 0 \\ -2 & 1 & 0 \\ -3 & 1 & 1 \end{pmatrix}.$$

Thus, in accordance with (12.60) we have

$$\begin{aligned} x &= P^{(R)} (U'')^{-1} Q \, L^{-1} \, b = \\ &= \begin{pmatrix} 0 & 0 & 1 \\ 1 & 0 & 0 \\ 0 & 1 & 0 \end{pmatrix} \begin{pmatrix} 1 & -1 & 4 \\ 0 & 1 & -11/4 \\ 0 & 0 & -1/4 \end{pmatrix} \times \\ &\quad \times \begin{pmatrix} 1 & 0 & 0 \\ 0 & 0 & 1 \\ 1 & 1 & 0 \end{pmatrix} \begin{pmatrix} 1 & 0 & 0 \\ -2 & 1 & 0 \\ -3 & 1 & 1 \end{pmatrix} \begin{pmatrix} 1 \\ 2 \\ 2 \end{pmatrix} = \\ &= (-1/4, 4, -7/4)^T. \end{aligned}$$

Closing this section we note the main advantage of the procedures discussed: when changing vectors in basis B all necessary updates of the corresponding LU-factors may be performed in the upper triangular matrix U without any changes in the lower triangular factor L.

2.3.3 The Forest-Tomlin Updating

In this section, we very briefly address the main ideas of the update method developed by Forrest and Tomlin, the so called *Forrest-Tomlin update* [64].

As in the previous section, after removing leaving vector A_r from basis B, shifting all the columns from the right of the leaving column-vector A_r to the left by one position, and entering the new column-vector A_k to the basis at the rightmost column, we obtain new basis $\tilde{B}$ (12.47) with almost upper triangular decomposition matrix U' (12.50) with sub-diagonal elements $u_{r+1,r+1}$, $u_{r+2,r+2}$, $\dots$, $u_{m,m}$. Forrest and Tomlin noted that for L and U nonsingular, all the sub-diagonal elements in rows $i = r+1, r+2, \dots, m$ are nonzero. Hence rows $r+1, r+2, \dots, m$, can be used to eliminate the elements in the columns $r, r+1, \dots, m-1$, of row r yielding a matrix of the form

$$
U'' = \begin{pmatrix}
u_{1,1} & \dots & u_{1,r-1} & u_{1,r+1} & u_{1,r+2} & \dots & u_{1,m} & f_1 \\
0 & \dots & u_{2,r-1} & u_{2,r+1} & u_{2,r+2} & \dots & u_{2,m} & f_2 \\
\vdots & \vdots & \vdots & \vdots & \vdots & \vdots & \vdots & \vdots \\
0 & \vdots & u_{r-1,r-1} & u_{r-1,r+1} & u_{r-1,r+2} & \dots & u_{r-1,m} & f_{r-1} \\
0 & \vdots & 0 & 0 & 0 & \dots & 0 & f'_r \\
0 & \vdots & 0 & u_{r+1,r+1} & u_{r+1,r+2} & \dots & u_{r+1,m} & f_{r+1} \\
0 & \vdots & 0 & 0 & u_{r+2,r+2} & \dots & u_{r+2,m} & f_{r+2} \\
\vdots & \vdots & \vdots & \vdots & \vdots & \dots & \vdots & \vdots \\
0 & 0 & 0 & 0 & 0 & \dots & u_{m,m} & f_m
\end{pmatrix}.
$$

Moving up all lower rows, i.e. rows $r+1, r+2 \dots, m,$ by one position and putting row r last, gives a matrix in the desired upper triangular form

$$
U''' = \begin{pmatrix}
u_{1,1} & \dots & u_{1,r-1} & u_{1,r+1} & u_{1,r+2} & \dots & u_{1,m} & f_1 \\
0 & \dots & u_{2,r-1} & u_{2,r+1} & u_{2,r+2} & \dots & u_{2,m} & f_2 \\
\vdots & \vdots & \vdots & \vdots & \vdots & \vdots & \vdots & \vdots \\
0 & \vdots & u_{r-1,r-1} & u_{r-1,r+1} & u_{r-1,r+2} & \dots & u_{r-1,m} & f_{r-1} \\
0 & \vdots & 0 & u_{r+1,r+1} & u_{r+1,r+2} & \dots & u_{r+1,m} & f_{r+1} \\
0 & \vdots & 0 & 0 & u_{r+2,r+2} & \dots & u_{r+2,m} & f_{r+2} \\
\vdots & \vdots & \vdots & \vdots & \vdots & \dots & \vdots & \vdots \\
0 & 0 & 0 & 0 & 0 & \dots & u_{m,m} & f_m \\
0 & \vdots & 0 & 0 & 0 & \dots & 0 & f'_r
\end{pmatrix}.
$$

This permutation is precisely the inverse of the permutation that shifted columns $r+1, r+2, \dots, m$ from right to left by one position and moved column r to the rightmost position.

Using matrix notation and denoting this permutation by Q, we obtain

$$Q^{-1}\tilde{M}L^{-1}\tilde{B} = U''', \tag{12.61}$$

where inverse $L^{-1} = M^{(m-1)} \dots M^{(2)}M^{(1)}$ is the Gaussian elimination that transformed the original basis B to the upper triangular form U, product matrix $\tilde{M} = \tilde{M}_{m-1} \dots \tilde{M}_{r+1}\tilde{M}_r,$ is the transformation matrix that eliminates elements $r, r+1, \dots, m-1$ of row r (and hence, produces matrix U''). Obviously, matrices $\tilde{M}_i$ are the transformation matrices that zero out elements i of row $r,\ i = r, r+1, \dots, m-1,$ each by one, i.e.

$$\tilde{M}_i = I - e_r g_i^T,$$

where column-vector e_r is rth column of the unit matrix of order m,

$$g_i^T = (\underbrace{0,0,\dots,0,}_{i} \mu_i,\ 0,0,\dots,0)^T$$

and $\mu_i = u_{r,i+1}/u_{i+1,i+1}$, $i = r, r+1, \dots, m-1$.

So, multiplier μ_i is positioned in column $i+1$ of row i, i.e.

$$\tilde{M}_i = \begin{pmatrix} 1 & \dots & 0 & 0 & 0 & 0 & \dots & 0 \\ \vdots & \vdots & \vdots & \vdots & \vdots & \vdots & \vdots & \vdots \\ 0 & \dots & 1 & 0 & 0 & 0 & \dots & 0 \\ 0 & \dots & 0 & 1 & -\mu_i & 0 & \dots & 0 \\ 0 & \dots & 0 & 0 & 1 & 0 & \dots & 0 \\ 0 & \dots & 0 & 0 & 0 & 1 & \dots & 0 \\ \vdots & \vdots & \vdots & \vdots & \vdots & \vdots & \dots & \vdots \\ 0 & \dots & 0 & 0 & 0 & 0 & \dots & 1 \end{pmatrix}, \quad i = r, r+1, \dots, m-1.$$

Finally, since Q is a permutation matrix, $Q^{-1} = Q^T$ and we can rewrite (12.61) as follows

$$Q^T\ \tilde{M}_{m-1} \dots \tilde{M}_{r+1} \tilde{M}_r\ M^{(m-1)} \dots M^{(2)} M^{(1)}\ \tilde{B} = U''', \qquad (12.62)$$

To illustrate how the method works, consider matrix

$$B = \begin{pmatrix} 1 & 1 & 1 \\ 2 & 1 & 1 \\ 1 & 2 & 3 \end{pmatrix}$$

from the example of section 2.2, page 341. There it was shown that

$$M^{(2)} M^{(1)} B = \begin{pmatrix} 1 & 0 & 0 \\ 0 & 1 & 0 \\ 0 & 1 & 1 \end{pmatrix} \begin{pmatrix} 1 & 0 & 0 \\ -2 & 1 & 0 \\ -1 & 0 & 1 \end{pmatrix} B =$$

$$= \begin{pmatrix} 1 & 1 & 1 \\ 0 & -1 & -1 \\ 0 & 0 & 1 \end{pmatrix} = U\,.$$

Assume that in the basis $B = (A_1, A_2, A_3)$ we have to replace column-vector $A_1 = (1,2,1)^T$ (that is $r = 1$) with some other column-vector $A_4 = (5,1,3)^T$. After shifting columns A_2 and A_3 to the left by one position and entering column

A_4 at the rightmost position, we obtain

$$\tilde{B} = \begin{pmatrix} 1 & 1 & 5 \\ 1 & 1 & 1 \\ 2 & 3 & 3 \end{pmatrix}.$$

The correspondent permutation matrix Q that shifts columns A_2 and A_3 to the left by one position and moves column A_1 to the rightmost position is

$$BQ = \begin{pmatrix} 1 & 1 & 1 \\ 2 & 1 & 1 \\ 1 & 2 & 3 \end{pmatrix} \begin{pmatrix} 0 & 0 & 1 \\ 1 & 0 & 0 \\ 0 & 1 & 0 \end{pmatrix} = \begin{pmatrix} 1 & 1 & 1 \\ 1 & 1 & 2 \\ 2 & 3 & 1 \end{pmatrix}.$$

Further, in accordance with (12.48) we produce vector

$$f = M^{(2)} M^{(1)} A_4 = \begin{pmatrix} 1 & 0 & 0 \\ 0 & 1 & 0 \\ 0 & 1 & 1 \end{pmatrix} \begin{pmatrix} 1 & 0 & 0 \\ -2 & 1 & 0 \\ -1 & 0 & 1 \end{pmatrix} \begin{pmatrix} 5 \\ 1 \\ 3 \end{pmatrix} = \begin{pmatrix} 5 \\ -9 \\ -11 \end{pmatrix}$$

and obtain matrix

$$U' = \begin{pmatrix} 1 & 1 & 5 \\ -1 & -1 & -9 \\ 0 & 1 & -11 \end{pmatrix},$$

see formulas (12.49) and (12.50). Now, we have to transform matrix U' to the form of U'' (see page 355) using sub-diagonal elements of rows 2 and 3 to eliminate elements in columns 1 and 2 of row $r = 1$. The correspondent matrices of the Gaussian transformations are as follows

$$\tilde{M}_1 = \begin{pmatrix} 1 & -\mu_1 & 0 \\ 0 & 1 & 0 \\ 0 & 0 & 1 \end{pmatrix} \quad \text{and} \quad \tilde{M}_2 = \begin{pmatrix} 1 & 0 & 0 \\ 0 & 1 & -\mu_2 \\ 0 & 0 & 1 \end{pmatrix}$$

where $\mu_1 = -1$ and $\mu_2 = 0$.

Observe that transformation $\tilde{M}_1$ zero outs both elements in columns 1 and 2 of row 1

$$\tilde{M}_1 U' = \begin{pmatrix} 1 & 1 & 0 \\ 0 & 1 & 0 \\ 0 & 0 & 1 \end{pmatrix} \begin{pmatrix} 1 & 1 & 5 \\ -1 & -1 & -9 \\ 0 & 1 & -11 \end{pmatrix} =$$

$$= \begin{pmatrix} 0 & 0 & -4 \\ -1 & -1 & -9 \\ 0 & 1 & -11 \end{pmatrix} = U'',$$

so transformation $\tilde{M}_2$ is unnecessary and for matrix $\tilde{M}$ we obtain

$$\tilde{M} = \tilde{M}_2 \tilde{M}_1 = I \tilde{M}_1 = \tilde{M}_1 = \begin{pmatrix} 1 & 1 & 0 \\ 0 & 1 & 0 \\ 0 & 0 & 1 \end{pmatrix}.$$

Finally, using permutation matrix Q^T we move up rows 2 and 3 of matrix U'' by one position and put row 1 last, and hence obtain the matrix in the desired upper triangular form

$$Q^T U'' = \begin{pmatrix} 0 & 1 & 0 \\ 0 & 0 & 1 \\ 1 & 0 & 0 \end{pmatrix} \begin{pmatrix} 0 & 0 & -4 \\ -1 & -1 & -9 \\ 0 & 1 & -11 \end{pmatrix} =$$

$$= \begin{pmatrix} -1 & -1 & -9 \\ 0 & 1 & -11 \\ 0 & 0 & -4 \end{pmatrix} = U''' .$$

Closing discussion of the Forrest-Tomlin update, we have to note that this method only deals with the upper triangular matrix U since lower triangular matrix L remains unchanged.

2.4 Other Types of Factorization

One of the most important tendencies in numerical linear algebra and linear and linear-fractional programming is that to improve the performance of computations we try (whenever it is possible) to get an advantage from any special features of matrix A, e.g. symmetry, orthogonality, etc. In this section we address the main ideas of some decompositions usually used in commercial codes for special LP problems. More detailed information on these topics may be found in the following books and articles: [57], [72], [79], [186].

2.4.1 Cholesky Decomposition

If the given square matrix $A = ||a_{ij}||_{m\times m}$ is symmetric positive definite, it is normal instead of general LU-factorization to use a special, more efficient, triangular decomposition

$$A = LL^T, \tag{12.63}$$

where L denotes a lower triangular matrix and L^T is an upper triangular one. This factorization (12.63) is called a *Cholesky* factorization. *Symmetric* means that $a_{ij} = a_{ji}$ for all $i = 1, 2, \ldots, m$; $j = 1, 2, \ldots, m$, i.e. $A = A^T$, while *positive definite* means $v^T Av > 0$ for all vectors $v \neq 0$.

Writing out equation (12.63) in components, we obtain the analogs of equations (12.29)-(12.30) formulated for LU-decomposition

$$l_{ii} = \begin{cases} a_{ii}^{1/2}, & \text{if } \ i = 1; \\ (a_{ii} - \sum\limits_{k=1}^{i-1} l_{ik}^2)^{1/2}, & \text{if } \ i > 1; \end{cases} \tag{12.64}$$

$$l_{ji} = (a_{ij} - \sum_{k=1}^{i-1} l_{ik}\, l_{jk})/l_{ii}, \quad j = i+1, i+2, \ldots, m; \tag{12.65}$$

for each $i = 1, 2, \ldots, m$. As in the case of Crout's algorithm for LU-decomposition (see section 2.1), we have to apply equations (12.64) and (12.65) successively in order $i = 1, 2, \ldots, m$. Performing these operations in the required order, we will see that those entries l_{ij} that occur on the right-hand side are already determined by the time they are needed.

The total operations count required for Cholesky factorization is about a factor 2 better than LU-decomposition of matrix A, where its symmetry would be ignored. Another advantage of this method is that because of the symmetry of A, the lower triangular matrix L (excluding its diagonal entries) may be stored in the lower triangular part of A. The only extra storage required for this method is a vector of length m to accommodate the diagonal of L.

For more general symmetric matrices, the factorization

$$A = LDL^T, \tag{12.66}$$

is more appropriate, where matrix D is a block diagonal matrix, and L denotes a unit lower triangular matrix.

Consider the following numeric example with symmetric and positive definite matrix

$$A = \begin{pmatrix} 4 & 2 & 2 \\ 2 & 5 & 3 \\ 2 & 3 & 6 \end{pmatrix}.$$

Using formulas (12.64) and (12.65) in consecutive order $i = 1, 2, 3$, we obtain

for $i = 1$:

$$\begin{array}{rrcccccl} & l_{11} & = & \sqrt{a_{11}} & = & \sqrt{4} & = & 2; \\ j=2: & l_{21} & = & a_{12}/l_{11} & = & 2/2 & = & 1; \\ j=3: & l_{31} & = & a_{13}/l_{11} & = & 2/2 & = & 1; \end{array}$$

for $i = 2$:

$$\begin{array}{rrcccccl} & l_{22} & = & \sqrt{a_{22} - l_{21}^2} & = & \sqrt{5 - 1^2} & = & 2; \\ j=3: & l_{32} & = & (a_{23} - l_{21} l_{31})/l_{22} & = & (3 - 1 * 1)/2 & = & 1; \end{array}$$

finally, for $i = 3$:

$$l_{33} = \sqrt{a_{33} - l_{31}^2 - l_{32}^2} = \sqrt{6 - 1^2 - 1^2} = 2\,.$$

So,

$$L = \begin{pmatrix} 2 & 0 & 0 \\ 1 & 2 & 0 \\ 1 & 1 & 2 \end{pmatrix} \quad \text{and} \quad L^T = \begin{pmatrix} 2 & 1 & 1 \\ 0 & 2 & 1 \\ 0 & 0 & 2 \end{pmatrix}.$$

We can check

$$LL^T = \begin{pmatrix} 2 & 0 & 0 \\ 1 & 2 & 0 \\ 1 & 1 & 2 \end{pmatrix} \begin{pmatrix} 2 & 1 & 1 \\ 0 & 2 & 1 \\ 0 & 0 & 2 \end{pmatrix} = \begin{pmatrix} 4 & 2 & 2 \\ 2 & 5 & 3 \\ 2 & 3 & 6 \end{pmatrix} = A\,.$$

Observe that formulas (12.64) and (12.65) refer only to componentsa_{ij} with $j \geq i$. Since A is symmetric, these formulas have enough information to complete the decomposition. In fact, formulas (12.64) and (12.65) give an efficient way to test whether a symmetric matrix is positive definite, see (12.64). Closing discussion of the Cholesky factorization, we have to note that this method is numerically highly stable and does not require any pivoting at all.

2.4.2 QR Decomposition

There is another matrix factorization that is sometimes very useful, the so-called *QR decomposition,*

$$A = QR,$$

where R is upper triangular and Q is orthogonal, i.e.

$$Q^TQ = I, \text{ and hence } Q^T = Q^{-1}.$$

Like the other decomposition methods we have considered above, theQR can be used to solve systems of linear equations

$$Ax = b.$$

Indeed, having factors Q and R we can rewrite this system as follows

$$QRx = b.$$

First, we solve system

$$Qy = b, \text{ where } Rx = y,$$

for unknown y and then solve $Rx = y$ for unknown x by backward substitution. Since $Q^{-1} = Q^T$, we have

$$Qy = b, \rightarrow Q^{-1}Qy = Q^{-1}b, \rightarrow y = Q^Tb.$$

This factorization can be applied to square or rectangular matrices but we restrict our considerations to the case of square matrix $A = ||a_{ij}||_{m \times m}$. Usually this method is a key algorithm for computing eigenvalues or least-square solutions. Since QR requires about twice as many operations as LU decomposition, and needs about factor 1.5 more memory than LU [82], it is less applied to find the solution of a square linear system. Nevertheless, there are several reasons why orthogonalization methods, such as QR, might be considered. The first of them, and the main, is that these methods usually do not require any pivoting and are numerically very stable, which is not the case for Gaussian elimination. Another advantage that might favor QR decomposition is the possibility of updating factors Q and R corresponding to a rank one modification of matrix A in $O(m^2)$ operations. Of course, it is also possible for LU factorization, however, the implementation of QR updating is much easier and simpler.

There are two standard algorithms for QR factorization - one of them involves the so-called *Householder transformations* or *Householder reflections* and the second one is based on the *Givens rotations*. There is another method used for QR factorization - usually referred to as *Gram-Schmidt orthogonalization* or *Gram-Schmidt process*. Since the "original" or "classic" Gram-Schmidt process is often numerically unstable, we do not consider this method (see [79] for a "modified" Gram-Schmidt method). The Householder transformations lead to algorithms involving less numerical operations than require Givens rotations and therefore are presented in the following.

DEFINITION 12.2 *If vector $v \neq 0$, then matrix*

$$H = I - 2\,\frac{v\,v^t}{v^t\,v}$$

is called the **Householder matrix** *or* **Householder reflection** *and vector v is called the* **Householder vector**.

The Householder matrix is symmetric and orthogonal. Applying an appropriate Householder vector v we can construct such a Householder matrix that, being used to multiply the given matrix A from the left, can annihilate all elements in a column of matrix A below a chosen element a_{ij}. So, for the elected element, say a_{11}, we can construct such a Householder matrix Q_1 that

$$Q_1 A = Q_1 \begin{pmatrix} a_{11} & a_{12} & \dots & a_{1m} \\ a_{21} & a_{22} & \dots & a_{2m} \\ \vdots & \vdots & \vdots & \vdots \\ a_{m1} & a_{m2} & \dots & a_{mm} \end{pmatrix} =$$

$$= \begin{pmatrix} \tilde{a}_{11} & \tilde{a}_{12} & \dots & \tilde{a}_{1m} \\ 0 & \tilde{a}_{22} & \dots & \tilde{a}_{2m} \\ \vdots & \vdots & \vdots & \vdots \\ 0 & \tilde{a}_{m2} & \dots & \tilde{a}_{mm} \end{pmatrix} = A^1.$$

Similarly, we can construct such a Householder matrix Q_2 that zero outs all elements below element $\tilde{a}_{22}$, and so on up to Q_{m-1}. So, we have

$$Q_{m-1} \dots Q_2 Q_1 A = R.$$

Using the orthogonality of matrices $Q_1 \dots Q_{m-1}$, we can easily establish that

$$Q = (Q_{m-1} \dots Q_2 Q_1)^{-1} = Q_1^T Q_2^T \dots Q_{m-1}^T.$$

Recalling that Householder matrices $Q_1 Q_2 \dots Q_{m-1}$ are symmetric, we can rewrite the last in the form of a product as follows

$$Q = Q_1 Q_2 \dots Q_{m-1}.$$

To produce matrix Q_1 we have to calculate vector

$$v_1 = A_1 + |A_1| \, e_1,$$

where column-vector $A_1 = (a_{11}, a_{21}, \dots, a_{m1})^T$ is column 1 of matrix A, $|A_1|$ denotes the length of vector A_1, i.e.

$$|A_1| = \sqrt{a_{11}^2 + a_{21}^2 + \dots + a_{m1}^2} \, ,$$

and e_1 is the leftmost column-vector of the unit matrix of order m. Then

$$Q_1 = I - 2 \, \frac{v_1 \, v_1^T}{v_1^T \, v_1}$$

and

$$Q_1 A = A^1.$$

We continue this process in a similar way on the matrix $\tilde{A}^1$ of order $(m - 1) \times (m - 1)$ where we have removed the first row and column, i.e.

$$\tilde{A}^1 = \begin{pmatrix} \tilde{a}_{22} & \tilde{a}_{23} & \dots & \tilde{a}_{2m} \\ \tilde{a}_{32} & \tilde{a}_{33} & \dots & \tilde{a}_{3m} \\ \vdots & \vdots & \vdots & \vdots \\ \tilde{a}_{m2} & \tilde{a}_{m3} & \dots & \tilde{a}_{mm} \end{pmatrix}.$$

The appropriate Householder vector will now be of dimension$(m-1)$, so we have to complete it with zero $v_2 = (0, \tilde{v}_2)^T$. We use vector-column $\tilde{v}_2$ to build

$$\tilde{Q}_2 = I - 2\,\frac{\tilde{v}_2\,\tilde{v}_2^T}{\tilde{v}_2^T\,\tilde{v}_2},$$

and then use matrix $\tilde{Q}_2$ to produce

$$Q_2 = \begin{pmatrix} 1 & 0 \\ 0 & \tilde{Q}_2 \end{pmatrix}.$$

After $m-1$ steps, we obtain required decomposition $A = QR$ with

$$R = Q_{m-1}\ldots Q_2 Q_1 A \quad \text{and} \quad Q = Q_1 Q_2 \ldots Q_{m-1}.$$

To illustrate how the method works, we consider the following example. Let

$$A = \begin{pmatrix} 1 & 6 & 0 \\ 2 & 1 & 2 \\ 2 & 2 & 1 \end{pmatrix}.$$

To find Q_1, we compute the according Householder vector

$$v_1 = \begin{pmatrix} 1 \\ 2 \\ 2 \end{pmatrix} + \sqrt{1^2+2^2+2^2}\begin{pmatrix} 1 \\ 0 \\ 0 \end{pmatrix} = \begin{pmatrix} 4 \\ 2 \\ 2 \end{pmatrix}.$$

Then,

$$Q_1 = I - 2\,\frac{\begin{pmatrix} 4 \\ 2 \\ 2 \end{pmatrix}(4,2,2)}{(4,2,2)\begin{pmatrix} 4 \\ 2 \\ 2 \end{pmatrix}} = I - \frac{1}{12}\begin{pmatrix} 16 & 8 & 8 \\ 8 & 4 & 4 \\ 8 & 4 & 4 \end{pmatrix} =$$

$$= \begin{pmatrix} -1/3 & -2/3 & -2/3 \\ -2/3 & 2/3 & -1/3 \\ -2/3 & -1/3 & 2/3 \end{pmatrix}.$$

So,

$$Q_1 A = \begin{pmatrix} -1/3 & -2/3 & -2/3 \\ -2/3 & 2/3 & -1/3 \\ -2/3 & -1/3 & 2/3 \end{pmatrix}\begin{pmatrix} 1 & 6 & 0 \\ 2 & 1 & 2 \\ 2 & 2 & 1 \end{pmatrix} =$$

$$= \begin{pmatrix} -3 & -4 & -2 \\ 0 & -4 & 1 \\ 0 & -3 & 0 \end{pmatrix} = A^1.$$

We have to continue with the matrix

$$\tilde{A}^1 = \begin{pmatrix} -4 & 1 \\ -3 & 0 \end{pmatrix}$$

and to calculate the appropriate vector

$$\tilde{v}_2 = \begin{pmatrix} -4 \\ -3 \end{pmatrix} + \sqrt{(-4)^2 + (-3)^2} \begin{pmatrix} 1 \\ 0 \end{pmatrix} = \begin{pmatrix} 1 \\ -3 \end{pmatrix}.$$

Then, we produce the next Householder vector

$$v_2 = \begin{pmatrix} 0 \\ 1 \\ -3 \end{pmatrix}$$

and matrix

$$\tilde{Q}_2 = I - 2 \frac{\begin{pmatrix} 1 \\ -3 \end{pmatrix} (1, -3)}{(1, -3) \begin{pmatrix} 1 \\ -3 \end{pmatrix}} = \begin{pmatrix} 4/5 & 3/5 \\ 3/5 & -4/5 \end{pmatrix}.$$

So,

$$Q_2 = \begin{pmatrix} 1 & 0 & 0 \\ 0 & 4/5 & 3/5 \\ 0 & 3/5 & -4/5 \end{pmatrix},$$

and

$$\begin{aligned} Q_2 A^1 &= \begin{pmatrix} 1 & 0 & 0 \\ 0 & 4/5 & 3/5 \\ 0 & 3/5 & -4/5 \end{pmatrix} \begin{pmatrix} -3 & -4 & -2 \\ 0 & -4 & 1 \\ 0 & -3 & 0 \end{pmatrix} = \\ &= \begin{pmatrix} -3 & -4 & -2 \\ 0 & -5 & 4/5 \\ 0 & 0 & 3/5 \end{pmatrix} = R. \end{aligned}$$

Finally,

$$\begin{aligned} Q = Q_1 Q_2 &= \begin{pmatrix} -1/3 & -2/3 & -2/3 \\ -2/3 & 2/3 & -1/3 \\ -2/3 & -1/3 & 2/3 \end{pmatrix} \begin{pmatrix} 1 & 0 & 0 \\ 0 & 4/5 & 3/5 \\ 0 & 3/5 & -4/5 \end{pmatrix} = \\ &= \begin{pmatrix} -1/3 & -14/15 & 2/15 \\ -2/3 & 1/3 & 2/3 \\ -2/3 & 2/15 & -11/15 \end{pmatrix}. \end{aligned}$$

The factors Q and R obtained are correct because if we check, we have

$$QR = \begin{pmatrix} -1/3 & -14/15 & 2/15 \\ -2/3 & 1/3 & 2/3 \\ -2/3 & 2/15 & -11/15 \end{pmatrix} \begin{pmatrix} -3 & -4 & -2 \\ 0 & -5 & 4/5 \\ 0 & 0 & 3/5 \end{pmatrix} =$$

$$= \begin{pmatrix} 1 & 6 & 0 \\ 2 & 1 & 2 \\ 2 & 2 & 1 \end{pmatrix} = A.$$

3. Re-using Basis

In many practical situations systems of linear equations do not occur in isolation but as a part of a sequence of related problems that change in some systematic way. For example, we may need to solve a sequence of linear systems $Ax = b$ having the same matrix A but different right-hand side vectors b, or conversely, having the same vector b and a slightly modified matrix A. In linear and linear-fractional programming such situations occur when using the simplex method we have to replace in the current basis some basic vector with some other (non-basic) vector. The techniques discussed in this section sometimes allow to avoid new factorization and to construct solution for the new system on the basis of the known solution for the original system.

LEMMA 12.1 (SHERMAN-MORRISON FORMULA) *If the given column-vectors $u = (u_1, u_2, \dots, u_m)^T$ and $v = (v_1, v_2, \dots, v_m)^T$ are such that*

$$1 + v^T A^{-1} u \neq 0,$$

then for inverse matrix $\tilde{A}^{-1}$ of $\tilde{A} = A + uv^T$, where $A = ||a_{ij}||_{m\times m}$, we have

$$\tilde{A}^{-1} = A^{-1} - \frac{A^{-1}uv^T A^{-1}}{1 + v^T A^{-1} u}. \tag{12.67}$$

Proof. The proof is trivial. We simply check if equality $AA^{-1} = I$ holds. So we simply multiply matrix A by its supposed inverse and check that we get the identity. We obtain

$$AA^{-1} = (A + uv^T)(A^{-1} - \frac{A^{-1}uv^T A^{-1}}{1 + v^T A^{-1} u}) =$$

$$= AA^{-1} - \frac{AA^{-1}uv^T A^{-1}}{1 + v^T A^{-1} u} + uv^T A^{-1} - \frac{uv^T A^{-1} uv^T A^{-1}}{1 + v^T u} =$$

$$= \; I - \frac{uv^TA^{-1}}{1+v^TA^{-1}u} + uv^TA^{-1} - \frac{uv^TA^{-1}\,uv^TA^{-1}}{1+v^TA^{-1}u} =$$

$$= \; I + uv^TA^{-1} - \frac{u\,(1+v^TA^{-1}u)\,v^TA^{-1}}{1+v^TA^{-1}u} =$$

$$= \; I + uv^TA^{-1} - uv^TA^{-1} \; = \; I.$$

For the modified system of linear equations $(A + uv^T)x = b$, the Sherman-Morrison formula gives the solution

$$\tilde{x} \; = \; (A+uv^T)^{-1}\,b \; = \; A^{-1}b - \frac{A^{-1}uv^TA^{-1}}{1+v^TA^{-1}u}\,b. \qquad (12.68)$$

Now, we can solve system $Az = u$ for known vector u and unknown z, and obtain $z = A^{-1}u$. If the solution $x = A^{-1}b$ for the original system $Ax = b$ is known, then from (12.68) we obtain

$$\tilde{x} \; = \; A^{-1}b - \frac{(A^{-1}u)\,v^T(A^{-1}b)}{1+v^T(A^{-1}u)} \; = \; x - \frac{z\,v^Tx}{1+v^Tz}\,. \qquad (12.69)$$

The following statement generalizes the Sherman-Morrison formula to a *rank-k* modification of matrix A.

LEMMA 12.2 (SHERMAN-MORRISON-WOODBURY FORMULA) *If the given matrices* $U = ||u_{ij}||_{m\times k}$ *and* $V = ||v_{ij}||_{m\times k}$ *are such that*

$$I + V^TA^{-1}U \neq 0,$$

then for inverse matrix $\tilde{A}^{-1}$ *of* $\tilde{A} = A + UV^T$ *we have*

$$\tilde{A}^{-1} \; = \; A^{-1} - \frac{A^{-1}UV^TA^{-1}}{I+V^TA^{-1}U}\,.$$

Let us apply the Sherman-Morrison formula to the basis update (12.43). We substitute column-vectors u and v in formula (12.67) with $A_k - A_r$ and e_j, respectively. Thus, in accordance with (12.67) we obtain

$$\tilde{B}^{-1} \;\; = \;\; B^{-1} - \frac{B^{-1}(A_k - A_r)\,e_j^TB^{-1}}{1+e_j^TB^{-1}(A_k - A_r)} \;\; =$$

$$= \;\; B^{-1} - \frac{(B^{-1}A_k - B^{-1}A_r)\,e_j^TB^{-1}}{1+e_j^TB^{-1}A_k - e_j^TB^{-1}A_r)}. \qquad (12.70)$$

Since vector A_r is in the current basis B in position j, we have $B^{-1}A_r = e_j$. Observe that $e_j e_j^T = e_j^T e_j = 1$. So we can continue (12.70) in the following way:

$$\begin{aligned}\tilde{B}^{-1} &= \ldots = \\ &= B^{-1} - \frac{(B^{-1}A_k - e_j)\, e_j^T B^{-1}}{1 + e_j^T B^{-1} A_k - 1} = \\ &= (I - \frac{(B^{-1}A_k - e_j)\, e_j^T}{e_j^T B^{-1} A_k})\, B^{-1} = \\ &= (I - \frac{B^{-1}A_k\, e_j^T - 1}{e_j^T\, B^{-1} A_k})\, B^{-1}.\end{aligned}$$

If solution $x = B^{-1}b$ for the current basis B is known, then the solution $\tilde{x}$ of updated system $\tilde{B}x = b$ is

$$\tilde{x} = (I - \frac{B^{-1}A_k\, e_j^T - 1}{e_j^T\, B^{-1} A_k})\, x\,. \tag{12.71}$$

Recall that $B^{-1}A_k$ gives coefficients x_{ik} of the decomposition of non-basis vector A_k in the current basis B, i.e.

$$\sum_{i=1}^{m} A_{s_i} x_{ik} = A_k, \quad k \in J_N,$$

so $B^{-1}A_k = (x_{1k}, x_{2k}, \ldots, x_{mk})^T$. Hence, the product $e_j^T\, B^{-1}A_k$ in the denominator of (12.71) is a scalar. Indeed,

$$e_j^T\, B^{-1}A_k = (0,0,\ldots,0,\underbrace{1,}_{j}0,\ldots,0)\begin{pmatrix} x_{1k} \\ x_{2k} \\ \vdots \\ x_{mk}\end{pmatrix} = x_{jk}.$$

Product $B^{-1}A_k\, e_j^T$ in the numerator of (12.71) gives us the following matrix

$$B^{-1}A_k\, e_j^T = \begin{pmatrix} x_{1k} \\ x_{2k} \\ \vdots \\ x_{mk}\end{pmatrix}(0,0,\ldots,0,\underbrace{1,}_{j}0,\ldots,0) =$$

$$= \begin{pmatrix} 0 & \dots & 0 & x_{1k} & 0 & \dots & 0 \\ 0 & \dots & 0 & x_{2k} & 0 & \dots & 0 \\ \vdots & & & \vdots & & & \vdots \\ 0 & \dots & 0 & x_{mk} & 0 & \dots & 0 \end{pmatrix}$$

with vector $(x_{1k}, x_{2k}, \dots, x_{mk})^T$ in column j. So the final expression (12.71) is utterly trivial even if it looks a little bit complicated.

To illustrate this method, we consider linear system

$$\begin{pmatrix} 1 & 1 & 0 \\ 3 & 1 & 2 \\ 4 & 2 & 1 \end{pmatrix} \begin{pmatrix} x_1 \\ x_2 \\ x_3 \end{pmatrix} = \begin{pmatrix} 15 \\ 20 \\ 25 \end{pmatrix},$$

where matrix A has the following LU-decomposition

$$A = \begin{pmatrix} 1 & 1 & 0 \\ 3 & 1 & 2 \\ 4 & 2 & 1 \end{pmatrix} = \begin{pmatrix} 1 & 0 & 0 \\ 3 & 1 & 0 \\ 4 & 1 & 1 \end{pmatrix} \begin{pmatrix} 1 & 1 & 0 \\ 0 & -2 & 2 \\ 0 & 0 & -1 \end{pmatrix} = LU.$$

This system has solution $x = (-7.5,\ 22.5,\ 10.0)^T$. Let us "rank-one" update matrix A in such a way that only $a_{32} = 2$ is changed from 2 to 6, that is $\tilde{a}_{32} = 6$. In accordance with (12.44) we write

$$\tilde{A} = A + 4\, e_3\, e_2^T = A + u\, v^T,$$

where the appropriate vectors u and v are as follows

$$u = e_3 = (0, 0, 1)^T, \quad v = 4\, e_2 = (0, 4, 0)^T.$$

So, the matrix of the modified system is

$$\begin{aligned} \tilde{A} = A + u\, v^T &= \begin{pmatrix} 1 & 1 & 0 \\ 3 & 1 & 2 \\ 4 & 2 & 1 \end{pmatrix} + \begin{pmatrix} 0 \\ 0 \\ 1 \end{pmatrix} (0, 4, 0) = \\ &= \begin{pmatrix} 1 & 1 & 0 \\ 3 & 1 & 2 \\ 4 & 2 & 1 \end{pmatrix} + \begin{pmatrix} 0 & 0 & 0 \\ 0 & 0 & 0 \\ 0 & 4 & 0 \end{pmatrix} = \begin{pmatrix} 1 & 1 & 0 \\ 3 & 1 & 2 \\ 4 & 6 & 1 \end{pmatrix}. \end{aligned}$$

If using known LU-factorization of matrix A we solve now system $Az = u$, we obtain $z = (1, -1, -1)^T$. Then, in accordance with (12.69) we have

$$\tilde{x} = x - \frac{z\, v^T x}{1 + v^T z} = \begin{pmatrix} -7.5 \\ 22.5 \\ 10.0 \end{pmatrix} - \frac{\begin{pmatrix} 1 \\ -1 \\ -1 \end{pmatrix} (0, 4, 0) \begin{pmatrix} -7.5 \\ 22.5 \\ 10.0 \end{pmatrix}}{1 + (0, 4, 0) \begin{pmatrix} 1 \\ -1 \\ -1 \end{pmatrix}} =$$

$$= \begin{pmatrix} -7.5 \\ 22.5 \\ 10.0 \end{pmatrix} - \frac{90}{-3} \begin{pmatrix} 1 \\ -1 \\ -1 \end{pmatrix} = \begin{pmatrix} 22.5 \\ -7.5 \\ -20.0 \end{pmatrix}.$$

We can check

$$\tilde{A}\tilde{x} = \begin{pmatrix} 1 & 1 & 0 \\ 3 & 1 & 2 \\ 4 & 6 & 1 \end{pmatrix} \begin{pmatrix} 22.5 \\ -7.5 \\ -20.0 \end{pmatrix} = \begin{pmatrix} 15 \\ 20 \\ 25 \end{pmatrix}.$$

Thus, we have found a solution of the modified system without having to re-factor the modified matrix.

4. Iterative Refinement of a Solution

For large systems of linear equations it is not too easy to obtain solutions with acceptable precision, or even with precision comparable with the computer's limit. The matter is that in the direct methods of linear algebra, roundoff errors typically have accumulative effect and hence, the closer is the matrix to be singular the larger errors we have. So we can easily lose several (sometimes two, three or even more) significant digits in the "solution" obtained. To restore full machine precision we can use the method called*iterative refinement* of the solution designed especially to improve the accuracy of a computed solution.

Let us suppose that we have computed an approximate solution$x^{(0)}$ to the linear system $Ax = b$, say using some form of LU-factorization. Let x denote the (unknown) exact solution of the system. Then we can compute the so called *residual* $r^{(0)}$

$$r^{(0)} = b - Ax^{(0)}.$$

If we multiply this formula from the left side by inverseA^{-1}, we obtain

$$A^{-1}r^{(0)} = A^{-1}b - A^{-1}Ax^{(0)}.$$

or

$$A^{-1}r^{(0)} = x - x^{(0)}.$$

Then we can rewrite the last in the following form

$$A\,\Delta x^{(0)} = r^{(0)}, \qquad (12.72)$$

where $\Delta x^{(0)}$ denotes correction $x - x^{(0)}$. Having solved system (12.72) and obtained its solution $\Delta x^{(0)}$, we then can take

$$x^{(1)} = x^{(0)} + \Delta x^{(0)}$$

as a new "improved" approximate solution, since

$$Ax^{(1)} = A(x^{(0)} + \Delta x^{(0)}) = Ax^{(0)} + A\Delta x^{(0)} = (b - r^{(0)}) + r^{(0)} = b.$$

So we can define an iterative process with steps

$$r^{(k)} = b - Ax^{(k)}, \qquad (12.73)$$

$$A\,\Delta x^{(k)} = r^{(k)}, \qquad (12.74)$$

$$x^{(k+1)} = x^{(k)} + \Delta x^{(k)}, \qquad (12.75)$$

for $k = 1, 2, \ldots$.

This iterative process may be repeated as many times as necessary and usually produces an approximate solution with a residual as small as we need.

Unfortunately, this iterative process is quite expensive, since it requires to compute the residual $r^{(k)}$ and to solve the subsequent system (12.74) in each kth iteration. Moreover, to produce precise corrections $\Delta x^{(k)}$ the calculations of residuals must be performed with higher machine precision that leads to extra computational costs. If we do not mind these extra expenses, iterative improvement is highly recommended ([147]). In many descriptions of this iterative method it is stressed that to reduce these extra computational costs only residuals may be computed with a higher precision and the rest of computations may be performed with standard precision.

Before closing this section, we just note that another analytical approach to improving approximate solutions is based on the fact that LU-decomposition used to solve subsequent systems (12.74) is itself not exact. Detailed information on this topic may be found in [147].

5. Sparse matrices

The aim of this section is to briefly overview data structures suitable for holding sparse matrices and vectors. The interest of considering sparse structures has many reasons: one of the most important of them is that the information can be stored in a much more compact way; second, because of avoiding redundant numerical operations involving zero entries the performance of computations may be improved dramatically. Moreover, this interest is not an optional one - when a sparse matrix of dimension $m \times m$ in a very large-scale programming problem contains only a few times m nonzero elements, it is often physically impossible to allocate in computer storage the room necessary for all m^2 elements. Several storage methods and corresponding structures exist for representation and performing manipulation with sparse matrices and vectors. But there is no one "best" method or data structure; most practical computer codes use different storage methods and structures at different stages since the choice usually depends on the nature of the manipulations, properties of the matrix, computer architecture, and programming languages used for implementation. There are many methods for storing the data (see for example [59] and [157]). Here we will discuss the storage of *Sparse Vectors, Coordinate Scheme, Collection of Sparse Vectors,* and *Linked List.*

5.1 Sparse Vectors

Independently of its sparseness any vector may be stored in a full-size storage. On the one hand, it is often used because of simplicity and speed which it may be manipulated with. On the other hand, in the case of very large size and high sparseness such method of storage is rather wasteful.

To use the memory of the computer more effectively, we should use such an approach that allows to store only non-zero entries of the vector and, at the same time, provides some special mechanism necessary to manipulate entries of the vector separately as well as the vector as a whole. Usually, non-zero elements of sparse vectors are stored in pairs $(Value,\ Position)$ organized in two arrays (each of the length of at least the number k of non-zero entries to be stored) as is shown in Table 12.1. Such a scheme of storing sparse vectors

$Entry\,No.$	1	2	$\ldots$	k
$Position$	j_1	j_2	$\ldots$	j_k
$Value$	a_{j_1}	a_{j_2}	$\ldots$	a_{j_k}

Table 12.1. Sparse vector storage.

is usually referred to as a *packed* form. It is very effective from the point of view of memory usage, but is not suitable for manipulation. This is why for performing such operations as addition and multiplication it is convenient first to *unpack* the vector from packed form to full-size and then perform the necessary operations. After completing all operations the result vector should be transformed back to packed form. Notice that a sequence of such operations may be performed efficiently if just one full-size vector for temporary storage is used. Moreover, it is very important in manipulating sparse data structures to avoid complete scans of full-size vectors.

5.2 Coordinate Scheme

One of the most widely and often used convenient ways to store a sparse matrix is to use a set of (un)ordered triplets $(Row,\ Column,\ Value)$. The following example best illustrates this storage scheme. Let us consider a sparse matrix of order 5×5 with 11 non-zero entries, for example

$$A = \begin{pmatrix} 1.0 & 0.6 & & & 0.2 \\ & 0.1 & & 1.2 & \\ 0.3 & & & & 0.7 \\ 3.3 & & 1.8 & & \\ & & 0.4 & 0.8 & \end{pmatrix} \tag{12.76}$$

In the case of the *Coordinate scheme for storing sparse matrices*, three arrays are used: two integer arrays for row and column indices, and a real array containing values of non-zero entries. For our matrix A we have the representation given in Table 12.2, where each non-zero entry of matrix A is represented by a triplet

Entry No.	1	2	3	4	5	6	7	8	9	10	11
Row	1	3	4	1	2	4	5	2	5	1	3
Col	1	1	1	2	2	3	3	4	4	5	5
Value	1.0	0.3	3.3	0.6	0.1	1.8	0.4	1.2	0.8	0.2	0.7

Table 12.2. Coordinate scheme for storing sparse matrices.

and corresponds to a column in the Table. In our example all these triplets are ordered by columns, but no ordering is actually necessary for this method. Usually, such a storage scheme needs less memory than full storage if the density (calculated as the ratio between its non-zero entries and its total number of entries) of the matrix to be stored is less than ≈ 0.55. For example, if in the computer we use to store a square matrix of order $m = 1000$, the high precision real data is stored in a 10 byte length *"Extended"* type and indices are stored in a 4 byte length *"Integer"* type, then for full-size storage of the matrix we have to allocate $m \times m \times 10 = 1000 \times 1000 \times 10 = 10,000,000 \approx 10MB$ (*megabytes*) of memory. The memory requirements for different densities of the matrix in the case of the coordinate scheme are given in Table 12.3. The Table shows that for matrices with density under 0.55, the coordinate scheme becomes more preferable than the full-size storage method. The insertion and deletion of elements when using this scheme are easy to perform, while the direct

Density	0.05	0.15	0.25	0.35	0.45	0.55	0.65	0.75	0.85
$\approx MBytes$	0.9	2.7	4.5	6.3	8.1	9.9	11.7	13.5	15.30

Table 12.3. Memory requirement for coordinate scheme.

access of elements is relatively expensive since it requires sequential scanning of the triplets until the entry needed is found. To improve the performance of direct access D.E.Knuth [112] suggested using the "entry pointers" for rows and columns that point to the first non-zero elements in each row and column, respectively. For our example, it means two additional arrays NR ("Next non-zero element in the same Row") and NC ("Next non-zero element in the same Column"), which consist of 11 elements each. These two additional arrays allow us to find easily the next non-zero element in the column or row we are staying in during scanning triplets. For further improvement of the performance

Entry No.	1	2	3	4	5	6	7	8	9	10	11
NR	4	11	6	10	8	0	9	0	0	0	0
NC	2	3	0	5	0	7	0	9	0	11	0

Table 12.4. Additional "next non-zero" pointers NR and NC.

of direct access we need two more arrays that would assist us in finding entries to the columns or rows. So we have to introduce two arrays each of length $m = 5$, say JR and JC (see Table 12.5). Finally, we have seven arrays: $Value$, Row, Col, NR, NC have the same length of 11 elements each, and arrays JR and JC have only 5 elements each.

Obviously, the additional memory requirement for arrays NR, NC, JR and JC makes this scheme of storage more expensive and complicated. Even so, as shows Table 12.6, for matrices with density under $\approx$ 0.38 (see a matrix example on page 372), the coordinate scheme requires less memory than the full-size storage method.

To illustrate how the scheme works, let us suppose that in matrix A we have to access entries of column 2. First of all, from array JC for column 2 we get

Entry No.	1	2	3	4	5
JR	1	5	2	3	7
JC	1	4	6	8	10

Table 12.5. Additional "entry" pointers JR and JC.

Density	0.32	0.34	0.36	0.38	0.40
$\approx MBytes$	8.322	8.842	9.362	9.882	10.402

Table 12.6. Full memory requirement for coordinate scheme.

$JC[2] = 4$. It means that starting from position 4 of array $Value$ we obtain value $Value[4] = 0.6$, which is in row 1 ($Row[4] = 1$). Then element 4 of array NC gives us $NC[4] = 5$ that points to the "next non-zero element in the same column", so $Value[5] = 0.1$. This entry is located in row 2, since $Row[5] = 2$. Further, since $NC[5] = 0$ it means that there are no more non-zero elements in column 2.

Analogically, if we need all non-zero elements of row 3, we have to start from $JR[3] = 2$ which points to the first non-zero entry of row 3 $Value[2] = 0.3$. This element is located in column 1, since $Col[2] = 1$. Then element 2 of array NR gives position of the "next non-zero element in the same row", i.e $NR[2] = 11$. So from $Value[11]$ we obtain entry 0.7 which is located in column 5 ($Col[11] = 5$). Further, $NR[1]] = 0$ means that we have reached the end of row 3 and there are no non-zero entries in this row.

There are several improved modifications of this scheme, suggested by I.S.Duff [57], M.H.E.Larcombe [124], and W.C.Rheinboldt and C.K.Mesztenyi [150]. Detailed information on these schemes may be found in [146].

5.3 Collection of Sparse Vectors

With this storage scheme, sparse matrix A is stored as the concatenation of the sparse vectors representing its columns or rows. Depending on the organization of the storage (by columns or rows) this storage scheme is referred to as *Compressed Column Storage* (*CCS*) or *Compressed Row Storage* (*CRS*).

In the case of CCS, all sparse column-vectors are stored in the same *real* array sequentially after one another. The components of each vector may be ordered or not. For each non-zero entry we have to have an *integer* index of corresponding row the entry is located in. A second *integer* array gives the locations of the first entries in each column. So, for our example (12.76) we have to allocate in the computer memory the following three arrays: *real* array $Value$ of length 11 for non-zero elements of matrix A to be stored, *integer* array Row of length 11 for row indices associated with corresponding non-zero elements of matrix A, and array JN of length 5 for indices that point the first elements of columns in array $Value$ (see Table 12.7).

Entry No.	1	2	3	4	5	6	7	8	9	10	11
Row	1	3	4	1	2	4	5	2	5	1	3
Value	1.0	0.3	3.3	0.6	0.1	1.8	0.4	1.2	0.8	0.2	0.7

Entry No.	1	2	3	4	5
JN	1	4	6	8	10

Table 12.7. Collection of sparse vectors – CCS.

For instance, if we have to scan all non-zero entries of column 3, we have to start from entry 3 of array JN, i.e. $JN[3] = 6$, which points to the first element of column 3 needed inside array $Value$. Further, from $JN[4] = 8$ we know that column 4 begins in the entry $Value[8]$, the last means that all elements of array $Value$ from 6 to $JN[4] - 1 = 8 - 1 = 7$ belong to column 3. So, we have the first non-zero entry of column 3 in $Value[6] = 1.8$. Entry $Row[6] = 4$ gives us information which row the given entry is located in, i.e. row 4. Then entry $Value[7]$ gives us the next non-zero entry of column 3, its row is $Row[7] = 5$. There are no more non-zero elements in given column 3 since in $Value[8]$ the first non-zero entry of column 4 is stored.

Obviously, this method of storing sparse matrices is very effective in memory usage. The memory requirements for different densities of matrix for such representation of the sparse matrix described on page 372 are given in Table 12.8.

This Table shows that for matrices with density under0.70, this scheme becomes more preferable than the full-size storage method.

Density	0.55	0.60	0.65	0.70	0.75	0.80	0.85
$\approx MBytes$	7.704	8.404	9.104	9.804	10.504	11.204	11.904

Table 12.8. Memory requirement for collection of sparse vectors.

This method has several serious disadvantages: first, it does not provide any data structure for direct access to rows of the matrix; second, the difficulty of inserting new entries.

The first disadvantages may be easily avoided if we transform this storage scheme into the form of CRS (see see Table 12.9). In this method, for each

Entry No.	1	2	3	4	5	6	7	8	9	10	11
Column	1	2	5	2	4	1	5	1	3	3	4
Value	1.0	0.6	0.2	0.1	1.2	0.3	0.7	3.3	1.8	0.4	0.8

Entry No.	1	2	3	4	5
IN	1	4	6	8	10

Table 12.9. Collection of sparse vectors – CRS.

non-zero entry we store an *integer* index of corresponding column the entry is located in. A second *integer* array gives the locations of the first entries in each row.

Some modifications of this method use one more array that contains the number of entries in each column (for method CCS) or row (for method CRS). For more information on this method see [87], [146], [148], etc.[4]

[4]Detailed information on the LU-factorization based on CCS and CRS may be found on Web site http://www.netlib.org/linalg/htmLtemplates/node89.html

Closing the discussion of "collection of sparse vectors" we illustrate the usage of this scheme for a matrix-vector product based on CRS representation. The matrix-product $y = Ax$ using CRS representation of sparse matrix $A = ||a_{ij}||_{m\times n}$ can be expressed in the usual way:

$$y_i = \sum_{j=1}^{n} a_{ij}\, x_j\,, \quad j = 1, 2, \ldots, n.$$

If vector x is stored as a conventional full-length array of size$1 \times n$, the algorithm may be implemented as shown in Figure 12.3.

CRS Sparse Matrix-Vector Product

```
For i := 1 To m Do                                {Loop over rows}
   Begin
   y[i] := 0;
   For j := IN[i] To IN[i + 1] - 1 Do             {Loop over columns}
      y[i] := y[i] + Value[j] * x[j];
   End
```

Figure 12.3. Algorithm – CRS Sparse Matrix-Vector Product.

5.4 The Linked List

Another alternative scheme which is used widely for storing sparse matrices is the so-called *linked list.* Its peculiarity is that we have to define a pointer (named *head*) to the first non-zero entry of each column (if the matrix is stored by columns). Each entry stored is associated with a pointer (named*link*) that points to the next non-zero entry in the same column or contains a null-pointer (named *nil*) if there are no more non-zero entries in the given column. Entries may be ordered or not. So, if the matrix is stored by columns, we have as many head pointers as there are columns in the matrix. Each non-zero entry is composed of two parts: the value of the entry itself, and the row index the entry is located in.

For our example-matrix presented in (12.76), we have the representation given in Table 12.10.

For instance, to retrieve the elements of column3, we begin to read $Head[3] = 6$. Then $Row[6] = 4$ gives the row index, the entry value is $Value[6] = 1.8$ and the pointer $Link[6] = 7$ gives the position of the next entry in the data structure.

Entry No.	1	2	3	4	5
Head	1	4	6	8	10

Entry No.	1	2	3	4	5	6	7	8	9	10	11
Row	1	3	4	1	2	4	5	2	5	1	3
Value	1.0	0.3	3.3	0.6	0.1	1.8	0.4	1.2	0.8	0.2	0.7
Link	2	3	*Nil*	5	*Nil*	7	*Nil*	9	*Nil*	11	*Nil*

Table 12.10. Linked list.

So, we obtain $Value[7] = 0.4$ in row 5 ($Row[7] = 5$). Since $Link[7] = Nil$, it indicates that $Value[7] = 0.4$ is the last non-zero entry in the column.

As is shown in Table 12.11, calculated for the matrix described on page 372, such representation becomes more preferable than the full-size storage method if the matrix to be stored has density under0.55.

Density	0.35	0.40	0.45	0.50	0.55	0.60	0.65
$\approx MBytes$	6.304	7.204	8.104	9.004	9.904	10.804	11.704

Table 12.11. Memory requirement for linked list.

The obvious advantage of this method is the ease with which we can find all entries inside an unordered column. This data structure is close to the method "Collection of Sparse Vectors", but does not require contiguous storing entries inside the columns (rows). Furthermore, to insert or delete elements we have simply to update the pointers to take care of the modification. In practice it is often necessary to reorganize this representation from a column-based form to a row-based one and conversely. Since such a transformation is quite complicated, sometimes it is more suitable to use simultaneously with column-oriented structures also the data structures which enable row-oriented manipulations. Obviously, the choice depends on the nature of manipulations we have to perform.

6. Discussion Questions and Exercises

12.1 For the given matrix

$$A = \begin{pmatrix} 3 & -5 & 4 \\ -8 & 4 & 1 \\ 5 & -6 & 2 \end{pmatrix}$$

find its LU-decomposition using Crout's method.

12.2 Using Gaussian elimination find the LU-decomposition of coefficient matrix A and then solve the system of equations $Ax = b$ given by

$$\begin{pmatrix} 3 & -5 & 4 \\ -8 & 4 & 1 \\ 5 & -6 & 2 \end{pmatrix} \begin{pmatrix} x_1 \\ x_2 \\ x_3 \end{pmatrix} = \begin{pmatrix} 2 \\ -3 \\ 1 \end{pmatrix}.$$

12.3 Find for certain permutations matrix P_r the LU decomposition of $P_r A$ if

$$\text{a) } A = \begin{pmatrix} 0 & 2 \\ 3 & 5 \end{pmatrix}, \quad \text{b) } A = \begin{pmatrix} 0 & 7 \\ 5 & 3 \end{pmatrix}, \quad \text{c) } A = \begin{pmatrix} 0 & 5 & 7 \\ 2 & 3 & 3 \\ 6 & 9 & 8 \end{pmatrix}.$$

12.4 For matrix A given in exercise 12.1 update its LU-factorization using the Bartels-Golub method, if column-vector $A_2 = (-5, 4, 6)^T$ in matrix A is replaced with column-vector $A_4 = (5, -4, -6)^T$.

12.5 For the given symmetric positive definite matrix

$$A = \begin{pmatrix} 1 & 1 & 4 & -1 \\ 1 & 5 & 0 & -1 \\ 4 & 0 & 21 & -4 \\ -1 & -1 & -4 & 10 \end{pmatrix}$$

find its Cholesky decomposition using method (12.64)-(12.65).

12.6 Find QR factorization using Householder transformations for matrix

$$A = \begin{pmatrix} 3 & 3 & 0 \\ 3 & 5 & 0 \\ 0 & 0 & 6 \end{pmatrix}.$$

12.7 Using the Sherman-Morrison formula recalculate solution for system

$$Ax = \begin{pmatrix} 3 & -5 & 4 \\ -8 & 4 & 1 \\ 5 & -6 & 2 \end{pmatrix} \begin{pmatrix} x_1 \\ x_2 \\ x_3 \end{pmatrix} = \begin{pmatrix} 60 \\ 32 \\ 30 \end{pmatrix} = b,$$

if in matrix A column-vector $A_2 = (-5, 4, 6)^T$ is replaced with column-vector $A_4 = (5, -4, -6)^T$.

12.8 For a given sparse matrix of order 6 construct its "Coordinate Scheme", if

$$A = \begin{pmatrix} 3 & & & & & 1 \\ & 4 & & 3 & 2 & \\ & & 8 & & & 2 \\ 2 & & & 9 & & \\ & 6 & & & 4 & \\ 5 & & 9 & & & 3 \end{pmatrix}.$$

12.9 For the sparse matrix given in the previous exercise construct its representations as a "Collection of Sparse Vectors" in format CCS and then in format CRS.

12.10 For the sparse matrix given in the previous exercise construct its representations as a "Linked List".

Chapter 13

THE WINGULF PACKAGE

WinGULF - is a **G**eneral, **U**ser-friendly **L**inear and linear-**F**ractional programming package for **Win**dows.

Originally, the package is a descendant of the linear programming package GULP developed by David J. Pannell[1] in the early 90's for MS-DOS computers, see [182]. In 1993 the author of this book, in cooperation with David J. Pannell, further developed the package and built into the package its linear-fractional facilities. Its version for MS-DOS was tested in EJOR by the *OR software group* and the results of the tests appeared in [183] in 1993. Later the author independently developed the WinGULF package, a version of GULF for Windows. Since 1998 the package has been under constant development, and in 2001 the package got a built-in *branch and bound* engine, so it is able now to solve pure and mixed integer problems with up to 500 unknown variables (with a maximum of 100 integers) and 500 main constraints. The Special Student Edition of the package is free of charge and may be downloaded from the Web-page of the author: *http:\\www.math.klte.hu\ ~bajalinov*.

The aim of the present chapter is to describe the package WinGULF and to show how people could use the package if they needed to solve LP or LFP problems with/without integer unknown variables and then investigate and utilize the solutions obtained.

[1]The University of Western Australia, School of Agriculture, Nedlands W.A. 6009 Australia

1. Program Overview and Background

WinGULF is a simple to use but powerful, menu driven linear programming and linear-fractional programming package for IBM compatible personal computers operating under MS Windows 95 or higher.

What distinguishes WinGULF from other available LP programs is first of all its ability to solve LFP problems as well as LP ones and, second, its ease and convenience of use, guards against mistaken input, informative error messages, a help command, ease of data entry, speed of calculation and the range of other built-in options. Data can be entered in a spreadsheet style editor within WinGULF or from a conventional[2] format text file.

The package is capable of solving any solvable LP and LFP problems with a non-empty feasible set and if the objective function is not unbounded on the feasible set.

The maximum problem size solvable by WinGULF is 500 columns by 500 rows with a maximum of 100 integer variables.

There is no minimum disk size required as WinGULF takes up only about 1Mb of disk space. WinGULF is not copy protected.

To solve an LP or LFP problem the package uses the well-known simplex algorithm [51], [52], [69], [131], [132], or Chapter 4 in this book. The user has the choice between two pivoting rules: the simple *steepest ascent* and the *highest step* pivot selection (see Chapter 4, Section 8). The second rule involves longer iterations but may result in fewer steps. The package includes an ideal feature for those who are learning about the simplex method: the facility to run the simplex procedure in *step-by-step mode*, and manually choose pivot columns and view the matrix after each iteration.

When you are ready to solve your continuous (without integer variables) problem (see Figure 13.1) click on the *Run* button (see Figure 13.2) or choose the *Run→Run* menu item, hot key *F8*. The Simplex procedure will start in *automatic mode* and it will result in the optimal solution found. During calculations the package displays the status of the current solution: phase number, iteration, objective value (see Figure 13.3). In case you need to inspect the Simplex process you have to click on the *Run by Step* button (see Figure 13.2) or choose the *Run→By Step* menu item, hot key *F9*. The Simplex procedure will start the *Step-by-Step* mode and you will see an initial Simplex tableau like the one shown in Figure 13.4. When running WinGULF in this mode the bottom

[2]MPS is an abbreviation of the Mathematical Programming System format commonly used by the Operations Research community to share mathematical programming problems. It is a text format, so one of the reasons for using it is the ability to port LP or IP problems from one computer to another.

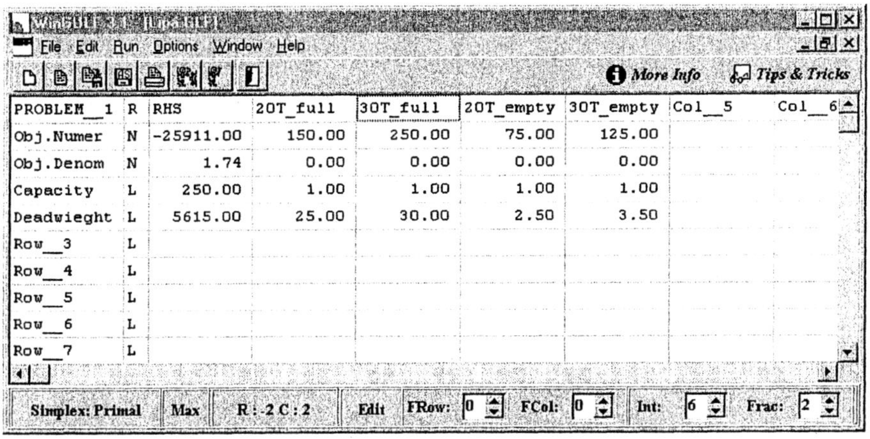

Figure 13.1. WinGULF – A continuous LFP problem.

Figure 13.2. WinGULF – Main functional buttons.

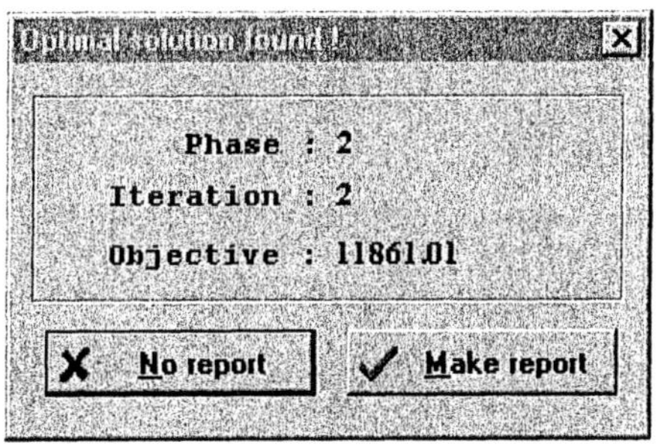

Figure 13.3. WinGULF – Status window.

row of its window contains the following controls (see Figure 13.4): *Cancel* button (it interrupts the procedure and returns control to the Editor grid), the *Iterate* <*K*> button (it performs <*K*>th Simplex iteration), the *User selected pivot* combo box (you can choose an alternative pivot column here). Also, the bottom row displays the name of the pivot column chosen by the algorithm automatically in accordance with the pivot rule selected in the*Defaults* dialog box (*Options→Defaults* menu item, *Methods* page). The dialog box *Defaults, Methods* page is shown in Figure 13.5. At the moment, only the Simplex method is

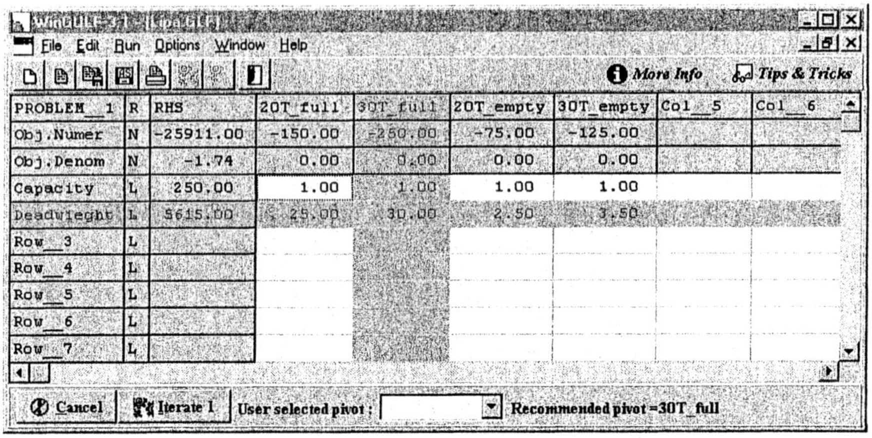

Figure 13.4. WinGULF – Step-by-Step mode.

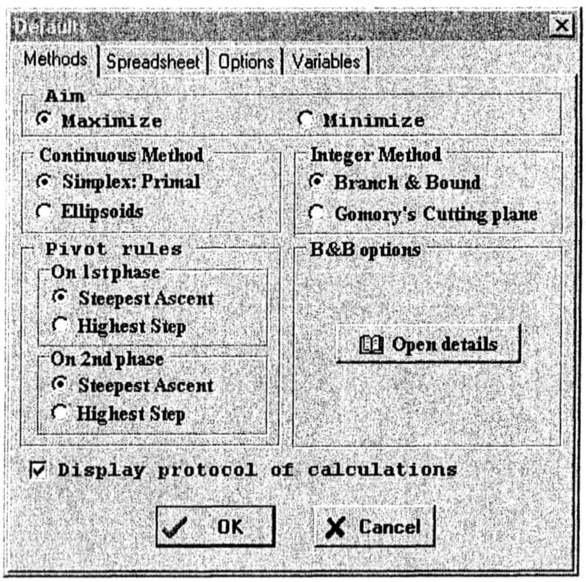

Figure 13.5. WinGULF – Defaults, Methods page.

implemented for continuous problems, and only the Branch-and-Bound method is available for integer problems.

2. The Editor

WinGULF is centered around a spreadsheet style editor which is used to enter a new problem or edit an existing problem. It operates similarly to an electronic spreadsheet program, such as Lotus 1-2-3[3], Quattro Pro[4] or Excel[5]. Coefficients of the problem matrix can be entered by moving around the spreadsheet and typing in values where required. Values may be entered by direct typing in the cell selected, or via a built-in pop-up calculator which appears if you right-click on the cell (see Figure 13.6).

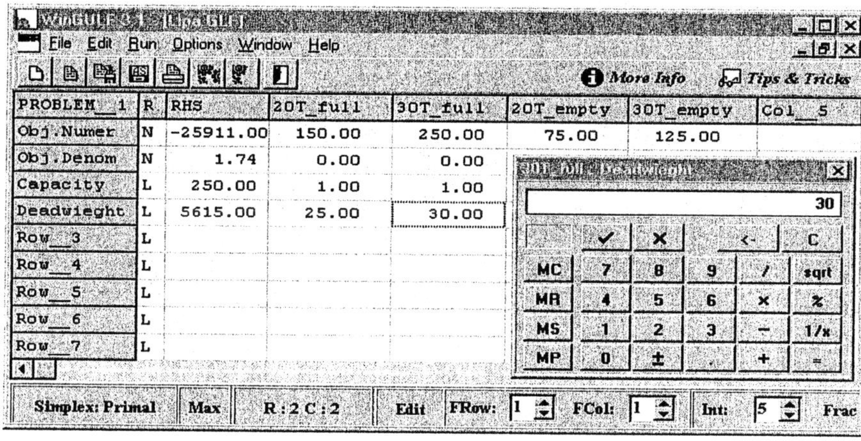

Figure 13.6. WinGULF – Built-in calculator.

At the stage of editing, the screen will show either the data from the file if it was found or a blank screen like that in Figure 13.7.

The upper-left position (available for editing if the spin buttons *FCol* and *FRow* for fixing columns and rows, respectively, are set to value 0, see Figure 13.1) is reserved for the problem name, which you may modify at will, as well as any of the spreadsheet positions. The number 1.00 in the "RHS" column of the "Obj.Denom" row and 0.00 in the other columns of the row (zeros are blanked) are the default values of the objective function denominator's constant term and coefficients respectively. If you retain these default values, WinGULF solves a standard LP problem using the objective function coefficients in the "Obj.Numer" row. To solve an LFP problem, the "RHS" value of the "Obj.Denom" row must be changed to a value other than 1.00 and/or other

[3]Lotus 1-2-3 is a registered trademark of Lotus Corporation
[4]Quattro Pro is a registered trademark of Borland Int.
[5]Excel is a registered trademark of Microsoft Ltd.

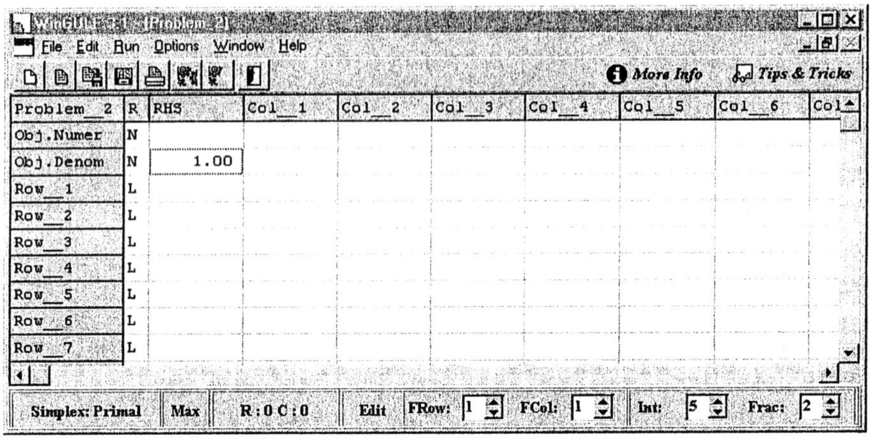

Figure 13.7. WinGULF – A new problem.

coefficients must be changed to values other than zero. When editing an LFP problem, coefficients of the problem are associated with the cells of the grid as follows:

$$\begin{array}{rcl} p_0 & \rightarrow & \text{cell: } Obj.Numer : RHS, \\ p_1, \ldots, p_n & \rightarrow & \text{cells from: } Obj.Numer : Col1 \text{ to: } Obj.Numer : ColN, \\ d_0 & \rightarrow & \text{cell: } Obj.Denom : RHS, \\ d_1, \ldots, d_n & \rightarrow & \text{cells from: } Obj.Denom : Col1 \text{ to: } Obj.Denom : ColN, \\ b_i & \rightarrow & \text{cells: } Rowi : RHS, \\ a_{ij} & \rightarrow & \text{cells: } Rowi : Colj. \end{array}$$

The leftmost column and the top row of the grid are reserved for row names and column names, respectively (see e.g. Figure 13.1).

If the problem is large, there may also be coefficients not displayed on the screen. Using corresponding scroll bars you will be able to move to the cells containing these coefficients to look at and, if necessary, change them. The bottom line in WinGULF's window is the status line. It tells you which method has been set as the current one, the aim of optimization, in which row and column the cursor is currently positioned, the mode of editing, two spin buttons for fixing rows and columns, and two spin buttons for formatting the numerical values displayed.

You can customize the spreadsheet of the Editor using the*Spreadsheet* page of the *Defaults* dialog box (see Figure 13.8).

If the problem has been solved and you choose the*Make Report* button (see Figure 13.3), the package generates a*Report* (discussed later in Section 3 and

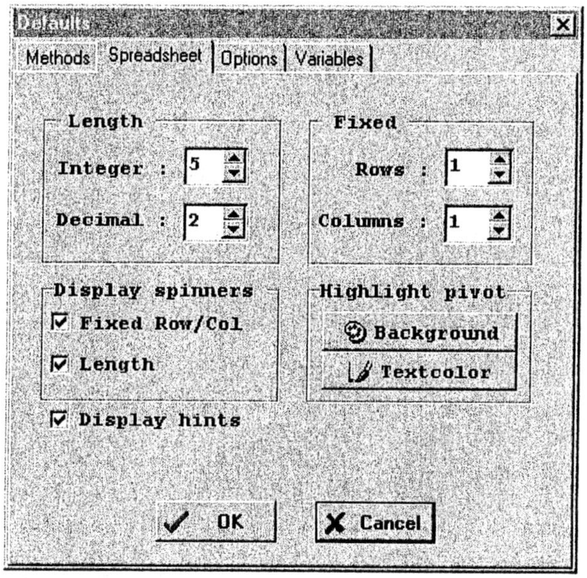

Figure 13.8. WinGULF – Defaults, the Spreadsheet page.

Section 4) which may consist of different parts depending on selected options shown in Figure13.9.

3. Problems with Continuous Variables

This section deals with continuous LP and LFP problems, i.e. such LP and LFP problems which do not include any integrality requirements for unknown variables.

3.1 Input and Main Options

As mentioned in the previous section, the package allows you to enter a new problem or edit an existing problem. Depending on your choice you have to use the corresponding button shown in Figure 13.2 or the appropriate menu item (*File→New* or *File→Open*) in the pull-down *File* menu list. To set the problem as a continuous one you have to choose the option*All variables are continuous* in the *Defaults* dialog box, the *Variables* page, as shown in Figure 13.10. After the problem has been entered and/or modified it may be saved as an MPS text file and/or printed on a printer.

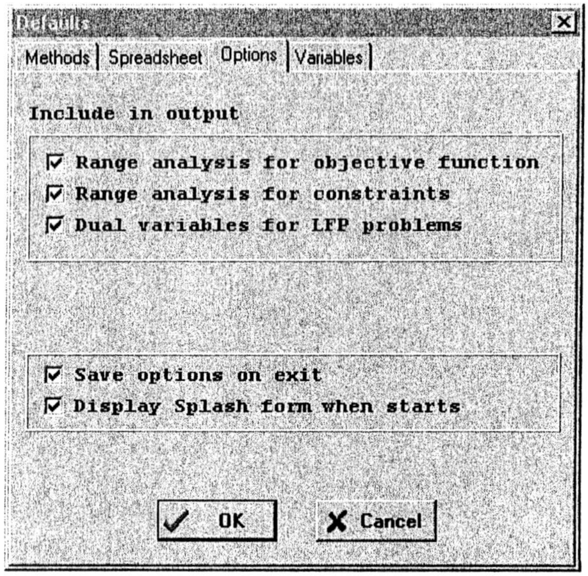

Figure 13.9. WinGULF – Defaults, the Options page.

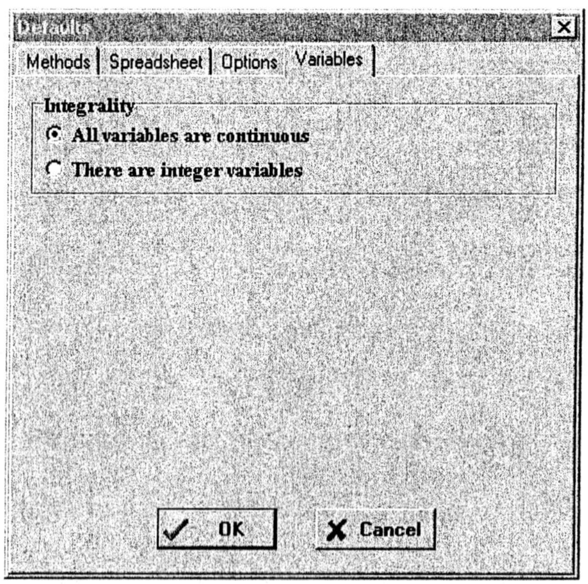

Figure 13.10. WinGULF – Defaults, the Variables page.

3.2 Output

If the problem has been solved the *Status window* dialog box will appear (see Figure 13.3). If you click on the *Make Report* button, the package generates an output report on the solution obtained (shown in Figure 13.11). The report

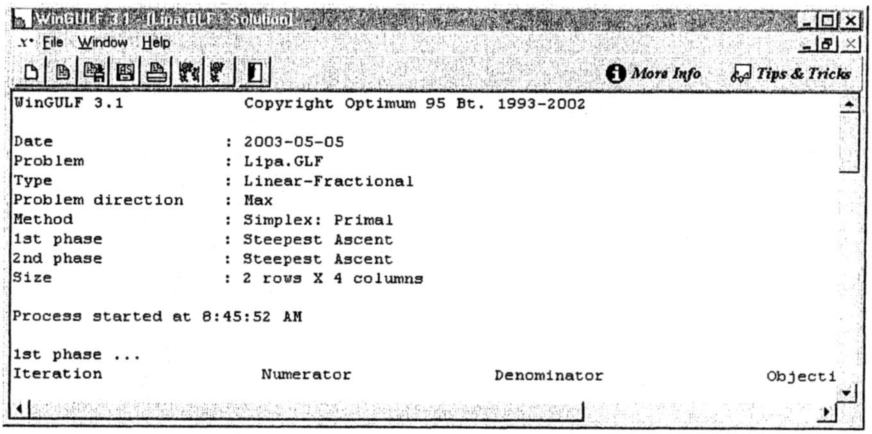

Figure 13.11. WinGULF – Continuous problem, report.

generated can be printed on a printer and/or can be saved to a text file on disk (default extension is '.SOL') for later viewing (see Figure 13.12).

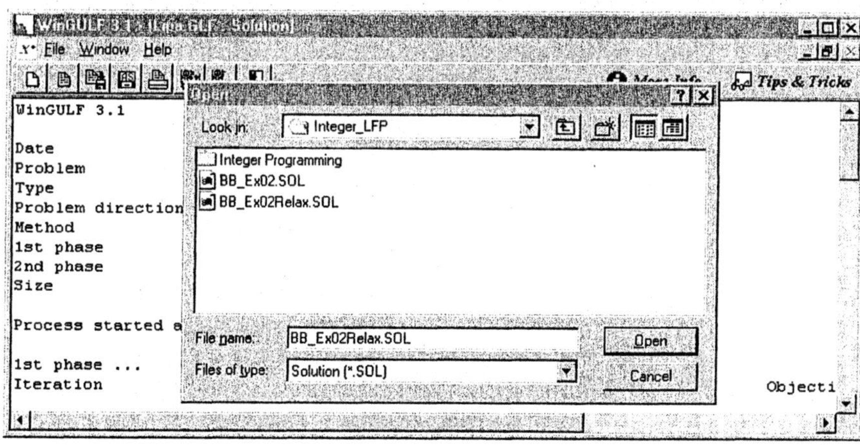

Figure 13.12. WinGULF – Opening the solution file for viewing.

After generating (or opening) a report, WinGULF displays it in a text window. Output (the report generated by WinGULF) includes some statistics on the problem (date, name, type, aim, method(s) used, etc.), levels (optimal values for objective function and unknown variables), slacks, shadow (i.e. reduced) costs, shadow prices and range analysis, each of which can optionally be suppressed.

Standard data format is used, so data can be exchanged with other LP packages on a mainframe or a microcomputer. It is possible to write your own data entry program which interfaces directly with WinGULF's solving algorithm, bypassing the data editor.

An output of the optimal solution for an LP problem consists of two tables: one each for columns and rows. For an LFP problem it consists of four tables: the same table as in the LP case but for the numerator and denominator separately. It is also possible to open a report for an optimal solution from within the editor, if it has been saved.

It is possible to select to print or not to print a range analysis for objective function and constraint limit coefficients (RHS column). It is also possible to choose whether or not to print the results for either columns (activities) or rows (constraints). Each of these options is set in the*Defaults* dialog box, the *Options* page (see Figure 13.9).

3.3 Interpreting an Optimal Solution

WinGULF provides a lot of information about the optimal solution found. There are several components of the optimal solution output. The first minor section gives the problem name, the problem direction (MAX or MIN), the number of iterations (irrelevant really, it just indicates how many feasible basic solutions WinGULF had to consider before finding the optimum) and the optimal objective function value (maximum profit or minimum cost) for an LP problem or optimal values of numerator, denominator and objective function (maximum efficiency or minimum specific cost) for an LFP problem. The other sections of the output are as follows.

3.3.1 Activity levels

These are the optimal levels of each activity, giving the maximum profit or the minimum cost for an LP problem or giving the maximum efficiency or the minimum specific cost for an LFP problem. In either case, the optimal levels are subject to the problem's constraints.

3.3.2 Activity status

WinGULF lists for each activity a single letter representing the activity's status. The letter could be any of **A**, **M**, **D** or **Z**. **A** for **A**ctive indicates that the activity forms a part of the optimal solution, i.e. it is selected at a level greater than zero. **Z** for **Z**ero indicates that the activity is not in the optimal basis. **M** for **M**ultiple solution indicates that although the activity is not in this optimal basis there is another equally profitable optimal solution which does include it. **D** for **D**egenerate indicates that the activity is included in the current basis but at zero level. This is due to there being a redundant constraint in the problem limiting the activity to zero level.

3.3.3 Shadow (or reduced) costs

For LP problems, the reduced cost of an activity has two possible interpretations:

1 if the activity is not part of the optimal solution, the shadow cost indicates by how much the objective function would be worsened by including one unit of the activity,

2 for all activities, the shadow cost indicates the amount by which the objective function value for an activity would have to be improved for that activity to be included in the optimal solution.

For LFP problems, there are three types of shadow (or reduced) costs: one of them for the numerator, the second one for the denominator of the objective function and the third one for the fractional objective function. All three types of reduced costs have interpretations similar to item (1) for LP. That is for variables (activities) not included in the optimal solution, they indicate the impact on the numerator or denominator or objective function, respectively, of including one unit of an activity which is not currently selected. In LFP problems reduced costs do not have an interpretation like the one described in the item (2) for LP: the amount by which the objective function value must improve to cause the activity to be selected. However, that information is often available in the range analysis output.

In LP, shadow (or reduced) costs are always positive, and indicate by how much the objective function would be worsened by selecting a sub-optimal activity. However, in LFP, shadow costs of the first two types can be positive or negative. A positive shadow cost indicates the amount by which the numerator or denominator (depending on which section of the output you are looking at) would be decreased by selection of a sub-optimal activity. A negative shadow cost indicates that the numerator or denominator would be increased by selecting the activity. This is true whether the objective is to maximize or minimize the objective function ratio. The shadow costs of the third type, i.e. shadow costs of objective functions, are always non-negative for maximization LFP problems, and non-positive for minimization LFP problems.

Because the over-riding objective of LFP is to optimize a ratio, it is possible for an activity which would improve, say, the numerator not to be included in the solution. This would occur if it also worsened the denominator by a proportionately greater amount.

If all activity shadow cost values for the numerator are non-negative (zero or greater), this indicates that the solution not only maximizes the ratio (e.g. rentability), but also the value of the numerator (e.g. profit).

Similarly, if all shadow costs for the denominator are zero or negative, the solution minimizes the denominator value (e.g. cost).

For both LP and LFP, activities which are already part of the optimal solution have a shadow cost of zero.

3.3.4 Constraint slacks

These give the difference between the use of a resource (or other constraint) and the constraint limit, i.e. the difference between the left-hand side and the right-hand side. For example, if a worker has 60 hours of labor available but the optimal solution uses only 40 hours, the labor constraint would have a slack of 20 hours. Alternatively, if the problem specified that there must be at least 500 units of an activity and the optimal solution includes 600 units, the slack for the constraint will be 100. Obviously, 'equal' constraints must have, by definition, no slack.

3.3.5 Shadow price

For constraints relating to limited physical resources (e.g. land) the shadow price may be interpreted as the value of acquiring one extra unit of that resource. For example, a shadow price of $100 for land in an LP problem indicates that an extra hectare of land would increase profit by $100 and so is worth leasing at any price less than $100. Analogously for an LFP problem a shadow price of $100 in the numerator output and a shadow price of $300 in the denominator output indicate that an extra hectare of land would increase profit by $100 and cost by $300. If $100/$300 = 0.33 is greater than the optimal value of the objective function then it is worth leasing land because it increases not just profit but rentability too.

For constraint types relating to personal preference or legal limits, the shadow price indicates the benefits from relaxing the preference or law by one unit. However, some constraints may be included in the problem solely to make it function correctly (e.g. in modelling marginal tax rates it is common to include a constraint per tax scale). For these constraints, the shadow price may have no interpretable meaning and care should be taken if assigning it one.

For 'less than' constraints in LP problems, shadow prices may be positive or, if the constraint is slack, zero. Positive shadow prices indicate that an increase in the constraint limit would increase profit.

'Greater than' constraints may have negative or zero shadow prices. Negatives indicate that a reduction in the constraint limit would increase profit and decrease cost.

In LFP problems shadow prices calculated separately for the numerator and the denominator may be positive as well as negative - independent of the constraint type. As in LP, shadow prices in the numerator output indicate that an increase in the constraint limit would increase or decrease profit subject to the sign of the shadow price. Shadow prices in denominator output have an analogous interpretation.

'Equal' constraints may have positive, negative or, very rarely, zero shadow prices. If an 'equal' constraint has a zero shadow price, it means that the constraint would have been met even if it had not been included in the matrix. Positive or negative shadow prices are interpreted as for 'less than' or 'greater than' constraints, i.e. the shadow price in LP indicates the change in total objective function if the constraint limit is increased by one unit. While in LFP the shadow prices in numerator or denominator output indicate the change in total objective function numerator or denominator respectively if the constraint limit is increased by one unit. A positive value indicates that an increase in the limit would increase the objective function (numerator or denominator respectively for LFP problems), while a negative value means that a higher limit gives a lower optimal value of objective function (numerator or denominator respectively for LFP problems).

3.3.6 Range analysis

Range analysis indicates the ranges within which objective function (separately numerator or denominator respectively for LFP problems) or constraint limit terms could be altered without affecting the composition of the optimal solution, i.e. without affecting which activities have positive values. It should be noted that within the indicated range for constraint limits, the optimal levels of activities selected are likely to vary.

Particular care needs to be taken in interpreting range analysis results. They should be regarded as "indicative" rather than "firm proof". To be certain of the effect of changing a particular constraint limit or objective function value, you should change it and re-solve the problem.

Range analysis output shows three columns: the original value of the objective function (numerator or denominator for LFP problems) or constraint limit; the lower limit and the upper limit. Note that in a maximization problem, the lower limit objective function and numerator in LFP problems for activities which are not part of the optimal solution is negative infinity. This indicates that it would be possible to reduce their objective function (the numerator for LFP problems) values indefinitely without affecting the solution. A little reflection will show that this is sensible. If an activity is not already part of the solution, making it less profitable is not going to cause it to enter the solution. On the other hand, making it more profitable is likely to bring it into the solution eventually. Analogously, the denominator upper limits for the same activities are positive infinity. This means that it would be possible to increase their cost indefinitely without affecting the solution. If an activity is not already part of the solution, making it more expensive is not going to cause it to enter the solution. On the other hand, making it less expensive is likely to bring it into the solution eventually. Range analysis shows how much more profitable or less expensive it needs to be for this to happen.

The upper limit constraint limits for 'less than' constraints which have some slack are positive infinity. If some resource is already being only partially used, increasing its availability will not affect the solution. Similarly, slack 'greater than' conditions have lower limits of negative infinity. Range analysis is discussed further in the next sections in which we look at examples.

3.4 An LP Example

In this section, we will work step by step through a sample LP problem.

Suppose that a pig farmer wishes to formulate a feed ration for his lactating sows. The sows have the nutritional requirements shown in Figure 13.13. The minimum requirements correspond to 'greater than' constraints, while the maximums are 'less than' constraints. The various available feeds are shown in Figure 13.14. The farmer wishes to produce 100 kg of a ration composed of a mixture of these feeds. The ration must satisfy the nutritional constraints outlined above and it should be as cheap to produce as possible.

This problem would be extremely difficult to resolve without using LP. No other easily usable technique is able to account simultaneously for the various constraints, the nutrient concentrations of different feeds and their costs. However, in LP this is a relatively simple problem. The problem is first re-organized into "matrix" form, from which it can be entered directly into the WinGULF editor. It is beyond the scope of this book to explain the process involved in translating a problem such as this into matrix form. Refer to elementary LP text

Crude protein (CP) %	min	16.00
Digestible energy (DE) MJ/kg	min	12.50
	max	13.50
Lysine (Ly) %	min	0.64
Phosphorus (P) %	min	0.54
Calcium (Ca) %	min	0.72
	max	2.00

Figure 13.13. WinGULF – Nutritional requirements of the sows.

	Wheat	lupins	Meat-meal	Dicalcium phosphate	Lime-stone	Lysine
Cost $/Ton	120.00	100.00	325.00	600.00	80.00	3800.0
CP %	11.00	28.00	50.00			
DE MJ/kg	14.60	14.20	11.50			
Lysine %	0.32	0.85	1.53		78.00	
Phosphorus %	0.26	0.29	4.80	22.30		
Calcium %	0.04	0.21	10.00	29.00	33.58	

Figure 13.14. WinGULF – Available feeds.

books for instructions in this area, for example [53], [91], [178], [187], [188]. A matrix for this example is presented in Figure 13.15.

LP example		Limit	Wheat	Lupins	MeatMl	Ca2P	LimeSt	Lysine
Cost	N		0.12	0.10	0.325	0.60	0.08	3.80
Obj.Denom	N	1.0						
CP min(kg)	G	16.0	0.11	0.28	0.50			
DE min(MJ)	G	1250.0	14.60	14.20	11.50			
DE max(MJ)	L	1350.0	14.60	14.20	11.50			
Ly min (g)	G	640.0	3.20	8.50	15.30			780.00
P min (g)	G	540.0	2.60	2.90	48.00	223.00		
Ca min (g)	G	720.0	0.40	2.10	100.00	290.00	335.80	
Ca max (g)	L	2000.0	0.40	2.10	100.00	290.00	335.80	
Mass (kg)	E	100.0	1.00	1.00	1.00	1.00	1.00	1.00

Figure 13.15. WinGULF – The matrix form of the problem.

After entering data[6] and solving the problem, you should have your screen looking like Figure 13.16.

[6]Note that this problem is included in the installation package of WinGULF. So, if WinGULF was installed properly you can find the problem in WinGULF's*Samples* sub-folder. The name of the file containing this problem is *Example1.GLF*

Optimal Solution

Problem name	: LP example
Problem direction	: MIN
Objective function value	: 10.509784
Number of iterations	: 4

Activities

No	Name		Level	Shad.Cost	LowerObj	Obj	UpperObj
1	Wheat	Z	0.0000	0.0206	0.10	0.120	INFINITY
2	Lupins	A	94.4722	0.0000	0.09	0.100	0.12
3	MeatMl	Z	0.0000	0.1241	0.20	0.325	INFINITY
4	Ca2P	A	1.1930	0.0000	0.08	0.600	1.20
5	LimeSt	A	4.3349	0.0000	-5.24	0.080	0.09
6	Lysine	Z	0.0000	3.7067	0.09	3.800	INFINITY

Constraints

No	Name		Slack	Shad.Price	LowerLim	Limit	UpperLim
1	CP min(kg)	G	10.4522	0.0000	-INFINITY	16.0	26.45
2	DE min(MJ)	G	91.5048	0.0000	-INFINITY	1250.0	1341.50
3	DE max(MJ)	L	8.4952	0.0000	1341.5048	1350.0	INFINITY
4	Ly min (g)	G	163.0134	0.0000	-INFINITY	640.0	803.01
5	P min (g)	G	0.0000	-0.0023	274.4441	540.0	1658.45
6	Ca min (g)	G	1280.0000	0.0000	-INFINITY	720.0	2000.00
7	Ca max (g)	L	0.0000	0.0000	1800.7186	2000.0	4146.52
8	Mass (kg)	E	0.0000	-0.0933	93.6077	100.0	100.59

Figure 13.16. WinGULF – Optimal solution output for an LP example.

Let us look briefly at what the different parts of this printout mean. First look at the 'Objective function value' at the top of the output. This is the cost of the cheapest ration which meets all the constraints specified. In this example the cheapest ration costs 10.5c per kilogram.

Now look at the 'Level' column under 'Activities'. This contains the optimal levels of the various feeds in the diet. In this case, to minimize feeding costs the farmer should mix a ration composed of 94.5% lupins, 1.2% Dicalcium Phosphate and 4.3% Limestone.

Now look at the shadow cost column. This indicates how far each feed is from entering the optimal solution. For example, if the cost of wheat fell by \$20.6 per tonne (2.06 c/kg) the least costly ration would change to include wheat. (As an experiment you could reduce the cost of wheat by \$21 and re-solve the problem to see what happens). The shadow cost of lupins is zero as lupins are already in the optimal solution.

We will discuss below the range analysis output (three columns at right of output). For now, consider the 'Slack' column under 'Constraints'. For 'less than' and 'greater than' constraints, the slack value indicates how far the constraint is from becoming limiting. For example, for digestible energy (DE) the minimum level allowed was 1250 MJ while the maximum was 1350 MJ. The diet actually selected includes 1341.5 MJ DE, so the minimum constraint is exceeded by 91.5 and the maximum constraint is undershot by 8.5. These are the values in the Slack column.

The Shadow price column shows the value of relaxing a constraint by one unit. For example, if it were decided that the level of phosphate in the diet could be reduced by 1 gram, the farmer could save 0.23 c on the cost of the 100 kg of ration. The minimum level of phosphate is 540 g. If this minimum level could be reduced to 500 g the shadow price of this constraint indicates that costs could be reduced by $40 \times \$0.0023 = \$0.093$. For constraints which are not currently limiting (i.e. have positive slacks) the value of relaxing them further is zero since they are not currently affecting the problem.

Now consider the range analysis output. This shows the ranges within which it is possible to change objective function and constraint limit values without changing the composition of the optimal solution. Objective function values of activities not currently in the optimal solution can be increased indefinitely without affecting the optimal solution. This is indicated in the printout by "INFINITY" in the upper limit column for objective function values (for wheat, meat meal and lysine). Note also that these ranges apply to changes in a single objective function value. If two or more changes are made at once, these ranges will not apply. Range analysis is thus a useful, but not definitive, guide.

The range analysis for constraint limits is very similar to that for objective function values except that within the upper and lower limits presented in the range analysis output, changes in constraint limit coefficients are likely to change the levels of activities in the optimal solution. To find out in what ways they change it is necessary to edit the problem and re-solve it. All that can be said without doing this is that within the upper and lower limits, levels of activities which are currently positive will not go to zero.

3.5 An LFP Example

In this section, we will work step by step through an example LFP problem.

Suppose that a refrigerator manufacturer is able to produce five types of refrigerator: Lehel 220, Lehel 120, Star 200, Star 160 and Star 250. The manufacturer has an order from dealers to produce 150, 70 and 290 units of Star 200, Star 160 and Star 250 respectively and 240 units without type detailing (that is, they can be of any type). The manufacturer wishes to formulate a production plan that satisfies given orders and maximizes profit per unit of cost. All necessary resources, excluding Freon 12 and TL 16, are not scarce. It is obvious that in the model we have to formulate it would be natural to require that all unknown variables associated with refrigerators should be integer. Even so, in this example we ignore the integrality restrictions and consider the problem as a pure continuous one. We re-consider this problem in integer form later, in Section 4.

So, the manufacturer has the following requirements and known data:

	Lehel 200	Lehel 120	Star 200	Star 160	Star 250
TL 16(l/unit)	0.20	0.13			
F 12(l/unit)			0.22	0.21	0.26
Price \$/unit	420.00	365.00	395.00	355.00	450.00
Cost \$/unit	320.00	290.00	300.00	280.00	340.00

This simple example shows how an LFP could be used when in an LP problem the linear objective function is replaced with the ratio of two linear functions. A matrix for this example is presented in Figure 13.17. After data entry[7] and the

LFP example		Limit	L 220	L 120	S 200	S 160	S 250
Profit \$	N		100.00	75.00	95.00	75.00	110.00
Cost \$	N		320.00	290.00	300.00	280.00	340.00
F12 (l)	L	125.00			0.22	0.21	0.26
TL16 (l)	L	80.00	0.20	0.13			
S200 min	G	150.00			1.00		
S160 min	G	70.00				1.00	
S250 min	G	290.00					1.00
Output	E	750.00	1.00	1.00	1.00	1.00	1.00

Figure 13.17. WinGULF – Matrix form for the LFP problem.

solution of the problem, your screen should look like Figures 13.18 and 13.19.

Let us look briefly at what the different parts of this printout mean. First look at the 'Objective function value' at the top of Figure 13.18. This is the profit and the cost of most rentable manufacturing plan which meets all the constraints specified. In this example the profit gained by the manufacturer from \$1 of expenditure is \$0.314280 .

[7]Note that this problem is included in the installation package of WinGULF. So, if WinGULF was installed properly you can find the problem in WinGULF's*Samples* sub-folder. The name of the file containing this problem is *Example2.GLF*

Optimal Solution

Problem name : LFP example
Problem direction : MAX
Objective function value : 75473.076923/240146.153846 = 0.314280
Number of iterations : 4

Activities - Numerator

No	Name		Level	Shadow Cost	Lower Obj	Obj	Upper Obj
1	L 220	A	232.69	0.00	84.2689	100.00	100.8255
2	L 120	Z	0.00	25.00	-INFINITY	75.00	90.5716
3	S 200	A	150.00	0.00	92.1506	95.00	96.8155
4	S 160	A	70.00	0.00	68.8942	75.00	90.1804
5	S 250	A	297.31	0.00	108.5624	110.00	529.2583

Activities - Denominator

No	Name		Level	Shadow Cost	Lower Obj	Obj	Upper Obj
1	L 220	A	232.69	0.00	317.3801	320.00	372.6060
2	L 120	Z	0.00	30.00	240.4530	290.00	INFINITY
3	S 200	A	150.00	0.00	294.2441	300.00	309.1179
4	S 160	A	70.00	0.00	232.3684	280.00	299.5385
5	S 250	A	297.31	0.00	-163.1100	340.00	344.6003

Figure 13.18. WinGULF – Optimal solution output for an LFP example, activities.

Constraints - Numerator

No	Name		Slack	Shadow Price	Lower Lim	Limit	Upper Lim
1	F12 (l)	L	0.00	38.46	123.1000	125.0	185.5000
2	TL16 (l)	L	33.46	0.00	46.5385	80.0	INFINITY
3	S200 min	G	0.00	-13.46	0.0000	150.0	158.6364
4	S160 min	G	0.00	-33.08	0.0000	70.0	79.0476
5	S250 min	G	7.31	0.00	-INFINITY	290.0	297.3077
6	Output	E	0.00	100.00	517.3077	750.0	917.3077

Constraints - Denominator

No	Name		Slack	Shadow Price	Lower Lim	Limit	Upper Lim
1	F12 (l)	L	0.00	76.92	123.1000	125.0	185.5000
2	TL16 (l)	L	33.46	0.00	46.5385	80.0	INFINITY
3	S200 min	G	0.00	-36.92	0.0000	150.0	158.6364
4	S160 min	G	0.00	-56.15	0.0000	70.0	79.0476
5	S250 min	G	7.31	0.00	-INFINITY	290.0	297.3077
6	Output	E	0.00	320.00	517.3077	750.0	917.3077

Figure 13.19. WinGULF – Optimal solution output for an LFP example, constraints.

Now look at the 'Level' column under 'Activities'. Both tables contain the optimal production levels of the various makes of refrigerator. In this case, to maximize profit gained per dollar of expenditure the manufacturer should produce 232.69 pieces of Lehel 220, 150 pieces of Star 200, 70 pieces of Star 160 and 297.31 pieces of Star 250. Lehel 120 should be excluded from manufacturing. Obviously, the optimal solution obtained cannot be utilized in the real-life application, since it contains non-integer values. We reconsider this problem in an integer form later, in Section 4.

Now look at the shadow cost column under 'Activities' in Figure 13.18. In the results for the numerator, all shadow costs, excluding Lehel 120 one, are zero as these refrigerators are in the optimal solution. Note that all shadow costs are non-negative. This means that the given plan maximizes the profit as well as rentability.

In the denominator results, all shadow costs are non-negative too. This means that the plan not only maximizes profit and rentability but also maximizes total cost.

The shadow price columns in Figure 13.19 show the value of relaxing a constraint by one unit. For example, if it were possible to curtail the production of Star 160 by 1 unit, the manufacturer could increase its profit by \$33.08 but would have to increase cost of production by \$56.15. The ratio $\$33.08/\$56.15 = 0.589$ is greater than the optimal value of the objective function (0.314280) so decreasing the number of Star 160 units would increase manufacturing efficiency. The shadow prices of Freon 12 are \$38.46 and \$76.92. If the volume of the resource could be increased by say 60 liters the profit and the cost would rise by $\$38.46 * 60 = \$2307.60$ and $\$76.92 * 60 = \$4615.20$, respectively.

Now consider the range analysis output. This shows the ranges within which it is possible to change objective function (numerator or denominator) and constraint limit values without changing the composition of the optimal solution. Objective function numerator values for activities which are not in the optimal solution can be decreased indefinitely without affecting the optimal solution. This is indicated in the printout by "-INFINITY" in the lower limit column of the numerator output (for Lehel 120). Similarly, the cost of this kind of production unit can be increased indefinitely without affecting the optimal solution. This is indicated in the printout by "INFINITY" in the upper limit column of the results for the denominator. Note also that these ranges apply to changes in a single objective function numerator or denominator value. If two or more changes are made at once, these ranges will not apply. Range analysis is thus a useful, but not a definitive, guide.

The range analysis for constraint limits is very similar to that for objective function values except that within the upper and lower limits presented in the range analysis output, changes in constraint limit coefficients are likely to change the levels of activities in the optimal solution. To find out in what ways they change it is necessary to edit the problem and re-solve it. All that can be said without doing this is that within the upper and lower limits, levels of activities which are currently greater than zero will not go to zero.

4. Problems with Integer Variables

In this section we deal with integer LP and LFP problems, i.e. such LP and LFP problems where one or more integrality requirements for unknown variables are included.

4.1 Input and Main Options

When we have to solve an LP or LFP problem with integer variables, similar to the case of continuous problems, we have to type in the problem or to open an existing problem from a disk file. Then, to set the problem as an integer one we have to choose the *There are integer variables* option in the *Defaults* dialog box, the *Variables* page, and mark those variables that have to be integers, as shown in Figure 13.20. After the problem has been entered and/or modified, it may be saved to an MPS text file and/or printed on a printer. If you leave option *Save integer variables in INT-files* checked, the package saves in a disk file (its name is <ProblemFileName>.INT) the information on variables - which one of them must be integer and which one may have a real (fractional) value.

At the moment only the Branch-and-Bound Method is implemented for integer problems. There are several customizable options for the method, see Figure 13.21. The package uses the depth-first search rule (see Chapter 8, Section 2), so you can choose which one of two branches (left or right) in the binary search tree must be examined first (*B&B strategy* option). Often it may be useful to switch on the *Preprocessor* option, since it allows you to avoid including in sub-problems such redundant branching restrictions as e.g.

$$x_1 \leq 12, \quad \text{and} \quad x_1 \leq 8.$$

If the *Preprocessor* is switched on it will result in only one 'resulting' restriction $x_1 \leq 8$ in the sub-problem. Obviously, using *Preprocessor* may sometimes significantly improve performance of the package.

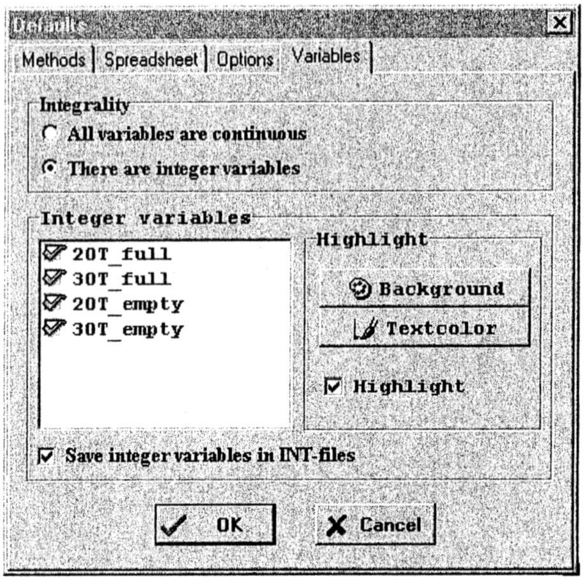

Figure 13.20. WinGULF – Defaults, the Variables page for integer problems.

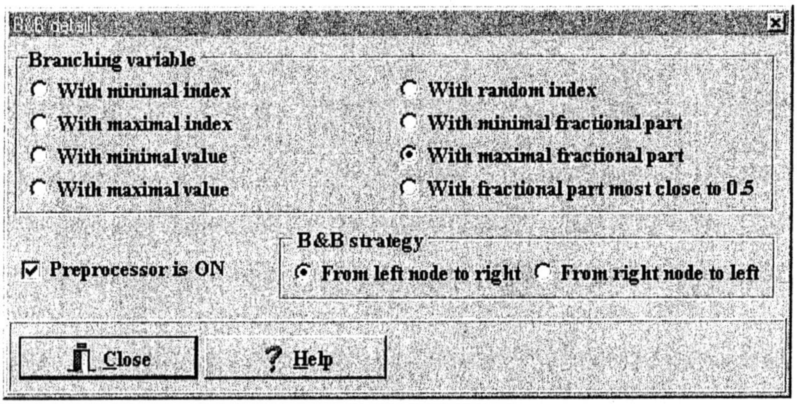

Figure 13.21. WinGULF – Branch-and-Bound Method, the Options dialog box.

4.2 Output

If you have finished editing the problem and have marked integer variables, to solve your integer problem you have to click on the *Run* button (note that the *Run by Step* button is not enabled for integer problems). The blank window as shown in Figure 13.22 will appear. Click on the *Run B&B* button to start

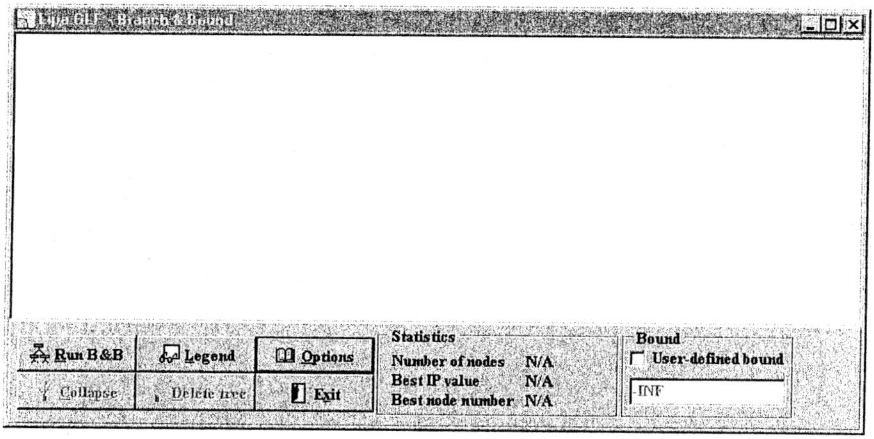

Figure 13.22. WinGULF – Branch-and-Bound Method, starting.

the Branch-and-Bound Method. When WinGULF has solved the problem it visualizes the binary tree as shown in Figure 13.23, and generates a report on

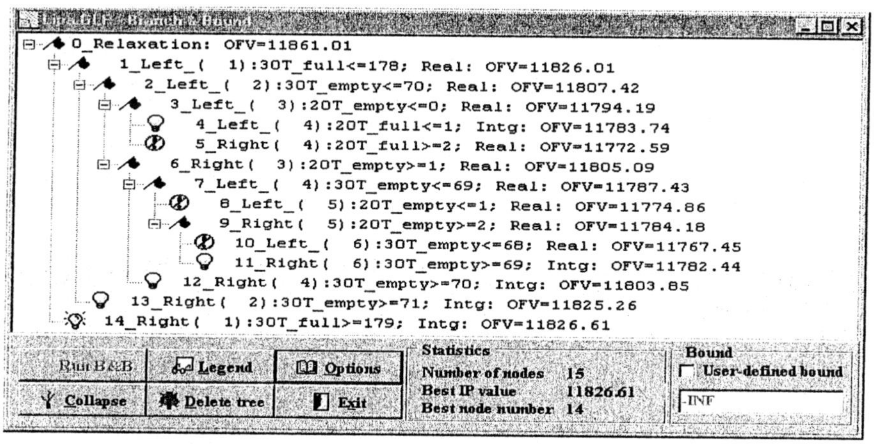

Figure 13.23. WinGULF – Branch-and-Bound Method, visualization.

the optimal solution found as shown in Figure 13.24. The report generated can be printed on a printer and/or can be saved to a text file on disk (default extension is '.SOL') for later viewing.

After generating (or opening) a report, WinGULF displays it in a text window. The first part of the output (the report generated by WinGULF) includes some statistics on the problem and information on the methods used to solve

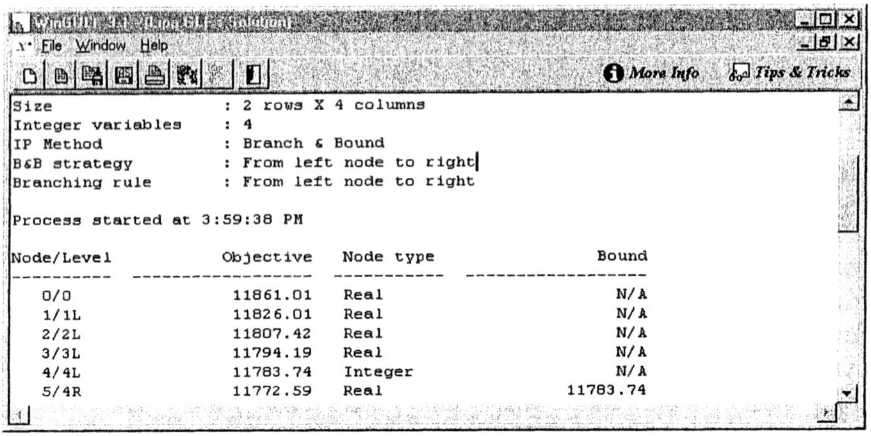

Size : 2 rows X 4 columns
Integer variables : 4
IP Method : Branch & Bound
B&B strategy : From left node to right
Branching rule : From left node to right

Process started at 3:59:38 PM

Node/Level	Objective	Node type	Bound
0/0	11861.01	Real	N/A
1/1L	11826.01	Real	N/A
2/2L	11807.42	Real	N/A
3/3L	11794.19	Real	N/A
4/4L	11783.74	Integer	N/A
5/4R	11772.59	Real	11783.74

Figure 13.24. WinGULF – Branch-and-Bound Method, report.

it. The second part of the report consists of the protocol of the calculations and describes the flow of the solution process. It includes four columns: the first one (*Node/Level*) gives information on the nodes (sub-problems) the package had to examine before it found an optimal integer solution, the second column (*Objective*) gives us objective values in associated node (sub-problem), while the *Node type* column describes the type of solution obtained in the given node (Real or Integer), finally, the *Bound* column contains the value of 'Bound' obtained in the current node. The last, third part of the output consists of two conventional portions: *Activities* for optimal values of unknown variables, and *Constraints* for the types of relations and for slacks.

4.3 An Integer Example

In this section, we reconsider the refrigerator manufacturer's LFP problem from Section 3.5, where we ignored the integrality requirements for unknown variables. As mentioned above, the problem may be opened from the *Example2.GLF* file (*Samples* sub-folder).

To set the problem as an integer one we have to open the *Defaults* dialog window, the *Variables* page, and mark all five variables L 220, L 120, S 200, S 160, and S 250, as integer. Solving this integer LFP problem by the Branch-and-Bound Method (the options used are: *Branching Variable= With fractional part most close to 0.5* and *B&B Strategy=From right node to left*) we obtain the visualization search tree shown in Figure 13.25. The report generated for the integer optimal solution found is shown in Figure 13.26. Let us look briefly at what the different parts of this printout mean. First look at the four columns

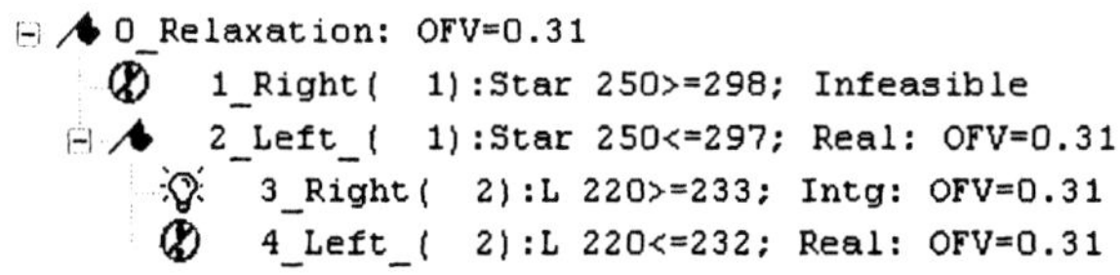

Figure 13.25. WinGULF – Search Tree for Integer LFP Example.

at the top of Figure 13.26. These columns describe the flow of looking for the optimal node (sub-problem). The *Node/Level* column gives information on the nodes (sub-problems) the package had to examine before it found an optimal integer solution. The first number gives the order number of the node, while the second one indicates the corresponding level and the branch (**R**ight or **L**eft) the node is located in. The *Objective* column gives objective values obtained in an associated node (sub-problem), while the *Node type* column indicates the type of solution obtained in the given node (Real, Integer, or N/A for infeasible node). The *Bound* column contains the value of Bound obtained in the current node.

Now look at the *Value* column under *Activities*. This column contains the integer optimal production levels of the various makes of refrigerator. In this case, to maximize the profit gained per dollar of expenditure, the manufacturer will have to produce 233 pieces of Lehel 220, 0 pieces of Lehel 120, 150 pieces of Star 200, 70 pieces of Star 160 and 297 pieces of Star 250. In this case the profit gained by the manufacturer from \$1 of expenditure is \$0.31427501.

The third part of the output in the *Slack* column contains differences between left-hand sides and right-hand sides of constraints. For example, consider value 0.08 in row 'F12 (l)'. The total amount of resource Freon 12 available was 125.00 liters. Since the type of the constraint is 'less than', slack 0.08 shows that if the manufacturer produces refrigerators in the quantities indicated above the necessity of Freon 12 is $125.00 - 0.08 = 124.92$ liters. Analogically, row 'S250 min' is associated with the constraint that establishes the low limit of 290 units for the production level of Star 250. Slack 7.00 in this row indicates that when producing refrigerators as it prescribes the optimal solution obtained, the manufacturer should produce 297 units of Star 250, i.e. the optimal value for Star 250 is greater than the low limit by 7.00 units.

Range analysis for integer programming problems is not available.

5. Future Developments

In this section we briefly overview the main directions of further developments in WinGULF intended for the near future. The main aim of these

```
Node/Level           Objective   Node type          Bound
----------  ------------------  -----------  -------------
   0/0                    0.31   Real                 N/A
   1/1R                    N/A   Infeasible           N/A
   2/1L                   0.31   Real                 N/A
   3/2R                   0.31   Integer              N/A
   4/2L                   0.31   Real                0.31

Optimal solution found at 11:02:33 AM in 00h.00m.00s.531ms.
Optimal value         : 0.31427501
Number of nodes       : 5
Best node number      : 3

Activities
----------
 No Name                          Value
--- ---------- - ------------------------
  1 L 220      A                 233.00
  2 L 120      Z                   0.00
  3 Star 200   A                 150.00
  4 Star 160   A                  70.00
  5 Star 250   A                 297.00

Constraints
-----------
 No Name                          Slack
--- ---------- - ------------------------
  1 F12 (1)    L                   0.08
  2 TL16 (1)   L                  33.40
  3 S200 min   G                   0.00
  4 S160 min   G                   0.00
  5 S250 min   G                   7.00
  6 Output     E                   0.00
```

Figure 13.26. WinGULF – Report for Integer LFP Example.

developments is to improve performance of the package, to make algorithms more stable and, of course, to make WinGULF more user-friendly.

Preprocessor

Most professional optimization packages have options that enable to preprocess the problem. It means that, for example, if the problem (LP or LFP) includes a constraint in the form $x_{25} = 15.5$, it would be preferable to avoid considering x_{25} as a variable. Instead, using simple algebraic operations we replace x_{25} with its value 15.5 everywhere it appears. Another possible case - the problem may include several redundant conditions like

$$x_3 \leq 15, \quad x_3 \leq 125, \quad \text{and} \quad x_3 \leq 12.$$

It is obvious that if we would like to improve the performance of the software package, we should try to avoid such redundancy, excluding the first two conditions from the problems and leaving the last one. There are also more sophisticated tests available that enable one to reduce the size of the LP or LFP problem to be solved. Generally, preprocessing is a good idea as it can reduce the time required to calculate solutions very dramatically.

The program package WinGULF, version 3.1 has such built-in facilities mainly used in branch-and-bound method. Our aim in future developments is to improve the performance of the preprocessor and to use more complicated and effective procedures to simplify the problems to be processed.

Scaling Problems
As discussed in Chapter 12, Section 1 the computer time required to solve LP or LFP problems as well as the correctness of the 'solution' obtained, can be affected by how the problem data is scaled. Most well-made packages have options to scale the data (including objective functions) automatically. At the moment, Version 3.1 of WinGULF freely downloadable from the Internet does not have such a facility. Only its professional version PGULF for UNIX/SOLARIS developed in ANSI C for high-performance parallel computers has such built-in automatic procedures, which may be parameterized to select one of two implemented algorithms for calculating scaling parameters. Hopefully, the next version of WinGULF will already have such facilities for automatic and manual scaling with corresponding options.

Re-starting
Sometimes we have to interrupt the process of solving a problem and have to re-start it later. Obviously, if we re-start the process from scratch we have to repeat all calculations performed earlier. So, it would be better to have such a procedure built in which would allow saving the best feasible solution obtained and re-using it during re-starting. Another reason to have such facilities is re-using a feasible or maybe an optimal solution of a problem when we have to solve some other problem. Having solved an LP or LFP problem and then saved its optimal solution we may wish to solve essentially the same problem but with just a few changes made (some data values altered, and/or some constraints

added/removed/modified). Generally speaking, re-staring from the previously saved solution, rather than starting from scratch, may dramatically reduce the time required to solve the new (modified) problem. This is why we consider this facility as one of the most important directions for further development.

Re-initialization

In Section 2, Chapter 12 we considered problems connected with the stability of the simplex algorithm. Most of these problems occur because of small numerical inaccuracies during pivot transformations. These small numerical errors may lead to a cumulative effect and hence, to large errors in calculations. One of the possible ways to make the simplex algorithm (or rather its implementation) more stable is to re-initialize the current simplex table using current basis and special methods of linear algebra (usually LU-factorization in general case, or Cholesky-factorization for symmetric matrices, or SV decomposition for singular matrices). Usually, optimization packages provide such options which allow you to set what type of decomposition you prefer to use and how often the procedure must re-initialize the basis. Another useful option may allow you to determine if the decomposition should be performed from scratch, or you prefer to re-use decomposition obtained earlier in the previous iterations and updated after each iteration performed. Since the WinGULF package first of all serves educational purposes, we intend in the next version to implement such facilities with a wide range of special tools for selecting preferred decomposition, its tuning and visualization of process.

Advanced Methods

A wide range of new methods, algorithms and improved computational techniques developed for LP may be relatively easily extended to the class of linear-fractional programming problems too. There are several effective special methods certainly developed directly for LFP. Theoretically, all these methods and techniques may be implemented and adapted for educational purposes and then may be incorporated into the next generations of WinGULF. However, in practice it means and requires many months (or even years) of a very hard job.

References

"Knowledge is of two kinds.
We know a subject ourselves,
or we know where we can find
information upon it"
– Samuel Johnson, 1775

[1] Abrham,J., Luthra,S., *"Comparison of Duality Models in Fractional Linear Programming"*, Zeitschrift für Operations Research, Vol.21, 1977, pp.125-130.

[2] Aggarwal,S.P., *"Analysis of the solution to a linear fractional functionals programming"*, Metrika, Vol.16, 1970, pp.9-26.

[3] Almogy,Y., Levin,O., *"A Class of Fractional Programming Problems"*, Operations Research, Vol.19, 1971, pp.57-67.

[4] Arora,S.R., Puri,M.C., *"Enumeration Technique for the Set Covering Problem with Linear Fractional Functional as its Objective Function"*, Zeitschrift für Angewandte Mathematik und Mechanik, Vol.56, 1977, pp.181-186.

[5] Arora,S.R., Puri,M.C., Swarup,K., *"The Set Covering Problem with Linear Fractional Functional"*, Indian Journal of Pure and Applied Mathematics, Vol.8, No.5, 1977, pp.578-588.

[6] Arvalo,M.T., Mrmol,A.M., Zapata,A. *"The Tolerance Approach in Multiobjective Linear Fractional Programming"*, Sociedad de Estadística e Investgación Operativa, Vol.5, No.2, 1997, pp.241-253.

[7] Ashton,D.J., Atkins,D.R., *"Multi-criteria Programming for Financial Planning"*, Journal of the Operational Research Society, Vol.30, 1979, pp.259-270.

[8] Bajalinov,E.B., *"Duality in Linear-Fractional Programming and its Applications"*, Ph.D. Thesis, Institute of Mathematics, Kirghiz Academy of Sciences, Frunze, Kirghiz Republic, 1984. (in Russian)

[9] Bajalinov,E.B., *"On the Economic Sense of Dual Variables in Linear-Fractional Programming"*, Ekonomika i matematicheskie metody, 3, Vol.24, 1988, pp.558-561. (in Russian)

[10] Bajalinov,E.B., *"On the Coincidence of the Optimal Solutions in Linear and Linear Fractional Programming Problems"*, Izvestia Akademii Nauk Kirgizskoi SSR, No.3, 1988. (in Russian)

[11] Bajalinov,E.B., *"On the System of Three Problems of Mathematical Programming"*, Kibernetika, No.6, 1989. (in Russian)

[12] Bajalinov,E.B., *"On Concordance of the Economic Interests"*, Izvestia Akademii Nauk Kirgizskoi SSR, No.3, 1990. (in Russian)

[13] Bajalinov,E.B., *"On an Approach to the Modelling of Problems Connected with Conflicting Economic Interests"*, European Journal of Operational Research, Vol.116, 1999, pp. 477-486.

[14] Bajalinov,E.B., Pannel,D.J., *"GULF : a General, User-friendly Linear and linear – Fractional programming package"*, Technical Report No. 93/86, Department of Mathematics, University of L.Kossuth, Debrecen, Hungary, 1993.

[15] Balas,E., *"An Additive Algorithm for Solving Linear Programs with Zero-One Variables"*, Operations Research, Vol.13, 1965, pp.517-546.

[16] Balas,E.,Ceria,S., Cornujols,G., *"A Lift-and-Project Cutting Plane Algorithm for Mixed 0/1 Programs"*, Mathematical Programming, Vol.58, 1993, pp.295-324.

[17] Balas,E.,Ceria,S., Cornujols,G., Natraj,N., *"Comory Cuts Revised"*, Operations Research Letters, Vol.19, 1996, pp.1-9.

[18] Barros,A.I., *"Discrete and Fractional Programming Techniques for Locations Models"*, Seria "Combinatorial Optimization", Vol. 3, Kluwer Academic Publishers, 1998.

[19] Barros,A.I., Frenk,J.B.G., Schaible,S., Zhang,S., *"A New Algorithm for Generalized Fractional Programs"*, Mathematical Programming, Vol.72, 1996, pp.147-173.

[20] Bartels,R.H., Golub,G.H., *"The Simplex Method of Linear Programming Using LU-Decomposition"*, Communication of the ACM, Vol.12, 1969, pp. 266-268 and 275-278.

[21] Beale,E.M.L., Small,R.E., *"Mixed Integer Programming by a Branch and Bound Technique"*, Proc. IFIP. Congr. 2, 1965, pp.450-451.

[22] Beasley,J.E., *"Advances in Linear and Integer Programming"*, Oxford Lecture Series in Mathematics and Its Applications, Vol.4, Oxfors University Press, 1996.

[23] Bector,C.R., *"Duality in Fractional and Indefinite Programming"*, Zeitschrift für Angewandte Mathematik und Mechanik, Vol.48, No.6, 1968, pp.418-420.

[24] Bector,C.R., *"Duality in Linear Fractional Programming"*, Utilitas Mathematica, Winnipeg, Vol.4, 1973, pp.155-168.

[25] Bector,C.R., *"Duality in Nonlinear Fractional Programming"*, Zeitschrift für Operations Research, Vol.17, 1973, pp.183-193.

[26] Bector,C.R., Chandra,S., Singh,C., *"Duality in Multiobjective Fractional Programming"*, in "International Workshop on Generalized Concavity, Fractional Programming and Economic Applications", University of Pisa, Italy, 1989.

[27] Belykh,V.M., Gavurin,M.K., *"An Algorithm for Minimizing a Fractional-Linear Function"*, Bulletin of the Leningrad State University, Vol.19, No.4, Oct.1980, pp.10-15. (in Russian)

[28] Bitran,G.R., Magnanti,T.L., *"Fractional Programming: Duality, Algorithms, Sensitivity Analysis and Applications"*, Technical Report No.92, Operations Research Center, Massachusets Institute of Technology, June 1974.

[29] Bitran,G.R., Magnanti,T.L., *"Duality and Sensitivity Analysis for Fractional Programs"*, Operations Research, No.4, Vol.24, 1976, pp.675-699.

[30] Bixby, R.E., *"Implementing the Simplex Method: The Initial Basis"*, ORSA J. on Computing, Vol.4, No.3, pp. 267-284, Summer, 1992.

[31] Bland,R., *"New Finite Pivoting Rules for the Simplex Method"*, Mathematics of Operations Research, Vol.2, 1977, pp. 103-107.

[32] Borde,J., Crouzeix,J.-P., *"Convergence of a Dinkelbach-type Algorithm in Generalized Fractional Programming"*, Zeitschrift für Operations Research, Vol.32, 1987, pp.A31-A54.

[33] Cambini,A., Martein,L., Schaible,S., *"On Maximizing a Sum of Ratios"*, Journal of Information and Optimization Sciences, Vol.10, 1989, pp.65-79.

[34] Ceria,S., Cornujols,G., Dawande,M. *"Combining and Strengthening Gomory Cuts"*, in Balas,E., Clausen,J., (eds.), Lecture Notes in Computer Science, Vol.920, Spinger-Verlag, 1995.

[35] Chandra,S, Chandramohan,M., *"An Improved Branch and Bound Method for Mixed Integer Linear Fractional Program"*, Zeitschrift für Angewandte Mathematik und Mechanik, Vol.59, No.10, 1979, pp.575-577.

[36] Chandra,S, Chandramohan,M., *"A Note on Integer Linear Fractional Program"*, Naval Research Logistics Quarterly, Vol.27, 1980, pp.171-174.

[37] Chandrasekaran,R., *"Minimal Ratio Spanning Trees"*, Networks, Vol.7, 1977, pp.335-342.

[38] Charnes,A., Cooper,W.W., *"Programming with Linear Fractional Functionals"*, Naval Res. Logistics Quart., Vol.9, No.3 and 4, 1962, pp.181-186.

[39] Chadha,S.S., *"Dual Fractional Program"*, ZAMM, Vol.51, 1971, pp.560-561.

[40] Chernov,Y.P., Lange,E.G., *"Problems of Nonlinear Programming with Fractional Economic Criteria. Methods and Applications"*, Kirghiz Academy of Science, Ilim, Frunze, 1978. (in Russian)

[41] Choo,E.U., Atkins,D.R., *"Bicriteria Linear Fractional Programming"*, Journal of Optimization Theory and Applications, Vol.36, 1982, pp.203-220.

[42] Chvátal,V., *"Linear Programming"*, Freeman, New York, 1983.

[43] Craven,B.D., Mond,B., *"The Dual of a Fractional Linear Program"*, Journal of Mathematical Analysis and Applications, Vol.42, 1973, pp. 507-512.

[44] Crouzeix,J.-P., Ferland,J.A., *"Algorithms for Generalized Fractional Programming"*, Mathematical Programming, Vol.52, 1991, pp.191-207.

[45] Crouzeix,J.-P., Ferland,J.A., Schaible,S., *"Duality in Generalized Linear Fractional Programming"*, Mathematical Programming, Vol.27, 1983, pp.342-354.

[46] Crouzeix,J.-P., Ferland,J.A., Schaible,S., *"An Algorithm for Generalized Fractional Programs"*, Journal of Optimization Theory and Applications, Vol.47, 1985, pp.35-49.

[47] Curtis,A.R., Reid,J.K., *"On the Automatic Scaling of Matrices for Gaussian Elimination"*, J. Inst. Maths. Applics., Vol.10, pp.118-124, 1972.

[48] Craven,B.D., *"Fractional Programming"*, Sigma Series in Applied Mathematics, Vol.4, Heldermann Verlag, Berlin, 1988.

[49] Dai,Y., Shi,J., *"A Conical Partition Algorithm for Maximizing the Sum of Several dc Ratios"*, Proceedings of the 5th International Conference on Optimization: Tech. Applications (ISOTA 2001), Hong Kong, 2001, pp.600-608.

[50] Dakin,R.,J., *"A Tree-Search Algorithm for Mixed Integer Programming Problems"*, Computer Journal, Vol.8, 1965, pp.250-255.

[51] Dantzig,G.B., *"Maximization of a Linear Function of Variables Subject to Linear Inequalities"*, In *Activity Analysis of Production and Allocation*, edited by T.C.Koopmans. New-York, John Wiley and Sons, 1951.

[52] Dantzig,G.B., *"Linear Programs and Extensions."* Princeton, New Jersey, Princeton University Press, 1963.

[53] Dent,J.B., Harrison,S.R. and Woodford,K.B., *"Farm Planning With Linear Programming: Concept and Practice"*, Butterworths, Sydney, 1986.

[54] Dinkelbach,W., *"Die Maximierung Eines Quotienten Zweier Linearer Funktionen Unter Linearen Nebenbedingungen"*, Wahrscheinlichkeitstheorie, Vol.1, 1962, pp.141-145.

[55] Dorn,W.S., *"Linear Fractional Programming"*, IBM Research Report RC-830, Yorktown Heights, New York, November 1962.

[56] Duff,I.S., *"A Survey of Sparse Matrix Research"*, Proc. IEEE 65, pp.500-535, 1977.

[57] Duff,I.S., Erisman,A.M., Reid,J.K., *"Direct Methods for Sparse Matrices"*, Clarendon Press, Oxford, 1986.

[58] Dutta,D., Rao,J.R., Tiwari,R.N., *"Fuzzy Approaches for Multiple Criteria Linera Fractional Optimization: a comment"*, Fuzzy Sets and Systems, Vol.54, 1993, pp. 347-349.

[59] Eijkhout,V., *"LAPACK Working Note 50: Distributed Sparse Data Structures for Linear Algebrs Operations"*, Technical Report CS 92-169, Computer Science Department, University of Tennessee, Knoxville, TN, 1992.

[60] Falk,J.E., Palocsay,S.W., *"Optimizing the Sum of Linear Fractional Functions"*, Recent Advances of Global Optimization, Princeton University Press, Princeton 1992, pp.221-258.

[61] Fletcher,R., Matthews,S.P.J., *"Stable Modification of Explicit LU Factors for Simplex Updates"*, Mathematical Programming, Vol.30, 1984, pp.267-284.

[62] Fletcher,R., Matthews,S.P.J., *"A Stable Algorithm for Updating Triangular Factors Under a Rank One Change"*, Mathematics of Computations, Vol.45, No.172, 1985, pp.471-485.

[63] Fletcher,R., *"Practical Methods of Optimization"*, Wiley-Interscience, 1987.

[64] Forrest,J.J.H., Tomlin,J.A., *"Updating Triangular Factors of the Basis Matrix to Maintane Sparsity in the Product Form Simplex Method"*, Mathematical Programming, Vol.2, 1972, pp.263-278.

[65] Freund,R.W., Jarre.F., *"An Interior-Point Method for Convex Fractional Programming"*, AT&T Numerical Analysis Manuscript, No.93-03, Bell Laboratories, Murray Hill, NJ, 1993.

[66] Freund,R.W., Jarre.F., *"An Interior-Point Method for Multi-Fractional Programs with Convex Constraints"*, AT&T Numerical Analysis Manuscript, No.93-07, Bell Laboratories, Murray Hill, NJ, 1993.

[67] Fukuda,K., Terlaky,T., *"Criss-Cross Method: a Fresh View on Pivot Algorithms"*, Mathematical Programming, B79, 1997, pp.369-395.

[68] Gavurin,M.K., *"Fractional-Linear Programming on an Unbounded Set"*, Bulletin of the Leningrad State University, Vol.19, No.4, Oct.1982, pp.12-16. (in Russian)

[69] Gass,S.I., *"Linear programming"*, McGraw-Hill, New-York, 1958.

[70] Gill,P.E., Golub,G.H., Murray,W., Saunders,M.A., *"Methods for Modifying Matrix Factorizations"*, Mathematics of Computations, Vol.28, 1974, pp. 505-535.

[71] Gill,P.E., Murray,W., Saunders,M.A., Wright,M.H., *"Maintaining LU Factors of a General Sparse Matrix"*, Linear Algebra and its Aplications, 1988, pp.239-270.

[72] Gill,P.E., Murray,W., Wright,M.H., *"Numerical Linear Algebra and Optimization"*, Addison-Wesley, 1991.

[73] Glover,F., *"A Multiphase-Dual Algorithm for the Zero-One Integer Programming Problem"*, Operations Research, Vol.13, 1965, pp.879-919.

[74] Glover,F., Laguna,M., *"Tabu Search"*, Kluwer Academic Publishers, 1997.

[75] Goedhart,M.H., Spronk, J., *"Financial Planning with Fractional Goals"*, European Journal of Operational Research, Vol.82, 1995, pp.111-123. North-Holland.

[76] Gol'stein,E.G., *"Dual Problems of Convex and Fractional-Convex Programming in Functional Spaces"*, Doklady Akademii Nauk SSSR, Vol.172, No.5, 1967, pp.1007-1010. (in Russian)

[77] Gol'stein,E.G., *"Duality Theory in Mathematical Programming and its Applications"*, Nauka, Moscow, 1971. (in Russian)

[78] Gol'stein,E.G., Yudin,D.B., *"Linear programming problems of transportation type"*, Nauka, Moscow, 1969. (in Russian)

[79] Golub,G.H., Van Loan,C.F., *"Matrix Computations"*, Baltimore, The Johns Hopkins University Press, 1996.

[80] Gomory,R., *"Outline of an Algorithm for Integer Solutions to Linear Programs"*, Bulletin of the American Mathematical Society, Vol.64, 1958, pp.275-278.

[81] Gomory,R., *"An Algorithm for Integer Solutions to Linear Programs"*, Recent Advances in Mathematical Programming, in Graves,R.L., and Wolfe,P. (eds.), McGraw-Hill, 1963, pp.269-302.

[82] Gondzio,J., *"Stable Algorithm for Updating Dense LU Factorization After Row or Column Exchange and Row and Column Addition or Deletion"*, Optimization, Vol.23, 1992, pp.7-26.

[83] Gondzio,J., *"Applying Schur Complements for Handling General Updates of a Large, Sparse, Unsymmetric Matrix"*, Technical Report ZTSW-2-G244/93, Systems Research Institute, Polish Academy of Sciences, 1995.

[84] Granot,D., Granot,F., *"On Integer and Mixed Integer Fractional Programming Problems"*, Annals of Discrete Mathematics 1, Studies in Integer Programming, (eds.) Hammer,P.L., North Holland Publishing Company, 1977, pp.221-231.

[85] Gupta,B., *"Finding the Set of all Efficient Solutions for the Linear Fractional Multiobjective Program with Zero-One Variables"*, Operations Research, Vol.18, 1981, pp.204-214.

[86] Gupta,R., Malhotra,R., *"Multi-Criteria Integer Linear Fractional Programming Problem"*, Optimization, Vol.35, 1995, pp.373-389.

[87] Gustavson,F.G., *"Some Basic Technigues for Solving Sprase Systems of Linear Equations"*, In "Sparse Matrices and thier Applications", Ed.: Rose,D.J., Willoughby,R.A., Proceedings of Symposium at IBM Research Center, NY, September 9-10, 1971.

[88] Hansen,P., De Aragão,M.V.P., Ribeiro,C.C., *"Hyperbolic* $0-1$ *Programming and Query Optimization in Information Retrieval"*, Mathematical Programming, Vol.52, 1991, pp.255-263.

[89] Hansen,P.,Pedrosa Filho,E.L., Ribeiro,C.C., *"Locations and Sizing of Offshore Platforms for Oil Exploration"*, European Journal of Operational Research, Vol.58(1), pp.202-214, 1992.

[90] Hansen,P.,Pedrosa Filho,E.L., Ribeiro,C.C., *"Modeling Location and Sizing of Offshore Platforms"*, European Journal of Operational Research, Vol.72(3), pp.602-605, 1994.

[91] Hardaker,J.B., *"Farm Planning by Computer", MAFF/ADAS, Reference Book 419*, Her Majesty's Stationery Office, London, 1980.

[92] Hartmann,K., *"Einige Aspekte der Ganzzahligen Linearen Quotientenoptimierung"*, Wiss Z.Tech.Hochsch. Chem.Leuna-Merseburg., Vol.15, No.4, 1973, pp.413-418.

[93] Hartmann,K., *"Zur Anwendung des Schnittverfahrens von Gomory auf Gemischt Ganzzahlige Lineare Quotientenoptimierungsprobleme"*, III Internat. Tagung "Mathematik und Kybernetik in der Ökonomie", 1973.

[94] Hartwig,H., *"Ein Simplexartigen Lösungsalgorithmus für Pseudolineare Optimierungsprobleme"*, Studia Sci. Math. Hungar., Vol.10, No.1-2, 1975, pp.213-236.

[95] Heath,M.T., *"Scientific Computing. An Introductory Survey"*, McGraw-Hill, 2002.

[96] Heesterman,A.R.G., *"Matrices and Simplex Algorithms"*, D.Reidel Publishing Company, 1983.

[97] Hirche,J., *"Optimizing of Sums and Products of Linear Fractional Functions Under Linear Constraints"*, Perprints Series, 95-03.

[98] Hoffman,A.J., Mannos,M., Sokolowsky,D., Weigmann,N., *"Computational Experience in Solving Linear Programms"*, Journal of the Society for Industrial and Applied Mathematics, Vol.1, No.1, 1953, pp.17-33.

[99] Illés,T., Szirmai,Á., Terlaky,T., *"The Finite Criss-cross Method for Hyperbolic Programming"*, European Journal of Operational Research, Vol.114, 1999, pp. 198-214.

[100] Ishii,H., Ibaraki,T., Mine,H., *"A Primal Cutting Plane Algorithm for Integer Fractional Programming Problems"*, Journal of the Operations Research Society of Japan, Vol.19, No.3, 1976, pp.228-244.

[101] Ishii,H., Ibaraki,T., Mine,H., *"Fractional Knapsack Problems"*, Mathematical Programming, Vol.13, 1976, pp.255-271.

[102] Ishii,H., Nishida,T., Daino.A., *"Fractional Set Covering Problems"*, Technical Report No. 1492, Osaka University, Vol. 29, 1979, pp.319-326.

[103] Isbell,J.R., Marlow,W.H., *"Attrition Games"*, Naval Research Logistics Quarterly, Vol.3, 1956, pp.71-94.

[104] Jagannathan,R., Schaible,S., *"Duality in Generalized Fractional Programming via Farkas' Lemma"*, Journal of Optimization Theory and Applications, Vol.41, 1983, pp.417-424.

[105] Jo,C.L., Kim,D.S., Lee,G.M., *"Duality for Multiobjective Fractional Programming Involving n-Set Functions"*, Optimization, Vol.29, 1994, pp.205-213.

[106] Jonathan,S., Kornbluth,H., Steuer,R.E., *"Multiple Objective Linear Fractional Programming"*, Management Science, Vol.27, No.9, 1981, pp.1024-1031.

[107] Kantorovich,L.V., *"Economic Accounting for the Best Utilization of Resources"*, AN SSSR, Moscow, 1960. (in Russian)

[108] Karmarkar,N, *"A new Polynomial-Time Algorithm For Linear Programming"*, Combinatorica, Vol.4, No.4, 1984, pp.373-395.

[109] Kaška,J., *"Duality in linear fractional programming"*, Ekonomicko-Matematicky Obzor, Vol.5, No.4, 1969, pp.442-453.

[110] Kaul,R.N., Bhatia,D., *"Generalized Linear Fractional Programming"*, Ekonomocko-Matematicky Obzor, Vol.10, No.3, 1974, pp.322-330.

[111] Khachian,L.G., *"A Polynomial Algorithm in Linear Programming"*, Doklady Akademii Nauk SSSR, Vol.244, 1979, pp.1093-1096.

[112] Knuth,D.E., *"The Art of Computer Programming"*, Vol.1:*"Fundamental Algorithms"*, Adisson-Wesley, 1968.

[113] Konno,H., Yajima,Y., *"Minimizing and Maximizing the Product of Linear Fractional Functions"*, Recent Advances of Global Optimization, Princeton University Press, Princeton 1992, pp.259-273.

[114] Kornbluth,J.S.H., *"A Survey of Goal Programming"*, OMEGA, Vol.1, 1973, pp.193-205.

[115] Kornbluth,J.S.H., Salkin,G.R., *"A Note on the Economic Interpretation of the Dual Variables in Linear Fractional Programming"*, ZAMM, 52, 1972.

[116] Kornbluth,J.S.H., Steuer,R.E., *"Multiple Objective Linear Fractional Programming"*, Management Science, Vol.9, No. 27, 1981, pp.1024-1039.

[117] Kornbluth,J.S.H., Vinso,J.D., *"Capital Structure and the Financing of Multinational Corporation: A Fractional Multiobjective Approach"*, Journal of Financial and Quantitative Amalysis, Vol.17, 1982, pp.147-178.

[118] Kotiah,T., Slater,N., *"On Two-Server Poisson Queues with Two Types of Customers"*, Operations Research, Vol.21, 1973, pp.597-603.

[119] Kotiah,T., Steinberg,D.I., *"Occurrence of Cycling and Other Phenomena Arising in a Class of Linear Programming Models"*, Commun. ACM, Vol.20, No.2, 1977, pp.102-112.

[120] Kuhn,H.W., *"The Hungarian Method for the Assignment Problem"*, Naval Research Logistics Quaterly, Vol.2, No.2-3, 1955, pp.83-97.

[121] Kuhn,H.W., Quandt,R.E., *"An Experimantal Study of the Simplex Method"*, Proceedings of the Symposia in Applied Mathemetics, Vol.15, American Mathematical Society, 1963.

[122] Kydland,F., *" Duality in fractional programming"*, Naval Research Logistics Quarterly, Vol.19, No.4, 1972, pp.691-697.

[123] Land,A.H., Doig,A.G., *"An Automatic Method of Solving Discrete Programming Problems"*, Econometrica, Vol.28, No.3, 1960, pp.497-520.

[124] Larcombe,M.H.E., *"A List Processing Approach to the Solution of Large Sparse Sets of Matrix Equations and the Factorizations of the Overall Matrices"*, In "Large Sparse Sets of Linear Eqiations", Ed.: Reid,J.K., Academic Press, London, 1971.

[125] Lasdon,L.S., *"Optimization Theory for Large Systems"*, Macmillan, Collier-MacMillan, London, 1970.

[126] Lawler,E.L., Lenstra,J.K., Rinnoy Kan,A.H.G., Shmoys,D.B., (eds.) *"The Traveling Salesman Problem"*, John Wiley & Sons, Ltd., 1985.

[127] Liebling,T.M., *"On the Number of Iterations of the Simplex Method"*, Methods of Operations Research, Vol.17, No.5, Oberwolfach-Tagung uber Operations Research, 13-19, 1977, pp. 248-264.

[128] Luhandjula,M.K., *"Fuzzy Approaches for Multiple Objective Linear Fractional Optimization"*, Fuzzy Sets and Systems, Vol.13, 1984, pp.11-23.

[129] Lutsma,F.A., *"Multi-Criteria Decision Analysis via Ratio and Difference Judgement"*, in Series of Applied Optimization, Vol.29., Kluwer Academic Publishers, 1999,

[130] Magee,T.M., Glover,F., *"Integer Programming: Mathematical Programming for Industrial Engineers"*, Avriel,M., and Golany,B.(eds.), Marcel Dekker, Inc. New York, 1995, pp.123-270.

[131] Martos,B., *"Hyperbolic Programming"*, Publ. Math. Inst., Hungarian Academy of Sciences, Vol.5, ser. B, 1960, pp.386-406.

[132] Martos,B., *"Hyperbolic Programming"*, Naval Research Logistics Quarterly, Vol.11, 1964, pp.135-155.

[133] Matsui, T., Saruwatari,Y., Shigeno,M., *"An Analysis of Dinkelbach's Algorithm for 0-1 Fractional Programming Problems"*, Technical Reports METR92-14, Department of Mathematical Engineering and Information Physics, University of Tokyo, 1992.

[134] Mukherjee,R.N., *"Generalizad Convex Duality for Multiobjective Fractional Programs"*, Journal of Mathematical Analysis and Applications, Vol.162, 1991, pp.309-316.

[135] Murty,K.G., *"Linear Programming"*, John Wiley and Sons, 1983.

[136] Myung,Y., Tcha,D., *"Return of Investment Analysis for Facility Location"*, Technical Reprt OR 251-91, Massachusetts Institute of Technology, 1991.

[137] Nemhauser,G.L., Wolsey,L.A., *"Integer and Combinatorial Optimization"*, John Wiley & Sons, New York, 1988.

[138] Nemirovskii,A.S., Nesterov,Y., *"An Interior-point Method for Generalized Linear-Fractional Programming"*, Mathematical Programming, Vol.69, 1995, pp.177-204.

[139] Nemirovskii,A.S., *"On Polynomiality of the Method of Analytical Centers for Fractional Programming"*, Mathematical Programming, Vol.73, 1996, pp.175-198.

[140] Nemirovskii,A.S., *"The Long-Step Method of Analytical Centers for Fractional Problems"*, Mathematical Programming, Vol.77, 1997, pp.191-224.

[141] Nesterov,Y., Nemirovskii,A.S., *"Interior Point Polynomial Algorithms in Convex Programming: Theory and Applications"*, SIAM, Philadelphia, 1994.

[142] Neumann von,J., *"A Model of General Economic Equilibrium"*, Review of Economic Studies, Vol.13, 1945, pp.1-9.

[143] Nykowski,I., Zolkiewski,Z., *"A Compromise Procedure for the Multiple Objective Linear Fractional Programming Problem"*, European Journal of Operational Research, Vol.19, 1985, pp.91-97.

[144] Orden,A., *"Computaional Investigation and Analysis of Probabilistic Parameters of Convergence of a Simplex Method"*, In Progress in Operations Research, Vol.2, Ed. Prekopa A., North-Holland, Amsterdam, The Netherlands, 1976, pp.705-715.

[145] Parker,G., Radin,R., *"Discrete Optimization"*, Academic Press, New York, 1988.

[146] Pissanetzky,S., *"Sparse Matrix Technology"*, Academic Press, 1984.

[147] Press,W.H., Teukolsky,S.A., Vetterling,W.T., Flannery,B.P., *"Numerical Recipes in C. The Art of Scientific Computing."*, Second Edition, The University of Cambridge, 1992.

[148] Reid,J.K., *"Fortran Subroutines for Handling Sparse Linear Programming Bases"*, HMSO, London, UK, Report AERE-R.8269, 1976.

[149] Reid,J.K., *"A Sparsity Exploiting Variant of the Bartels-Golub decomposition for Linear Programming Bases"*, Mathematical Programming, Vol.24, 1982, pp.55-69.

[150] Rheinboldt,W.C., Mesztenyi,C.K., *"Programs for the Solution of Large Sparse Matrix Problems Based on the Arc-graph Structures"*, Computer Science Center, University of Maryland, College Park, Technical Report TR-262, 1973.

[151] Ritter,K., *"A Parametric Method for Solving Certain Nonconcave Maximization Problems"*, Journal of Computer System Science, Vol.1, 1967, pp.44-54.

[152] Robillard,P., *"(0/1) Hyperbolic Programming Problems"*, Naval Research Logistics Quarterly, Vol.18, 1971, pp.47-57.

[153] Roos,C., Terlaky,T., Vial,J.-Ph., *"Theory and Algorithms for Linear Optimization"*, John Wiley & Sons, 1997.

[154] Rosing,K.E., *"Considering offshore production platforms"*, European Journal of Operational Research, Vol.72(1), pp.204-206, 1994.

[155] Rothenberg,R.I., *"Linear programming"*, North-Holland, 1979.

[156] Saad,O.M., *"An Algorithm for Solving the Linear Fractional Programs"*, Journal of Information & Optimization Science, Vol.14, No.1, 1993, pp.87-93.

[157] Saad,Y., *"SPARSKIT: A Basic Tool Kit for Sparse Matrix Computation"*, Technical Report CSRD TR 1029, CSRD, University of Illinois, Urbana, IL, 1990.

[158] Saunders,M.A. *"A Fast, Stable Implementation of the Simplex Method Using Bartels-Golub Updating"*, In Sparse Matrix Computations, Ed. Bunch,J.R., Rose,D.J., Academis Press, 1976, pp.213-226

[159] Savelsbergh,M.W.P., *"Preprocessing and Probing Techniques for Mixed Integer Programming Problems"*, ORSA Journal on Computing, Vol.6, No.4, 1994, pp.445-454.

[160] Saxena,P.C., Patkar,V.N., Parkash,O., *"A Note on an Algorithm for Integer Solution to Linear and Piecewise Linear Programs"*, Pure and Applied MAthematic Science, Vol.9, No.1-2, 1979, pp.31-36.

[161] Schaible,S., *"Fractional Programming: Transformations, Duality and Algorithmic Aspects"*, Technical Report Vol.73-9, Department of Operations Research, Stanford University, November 1973.

[162] Schaible,S., *"Duality in Fractional Programming: A Unified Approach"*, Operations Research, Vol.24, No.3, 1976, pp.452-461.

[163] Schaible,S., *"Fractional Programming: Applications and Algorithms"*, European Journal of Operations Research, Vol.7, 1981, pp.111-120.

[164] Schaible,S., *"Fractional Programming with Sums of Ratios"*, Scala and Vector Optimization in Economic and Financial Problems, Procedding of the Italian Workshop, (eds.) Castagnoli,E., Giorgi,E., Stampato da Elioprint, 1996, pp.163-175.

[165] Scott,C.H., Jefferson, T.R., *"Fractional Programming Duality via Geometric Programming Duality"*, Journal of Australian Mathematical Society, Sseies B, Vol.21, 1980, pp.398-401.

[166] Seshan,C.R., *"On duality in Linear Fractional Programming"*, Proc. Indian Acad. Sci., Sect. A. Math.Sci., Vol.89, 1980, pp. 35-42.

[167] Seshan,C.R., Tikekar,V.G., *"Algorithms for Integer Fractional Programming"*, Journal of Indian Institute of Science, Vol.62, No.2, 1980, pp.9-16.

[168] Sharma,I.C., Swarup,K., *"On Duality in Linear Fractional Functionals Programming"*, Zeitschrift für Operations Research, Vol.16, 1972, pp.91-100.

[169] Shigeno,M., Saruwatari,Y., Matsui,T., *"An Algorithm for Fractional Assignment Problems"*, DAMATH: Discerete Applied Mathematics and Combinatorial Operations Research and Computer Science, Vol.56, 1995.

[170] Shor,N.Z., Solomon,D.I., *"Decomposition Methods in Linear Fractional Programming"*, Chişinău, "Ştiinţa", 1989.

[171] Skeel,R.D., *"Scaling for Stability in Gaussian Elimination"*, J. Assoc. Comput. Mach., Vol.26, pp.494-526, 1979.

[172] Sniedovich,M., *"Fractional Programming Revised"*, European Journal of Operations Research, Vol.33, pp.334-341, 1988.

[173] Stancu-Minasian,I.M., *"Fractional Programming: Theory, Methods and Applications"*, Kluwer Academic Publishers, 1997.

[174] Suhl,U.H., Suhl,L.M., *"Computing Sparse LU Factorization for Large-scale Linear Programming Bases"*, ORSA Journal of Computing, Vol.2, 1990, pp.325-335.

[175] Swarup,K., *"Some Aspects of Duality for Linear Fractional Functional Programming"*, Zeitschrift für Angewandte Mathematik und Mechanik, Vol.47, No.3, 1967, pp.204-205.

[176] Swarup,K., *"Duality in Fractional Programming"*, Unternehmensforshung, Vol.12, No.2, 1968, pp.106-112.

[177] Taha,H.A., *"Integer Programming. Theory, Applications, and Computations"*, Academic Press, 1975.

[178] Taha,H.A., *"Operations Research, An Introduction"*, Collier MacMillan, New York, 1976.

[179] Terlaky,T., *"A New, Finite Criss-Cross Method for Solving Linear Programming Problems"*, Alkalmazott Matematikai Lapok, Vol.10, 1984, pp.289-296. (in Hungarian).

[180] Terlaky,T., *"A Convergent Criss-Cross Method"*, Math. Oper. und. Statist. Ser. Optim. Vol.16, No.5, 1985, pp.683-690.

[181] Terlaky,T., Zhang,S., *"Pivot Rules for Linear Programming: a Survey on Recent Theoretical Developments"*, Annals of Operations Research, Vol.46, 1993, pp.203-233.

[182] Thiriez,H., *"GULP (version 4.1)"*, in European Journal of Operational Research, Vol.39, 1989, pp.345-346. North-Holland.

[183] Thiriez,H., *"GULF (version 2.2)"*, in European Journal of Operational Research, Vol.67, 1993, pp.295-296. North-Holland.

[184] Vanderbei,R.J., *"Linear Programming. Foundations and Extensions"*, International Series in Operations Research & Management Science, Kluwer Academic Publishers, 1996.

[185] Verma,V., Bakshi,H.C., Puri,M.C., *"Ranking in Integer Linear Fractional Programming Problems"*, Zeitschrift für Operations Research, Vol.34, 1990, pp.325-334.

[186] Wilkinson,J.H., Reinsch,C., *"Linear Algebra"*, Vol.2 of *"Handbook for Automatic Computaion"*, New York, Springer-Verlag, 1971.

[187] Williams,H.P., *"Model Building in Mathematical Programming"*, John Wiley & Sons, A Wiley - Interscience Publication, 1985.

[188] Winston,W.L., *"Indtroduction to Mathematical Programming. Applications & Algorithms"*, PWS-Kent Publishing Company, Boston, 1991.

[189] Wolfe,P., Cutler,L., *"Experiments in Linear Programming"*, In Recent Advances in Mathematical Programming, Ed. Graves,R.L., and Wolfe,P., McGraw-Hill, New York, 1963, pp. 177-200.

[190] Zolkiewski,Z., *"A Multicriteria Linear Programming Model with Linear Fractional Objective Functions"*, Ph.D. Thesis, Central School of Planning and Statistics, Warsaw, Poland, 1983. (in Polish)

[191] Yudin,D.B., Gol'stein,E.G., *"Linear programming (Theory, Methods and Applications)"*, Nauka, Moscow, 1969. (in Russian)

Index